The Monochord in Ancient Greek Harmonic Science

Among the many instruments devised by students of mathematical sciences in ancient Greece, the monochord provides one of the best opportunities to examine the methodologies of those who employed it in their investigations. Consisting of a single string which could be divided at measured points by means of movable bridges, it was used to demonstrate theorems about the arithmetical relationships between pitched sounds in music. This book traces the history of the monochord and its multiple uses down to Ptolemy, bringing together all the relevant evidence in one comprehensive study. By comparing the monochord with a number of other ancient scientific instruments and their uses, David Creese shows how the investigation of music in ancient Greece not only shares in the patterns of demonstrative and argumentative instrument use common to other sciences, but also goes beyond them in offering the possibility of a rigorous empiricism unparalleled in Greek science.

DAVID CREESE is Assistant Professor of Greek and Latin Literature at the University of British Columbia.

CAMBRIDGE CLASSICAL STUDIES

THE MONOCHORD IN ANCIENT GREEK HARMONIC SCIENCE

DAVID CREESE

CAMBRIDGE UNIVERSITY PRESS
Cambridge, New York, Melbourne, Madrid, Cape Town, Singapore,
São Paulo, Delhi, Dubai, Tokyo

Cambridge University Press
The Edinburgh Building, Cambridge CB2 8RU, UK

Published in the United States of America by Cambridge University Press, New York

www.cambridge.org
Information on this title: www.cambridge.org/9780521843249

First published 2010

Printed in the United Kingdom at the University Press, Cambridge

A catalogue record for this publication is available from the British Library

Library of Congress Cataloguing in Publication data
Creese, David E. (David Evan), 1972–
The monochord in ancient Greek harmonic science / David Creese.
p. cm. – (Cambridge classical studies)
Includes bibliographical references and index.
ISBN 978-0-521-84324-9 (hardback)
1. Monochord. 2. Musical intervals and scales – Greece – History – To 500.
3. Music theory – Greece – History – To 500. I. Title. II. Series.
ML3809.C87 2009
787.7 – dc22 2009040103

ISBN 978-0-521-84324-9 Hardback

CONTENTS

PREFACE

This book is about an instrument that shaped the way scientists thought about music in Greek antiquity. It is also about the way Greek science shaped the use of a musical instrument. With its single string, movable bridge and graduated rule, the monochord (*kanōn*) straddled the gap between notes and numbers, intervals and ratios, sense-perception and mathematical reason. By representing musical sounds as visible, measurable distances (lengths of string), and by representing numbers audibly to the musical ear, it offered a way to study music as an arithmetical science through the medium of geometry. As an audible extension of the lettered diagram of Greek mathematics, what it offered the harmonic scientist, most importantly, was a means of attempting to extend the irrefutability of mathematical argument into the auditory realm, even when whether or not it succeeded in doing so was controversial.

It gives me great pleasure to record my thanks to the many people who have helped me in the writing of this book. My greatest debt is to Andrew Barker, who supervised the doctoral thesis on which it is based, and whose continued interest, encouragement and critical engagement were immensely helpful not only during the initial research but also during the final revisions of the book, and at countless points in between. I also benefited greatly from the comments and advice of my doctoral examiners, Charles Burnett and Niall Livingstone, many of whose questions and ideas have caused me to rethink my interpretations of the ancient evidence and reshape my arguments about them in this book. G. E. R. Lloyd was especially generous in reading the entire manuscript twice in different versions and offering detailed and challenging suggestions for its improvement, particularly in the development of comparisons between harmonics and other ancient sciences. I am also grateful to Alan C. Bowen, Peter Merten, Nathan Sidoli and Malcolm

Wilson for reading sections of the book in draft and for correcting some of my more egregious misunderstandings of astronomy and physics. For their invitations to present papers related to the research for this book, I should like to thank John Beatty, Sylvia Berryman, Robert Daum, Ingrid Holmberg, Donatella Restani and Gordon Shrimpton. Maya Yazigi answered my questions about the Arabic text of Diocles' *On Burning Mirrors*; Colin Gough helped me to understand the physics of sympathetic vibration; and Leonid Zhmud corresponded with me about Duris of Samos and the early history of the monochord. I have also benefited from discussions with Sylvia Berryman, Lisa Cooper, C. W. Marshall, Hallie Marshall, Francesco Pelosi, Massimo Raffa, Eleonora Rocconi and Robert B. Todd.

Final revisions to the manuscript were carried out during a period of research leave generously granted by the University of British Columbia, part of which was spent at the Scuola Normale Superiore in Pisa. I am especially grateful to Chiara Martinelli and Glenn Most for their invitation to spend what was a very productive time there, and also to Adrian Hasler for providing a similarly ideal (if very different) research environment during the previous summer. For their patience and assistance I thank Michael Sharp and Elizabeth Hanlon at Cambridge University Press, and for her care in copy-editing a demanding typescript I thank Linda Woodward, whose knowledge of the subject was as helpful as her attention to detail. I have received support of many kinds from my family, especially my parents, who read and commented on parts of the book, and above all my wife, Maggi, whose readings of many drafts saved me from errors too numerous to mention, and without whose constant interest, affection and assistance the book could not have been written at all.

ABBREVIATIONS AND EDITIONS

Ael. *VH*	Aelianus, *Varia Historia*
Archim. *Method*	Archimedes, *Method of Mechanical Problems*
Archyt.	Archytas, ed. Huffman 2005
Arist.	Aristotle
An. post.	*Analytica posteriora*, ed. Ross 1964
An. pr.	*Analytica priora*
De an.	*De anima*
Eth. Nic.	*Ethica Nicomachea*
Gen. an.	*De generatione animalium*
Metaph.	*Metaphysica*
Ph.	*Physica*
Pol.	*Politica*
Sens.	*De sensu*
Top.	*Topica*
[Arist.]	[Aristotle]
De audib.	*De audibilibus* (ap. Ptol. *In Harm.*), ed. Düring 1932
Pr.	*Problemata*, ed. Hett 1936
Aristid. Quint.	Aristides Quintilianus, *De musica*, ed. Winnington-Ingram 1963, cited either by book and chapter (e.g. III.3) or by page and line number (e.g. 99.9–12)
Aristox.	Aristoxenus
El. harm.	*Elementa harmonica*, ed. da Rios 1954
fr.	Fragmenta, ed. Wehrli 1945
Ath.	Athenaeus, *Deipnosophistae*
Bacch.	Bacchius, *Isagoge artis musicae*, ed. Jan 1895
Boeth. *Mus.*	Boethius, *De institutione musica*, ed. Friedlein 1867

Chrysipp.	Chrysippus, ed. von Arnim 1903–24
Cic. *Att.*	Cicero, *Epistulae ad Atticum*
De audib.	*see* [Arist.] *De audib.*
Diog. Laert.	Diogenes Laertius
DK	Diels and Kranz 1951
Eratosth.	Eratosthenes
Euc. *El.*	Euclid, *Elementa*, ed. Heiberg 1883–8
[Euc.] *Sect. can.*	[Euclid], *Sectio canonis*, ed. Jan 1895
Eutoc. *In Sph. Cyl.*	Eutocius, *On Archimedes' Sphere and Cylinder*, ed. Heiberg and Stamatis 1915
Exc. ex Nicom.	*Excerpta ex Nicomacho*, ed. Jan 1895
Exc. Neap.	*Excerpta Neapolitana*, ed. Jan 1895
FGrH	*Fragmente der Griechischen Historiker*, Jacoby 1923–58
FHG	*Fragmenta Historicorum Graecorum*, Müller 1841–70
Gal. *De plac. Hipp. et Plat.*	Galen, *De placitis Hippocratis et Platonis*
Gaud.	Gaudentius, *Isagoge harmonica*, ed. Jan 1895, cited either by chapter (e.g. 11) or by page and line number (e.g. 341.13–17)
Gell. *NA*	Aulus Gellius, *Noctes Atticae*
Gem. *Intro. astr.*	Geminus, *Introductio astronomiae*
Hdt.	Herodotus
Hero	Heron of Alexandria
Deff.	*Definitiones*, ed. Heiberg 1903
Stereom.	*Stereometrica*
Hippoc.	Hippocrates
Off.	*De officina medici*
Vict.	*On Regimen*, ed. Joly 1967
Hom.	Homer
Il.	*Iliad*
Od.	*Odyssey*
Hor. *Carm.*	Horace, *Carmina*
Iambl.	Iamblichus
Comm. math.	*De communi mathematica scientia*

In Nic.	*In Nicomachi Arithmeticam introductionem*
VP	*De vita Pythagorica*
Isoc.	Isocrates
LSJ	H. G. Liddell, and R. Scott, 1843, *A Greek–English Lexicon*, 9th edn, ed. H. S. Jones (Oxford, 1940), with a revised supplement, ed. P. G. W. Glare (Oxford, 1996)
Macrob. *In Somn.*	Macrobius, *Commentarius ex Cicerone in Somnium Scipionis*
Nicom.	Nicomachus of Gerasa
Ar.	*Arithmetica introductio*
Harm.	*Harmonicum enchiridium*, ed. Jan 1895, cited either by chapter (e.g. 6) or by page and line number (e.g. 246.1–5)
Papp.	Pappus of Alexandria, *Collectio*, ed. Hultsch 1876–8
PBerol.	*Berlin Papyri*
PHerc.	*Papyri Herculanenses*
Pherecr.	Pherecrates
PHib.	*Hibeh Papyri*, ed. Grenfell and Hunt 1906
Philo, *Opif.*	Philo of Alexandria, *De opificio mundi*
Philolaus	ed. Huffman 1993
Phld. *De mus.*	Philodemus, *De musica* IV, ed. Delattre 2007
Pl.	Plato
Euthd.	*Euthydemus*
Grg.	*Gorgias*
Lach.	*Laches*
Leg.	*Leges*
Men.	*Meno*
Phd.	*Phaedo*
Phdr.	*Phaedrus*
Phlb.	*Philebus*

Resp.	*Respublica*, ed. Slings 2003
Ti.	*Timaeus*, ed. Burnet 1902
[Pl.] *Epin.*	[Plato] *Epinomis*
Plut.	Plutarch
Alc.	*Alcibiades*
De an. procr.	*De animae procreatione in Timaeo*
Non posse	*Non posse suaviter vivi secundum Epicurum*, ed. Einarson and De Lacy 1967
Pericl.	*Pericles*
Quaest. conv.	*Quaestiones convivales*
Quaest. Plat.	*Quaestiones Platonicae*
Sol.	*Solon*
PMG	*Poetae Melici Graeci*, Page 1962
Poll. *Onom.*	Pollux, *Onomasticon*, ed. Bethe 1900–37
Porph.	Porphyry
In Harm.	*In Ptolemaei Harmonica commentarium*, ed. Düring 1932
VP	*Vita Pythagorae*
POxy.	*Oxyrhynchus Papyri*
Procl.	Proclus
Hypotyp.	*Hypotyposis*
In Euc.	*In primum Euclidis elementorum librum commentarii*
In Ti.	*In Platonis Timaeum commentaria*, ed. Diehl 1904
Ptol.	Ptolemy
Alm.	*Almagest* (= *Syntaxis mathematica*), ed. Heiberg 1898–1903, cited either by book and chapter (e.g. V.1) or by page and line number (e.g. 351.12–19)
Harm.	*Harmonica*, ed. Düring 1930, cited either by book and chapter (e.g. I.10) or by page and line number (e.g. 23.21–24.1)
Judic.	*De iudicandi facultate et animi principatu* (Περὶ κριτηρίου καὶ ἡγεμονικοῦ), ed. Huby and Neal 1989

Sect. can.	*see* [Euc.] *Sect. can.*
Sent. Vat.	*Vatican Sayings* (= *Gnomologium Vaticanum*)
Sext. Emp. *Math.*	Sextus Empiricus, *Adversus mathematicos*
Simpl. *In Phys.*	Simplicius, *In Aristotelis physicorum libros commentaria*, ed. Diels 1882–95
Stob. *Flor.*	Stobaeus, Ἀνθολόγιον
Suppl. Hell.	*Supplementum Hellenisticum*, ed. Lloyd-Jones and Parsons 1983
SVF	*Stoicorum veterum fragmenta*, ed. von Arnim 1903–24
Theo. Sm.	Theon of Smyrna, *Expositio rerum mathematicarum ad legendum Platonem utilium*, ed. Hiller 1878
Theophr.	Theophrastus
Char.	*Characteres*, ed. Diggle 2004
fr.	Fragmenta, ed. Fortenbaugh *et al.* 1992
Thuc.	Thucydides
Ti. Locr.	[Timaeus Locrus], *De universi natura* (Περὶ φύσιος κόσμω καὶ ψυχᾶς), ed. Thesleff 1965, cited by chapter and, where necessary, by page and line number; chapters 21–3 correspond to 96b–c
TLG	*Thesaurus Linguae Graecae*, http://www.tlg.uci.edu/

FIGURES

TABLES

INTRODUCTION: THE GEOMETRY OF SOUND

It is the very same taste which relishes a demonstration in geometry, that is pleased with the resemblance of a picture to an original and touched with the harmony of music. All these have unalterable and fixed foundations in nature, and are therefore equally investigated by reason and known by study; some with more, some with less clearness, but all exactly in the same way.[1]

Musical sounds are elusive things. They are among the most ephemeral of the objects of human perception and the productions of human art. Not even a single musical phrase, let alone an entire melody, is accessible to the ear all at once, as a painting is to the eye. But this is only one of the difficulties. Even when musical sounds are considered in isolation, they remain extraordinarily resistant to analysis. Each of us, as much today as in antiquity, recognises some sounds, and some combinations or sequences of sounds, as more musical, more beautiful, more concordant than others, but we may also find it difficult to say precisely what makes them so, or to define these categories in a way that will account for differences of individual taste, culture, or the age in which we live. Determining to what extent categories such as the 'musical', the 'beautiful' and the 'concordant' overlap presents even greater challenges.

One approach to the problem of defining and explaining musical beauty is to assume that the realm of music is not unique or self-contained, and that when we judge sounds as beautiful, we do so on the basis of a broader definition of beauty which applies to other perceptible things as well. Claudius Ptolemy (*fl.* AD 146–c. 170), one of the best representatives of this view among those who wrote on harmonics in Greek antiquity, argued that of all the senses, hearing and sight are the most closely connected to the faculty of reason, and that this accounts for the fact that while

[1] Joshua Reynolds, *Discourses on Art*, discourse VII, delivered to the Students of the Royal Academy, on the Distribution of the Prizes, 10 December 1776.

other senses take pleasure from their objects, sight and hearing alone find beauty in them: a smell may be fragrant, a taste may be delicious, a thing may be soft to the touch, but only a sight or a sound can be beautiful.[2] Thus when the fifth-century BC sculptor Polyclitus locates beauty in proportionality (*symmetria*), by which he means the proportional relationship of parts to one another and to the whole,[3] the student of music who accepts this thesis will say that beauty in music must arise from some sort of proportionality between sounds. If our student of music also accepts Ptolemy's view, he will say that we can appreciate beauty in music because hearing, like sight, communicates more directly with the faculty of reason than the other senses. Beauty, then, will be a kind of rational judgement made through our two most rational senses. And one way to extend the argument into the realm of music is to suggest that musical sounds are more beautiful than non-musical ones.

Without this set of preliminary assumptions, the thesis that musical sounds are musical in virtue of being proportional seems almost laughable. One reason for this is that proportion, in Polyclitus' sense at least, is something we identify and assess primarily with our eyes: we can look at a polygon drawn on a board and say at a glance whether it is a square or a rectangle (a proportional judgement, and readily definable as such). But the gap between visual and aural judgement seems unbridgeable, because there are no sounds which exhibit proportion in the way we know it from sight. Sounds can differ in timbre, in pitch, in volume, in duration – all of which are perceived as qualities except the last. The quantitative differences of duration constitute musical rhythm, which is the only proportionality in music directly accessible to perception. Since the same verse can be either spoken or sung in identical rhythm, however, and since only the latter would be counted fully musical by most people, the study of rhythm alone does not answer the question of what makes sounds musical. So we require a new starting point.

In spring, before the leaves come out in the hardwood forests of south-western Ontario, the ground is blanketed with the white

2 Ptol. *Harm.* 5.19–24, 93.11–94.1.

3 For example, of fingers to hand, hand to forearm, forearm to arm, and so on: Polyclitus ap. Gal. *De plac. Hipp. et Plat.* V.3 (= DK 40 A 3).

three-petalled wildflower called the trillium. It is a beautiful sight, but saying precisely why can be as difficult as explaining why a melody is beautiful. We could perhaps list some of the visible aspects of trilliums that must contribute to their beauty: their shape, their colour, their texture, the curl of their petals. But it is hard to get further, partly because the beauty we assess with our eyes, like the beauty we assess with our ears, seems to arise from many factors in combination. One approach, then, might be to narrow the focus and consider one factor in isolation. Suppose for the moment that we select a single trillium, and ignore all the things that make it beautiful that have nothing to do with its shape. Doing this requires restricting our vision: we might contemplate a black-and-white photograph of a pressed trillium, and find that it is still beautiful. Why? Ptolemy would probably say because of its various manifestations of *symmetria* (commensurability, proportion): literally, the ways in which its parts measure each other.[4] The idealised trillium (more perfect than any individual specimen we might obtain) has three petals of identical size and shape, whose points trace the outline of an equilateral triangle, the simplest and most symmetrical polygon. Polyclitus would probably say that this sort of basic mathematical structure is what makes works of art beautiful too. The hunch was that music, when it is beautiful, is so because it participates similarly in the mathematics of nature. But because proportion is a visual concept, sounds must somehow be rendered visible in order for us to investigate their proportionality. This can only be done indirectly, by assessing the dimensions of the physical objects that resonate when sounds occur. Some of these objects do not immediately appear to possess proportionality (the human vocal organs, for instance); others do (panpipes). The key, then, is to remove all the factors which cannot lead to an investigation of proportionality, like colour and texture for the trillium: we need an instrument that will provide a black-and-white photograph of pressed sounds. This is the monochord.

In Greek the instrument was called, simply, the 'measuring-rod' (*kanōn*). It consisted of a single string stretched over a soundbox

[4] See e.g. Ptol. *Harm.* 92.27–30.

whose surface could be marked with measurements, like a ruler. Fixed bridges at either end raised the string above the ruler, and a movable bridge allowed the string to be divided at any point in between. As long as the string was uniform, the only factor which could now contribute to its pitch was the length of the plucked section: thickness, tension and linear density were controlled. Divide the string in half by placing a movable bridge at its mid-point, and the half-length will sound a note an octave above that produced by the whole length. Here are proportion and musical beauty in one place, for the octave is (and was also for the Greeks) a privileged interval in music, and no matter where we construct it on the string, the lower note will always take twice as much length as the upper one. Thus we could say that the ratio of the octave is 2:1. Likewise, a division into two thirds generates the fifth (3:2), and one into three quarters produces the fourth (4:3). Musical relationships are now quantifiable, and just as the musician can say that a fourth and a fifth together make an octave, the mathematician can say that $4:3 \times 3:2 = 2:1$. The proportions which appear to underlie these three fundamental concords are all to be found in the first four numbers, 4:3:2:1. The special character of certain musical intervals, otherwise accessible to perception only as qualities, now opens itself to enquiry in the realm of arithmetic: one can attempt to define what makes musical concords concordant on the basis of their mathematical properties alone. This was the approach taken by a number of Greek musical writers, the earliest of whom were associated with the Pythagoreans.

In order to investigate music by means of ratio and proportion, therefore, the scientist needs to make sacrifices. Questions about how timbre and volume contribute to what is beautiful in music cannot be addressed; the causes of these attributes will be puzzled out in the science of acoustics. Questions of rhythm, too, will constitute a separate branch of investigation. The field of enquiry is narrowed to questions about the relationships between pitches in music, and because the questions have been framed in terms of proportion, these relationships will be further limited to those which can be expressed as a ratio of numbers. Indeed, those which cannot be so expressed will be considered unmusical: proportionality itself then becomes a condition of the musical. One of those who took this view was Adrastus of Aphrodisias, who wrote on music in the second century AD:

ὑπὸ μὲν οὖν τῶν ἀλόγων ἄλογοι καὶ ἐκμελεῖς γίνονται ψόφοι, οὓς οὐδὲ φθόγγους χρὴ καλεῖν κυρίως, ἤχους δὲ μόνον, ὑπὸ δὲ τῶν ἐν λόγοις τισὶ πρὸς ἀλλήλους πολλαπλασίοις ἢ ἐπιμορίοις ἢ ἁπλῶς ἀριθμοῦ πρὸς ἀριθμὸν ἐμμελεῖς καὶ κυρίως καὶ ἰδίως φθόγγοι.

Under irrational relations noises are irrational and unmelodic, and should not strictly even be called notes, but only sounds; but under relations that place them in certain ratios to one another, the multiple or the epimoric or simply that of number to number,[5] they are melodic, and are strictly and properly notes.[6]

Multiple ratios are those in which one term is a multiple of the other (*mn*:*n*); the ratios of the octave (2:1), octave plus fifth (3:1) and double octave (4:1) are of this form.[7] Epimoric ratios are those which have a 'part (*morion*) in addition': that is, the greater term exceeds the smaller by a simple part of the smaller ((n + 1):*n*); the ratios of the fifth (3:2), fourth (4:3) and tone (9:8) are of this form.[8] 'Number to number' ratios are those whose terms have no special relationship; the most common example is the interval left over when two tones are taken away from a fourth. This is the so-called *leimma* ('leftover'), an interval slightly smaller than half of a tone, whose ratio is 256:243.[9]

Adrastus calls this last category 'number to number' because this is the way such ratios are referred to in Greek: the *leimma* is 'the ⟨ratio⟩ of 256 to 243' (ὁ τῶν σνϛʹ πρὸς τὰ σμγʹ ⟨λόγος⟩).[10]

[5] An *arithmos* is something slightly different from what we mean by the term 'number': in Greek terms, it is 'a plurality (*plēthos*) composed of units' (Euc. *El.* VII def. 2); in our terms, this means a positive integer greater than one. Thus 'ratio of numbers' in the Greek sense excludes a quantity such as π (the ratio of the circumference of a circle to its diameter, two incommensurable magnitudes).

[6] Adrastus ap. Theo. Sm. 50.14–19, trans. Barker 1989: 214.

[7] Octave plus fifth: $2:1 \times 3:2 = 3:1$. Double octave: $(2:1)^2 = 4:1$. These are often referred to as compound ratios. Here and throughout the book I avoid the modern names for intervals greater than the octave in favour of those used by Greek authors (I write 'octave plus fifth', for example, rather than 'twelfth').

[8] Just as the tone is the interval by which a fifth exceeds a fourth, so too $9:8 = 3:2 \div 4:3$. Note that the definition is more exact than the expression (n + 1):*n* in that it excludes 2:1, which is not an epimoric ratio but a multiple: see Theo. Sm. 76.21–77.2. The Latin writers translate the term *superparticularis*, 'superparticular'.

[9] $4:3 \div (9:8)^2 = 256:243$.

[10] Number to number ratios are also called 'epimeric' (ἐπιμερής): see e.g. Theo. Sm. 78.6. The Latin equivalent is *superpartiens*, 'superpartient'.

This marks an important difference between Greek usage and ours, and it is a linguistic difference with ideological consequences. The privileged status of multiple and epimoric ratios is reflected in the fact that they can be expressed in a single word. Just as we can say 'duple' for 2:1, 'triple' for 3:1 and 'quadruple' for 4:1, Greek authors could also say 'hemiolic' (literally 'half-and-whole') for 3:2, 'epitritic' (literally 'a quarter in addition') for 4:3, 'epogdoic' for 9:8. This usage was not limited to ratios with smaller terms: the 27:1 ratio, for example, is ἑπτακαιεικοσιπλάσιος; the 17:16 ratio is ἐφεκκαιδέκατος.[11] The fact that the proportions found in the first four numbers (4:3:2:1) correspond to intervals which Greek musicians unanimously identified as concords was thus taken as an indication that the special status of these intervals reflected a broader principle which could be seen in the simplicity of the ratios and of the form of their expression.

Investigating music within these parameters means studying the different combinations and arrangements of intervals which arise in music with the ratios always in view. The science which pursued this investigation was concerned with the mathematical 'fitting-together' (*harmonia*) of the constituent notes and intervals of music. Aristotle called it 'mathematical harmonics', to distinguish it from 'hearing-based harmonics',[12] for the thesis that musical intervals acquire their particular qualities through the ratios to which they seem to correspond was not uncontroversial in antiquity. Those who had investigated music before Aristotle had done so in a variety of ways, not all of which are clear to us now from the surviving remnants of their work. But for Aristotle this variety could be condensed into a single dichotomy: some prioritised the mathematical aspects of the study of *harmonia*, and others prioritised the audible aspects. Aristotle called the former 'mathematical harmonicists', or more literally, 'those who investigate harmonics

[11] ἑπτακαιεικοσιπλάσιος: Nicom. *Harm.* 260.17; ἐφεκκαιδέκατος: Adrastus ap. Theo. Sm. 69.15. The more common way of writing epimorics with larger terms is just as brief: e.g. ὁ ἐπὶ ιϛʹ for 17:16 (Ptol. *Harm.* 24.12). Readers familiar with Boethius and the Latin tradition will recognise 'hemiolic' as 'sesquialter', 'epitritic' as 'sesquitertian' (*sesquitertius*) and 'epogdoic' as 'sesquioctave' (*sesquioctauus*).

[12] ἁρμονικὴ ἥ τε μαθηματικὴ καὶ ἡ κατὰ τὴν ἀκοήν, *An. post.* 79a1–2; with the former cf. *Metaph.* 997b21.

according to numbers'.[13] The hearing-based harmonicists, he says, know 'the fact that' (τὸ ὅτι), but the mathematical harmonicists know 'the reason why' (τὸ διότι), for the mathematical scientists 'are in possession of the demonstrations of the causes'.[14]

Later in the same text (the *Posterior Analytics*), Aristotle lists some of the concerns of mathematical harmonics:

τί ἐστι συμφωνία; λόγος ἀριθμῶν ἐν ὀξεῖ καὶ βαρεῖ. διὰ τί συμφωνεῖ τὸ ὀξὺ τῷ βαρεῖ; διὰ τὸ λόγον ἔχειν ἀριθμῶν τὸ ὀξὺ καὶ τὸ βαρύ. ἆρ' ἔστι συμφωνεῖν τὸ ὀξὺ καὶ τὸ βαρύ; ἆρ' ἐστὶν ἐν ἀριθμοῖς ὁ λόγος αὐτῶν; λαβόντες δ' ὅτι ἔστι, τίς οὖν ἐστιν ὁ λόγος;

What is concord? – a ratio of numbers between the high-pitched and the low-pitched. Why does the high-pitched form a concord with the low-pitched? – because the high-pitched and the low-pitched stand in a ratio of numbers. Does there exist a concord between the high-pitched and the low-pitched? – Is their ratio in numbers? Granted that it is, what then is the ratio?[15]

This book is about the monochord and its use within the tradition of mathematical harmonics, from the instrument's first appearance to the *Harmonics* of Claudius Ptolemy. The chronological scope of the book is defined in two ways. Firstly, it is in Ptolemy's work that the monochord receives its fullest, most detailed, most creative and methodologically rigorous treatment in antiquity. There are a number of important ancient witnesses who followed him: his earliest commentator, Porphyry, for one, and Boethius, whose *De institutione musica* transmitted the instrument and its use to the Latin West. But Porphyry is in some ways more helpful for his testimony about those whose monochord-informed harmonics preceded Ptolemy, and Boethius contributes little to the subject that is new.[16] Secondly, the history of the monochord in the Middle Ages has been written by others. Boethius' treatise has been studied both from the point of view of mediaeval music theory and from that of its Neoplatonic background,[17] and the uses of the

[13] οἱ κατὰ τοὺς ἀριθμοὺς ἁρμονικοί, *Top.* 107a15–16. [14] *An. post.* 79a2–4.

[15] *An. post.* 90a18–23, trans. Barker 1989: 70–1, following Ross' reading (ἐν ὀξεῖ καὶ βαρεῖ) rather than Bekker's (ἐν ὀξεῖ ἢ βαρεῖ) in 90a19.

[16] The instrument also appears in the work of Aristides Quintilianus (possibly third century AD), Gaudentius (possibly fourth century AD) and many Latin authors of late antiquity.

[17] See especially Bower 1989, Heilmann 2007. Mathiesen (1999: 629–36) gives an overview of the work. Barbera (1991) examines Boethius' transmission of a version

monochord in the Latin West from Boethius to 1500 have been exhaustively treated.[18] An area which is still in need of further work is the Arabic tradition. The most important items among the Greek literature on the monochord were known to Arabic music theorists of the ninth and tenth centuries, who added to the tradition by adapting the instrument to a new theoretical context, but this important aspect of its mediaeval legacy has been little studied.[19]

The aim of the book is to contextualise the monochord and its use within this chronological scope on four levels. The first, and narrowest, is mathematical harmonics: I shall attempt to establish when the instrument first came into use (toward the end of the fourth century BC, I shall suggest, although we cannot be certain); who among early mathematical harmonicists used it, and who did not; what mathematical harmonics could be done without it; and what it contributed to the science when it first appeared.

The second level is Greek harmonics more broadly. Aristotle's statement that it is the mathematical scientists who are in possession of the demonstrations of the causes, and the implied superiority of the mathematical approach to harmonics which follows from this, was challenged by two of his students, Aristoxenus and Theophrastus. Aristoxenus did not go so far as to deny that certain ratios can be found in the physical dimensions of instruments when they produce certain intervals, but he denied absolutely the value of such observations for the study of musical theory. We do not perceive music quantitatively, he argued, and so our science must be carried out in the realm of what we do perceive. Furthermore, pursuing the study of music on the authority of perception does not require relegating reason to a negligible role, nor does

of a short Hellenistic treatise on monochord division (the *Sectio canonis*); related issues (including Boethius' transmission of Ptolemy) are considered by Bowen and Bowen 1997.

[18] The first strides forward were made by Wantzloeben (1911); more work was done by Adkins (1963, 1967). Other contributions followed: e.g. Hughes 1969, Lindley 1980, Brockett 1981, Herlinger 1987, Pesce 1999 and especially Meyer 1996. See Herlinger 2002 for a summary of the literature on the monochord in the Middle Ages.

[19] The most important theorists in this tradition are Al-Kindī (ninth century) and Al-Fārābī (tenth century), who knew many of the important Greek harmonic texts which have survived to modern times, including Ptolemy's *Harmonics*. See Barbera 1991: 7–8 and Mathiesen 1999: 610–11 for brief accounts; the sources are listed by Shiloah (1979 and 2003).

it necessarily fail to demonstrate the causes of things to do with music: the third book of what survives as his *Elementa harmonica* is concerned almost exclusively with giving demonstrations, in Aristotle's sense of the word, of things to do with melodic succession.[20] The reason Aristoxenus' non-rational approach to harmonics (by which I mean merely 'not thinking about intervals as ratios', rather than 'unreasoned') is so important to the history of the monochord is that it provoked a counter-attack: this came in the form of a very short treatise called *The Division of the Monochord* (usually cited by its Latin title, *Sectio canonis*), in which the instrument made its first appearance in Greek literature. It is attributed, insecurely, to Euclid, but is probably to be dated to his generation (c. 300 BC).[21] It consists almost entirely of demonstrations of the primary theorems of mathematical harmonics, some of which are formulated in such a way as to refute specific rival arguments of Aristoxenus. That the vindication of the mathematical approach brought the monochord into the literature of music theory for the first time is significant: from its first appearance it was a polemical instrument as much as a musical one. And yet in later centuries, as harmonic theorists sought in different ways to bring the rival traditions together and combine their inheritances, the monochord also appears in a mediating role.

The third level is Greek mathematics. The arguments of the *Sectio canonis* are framed in the formulaic language of Euclidean arithmetic: they rely on theorems demonstrated in the arithmetical books of Euclid's *Elements* (V, VII–IX), and they employ a similar style of presentation. More specifically, they prove propositions in arithmetic through constructions in geometry: simple constructions, in which numbers are represented as line segments, and ratios as the relationships between their lengths. The monochord comes to participate in this presentation, for it makes the

[20] *El. harm.* III contains twenty-three theorems set out as 'proofs' or 'demonstrations' (*apodeixeis*). It is now generally accepted that the text of the *Elementa harmonica* as we have it is a combination of more than one original work: see Mathiesen 1999: 294–334, Gibson 2005: 39–75. For a recent attempt to delineate its ancient components and their relationships, see Barker 2007, ch. 5; for the unitarian position, see Bélis 1986, ch. 1.

[21] Besides the authorship of the treatise, its date and unity have been the subject of significant debate. These issues will be discussed in chapter 3. The monochord is attested independently in a fragment of Duris of Samos (c. 300 BC); this will be examined in chapter 2.

connection between ratios and intervals by means of simple geometrical constructions. A ratio can only be heard as an interval if it can first be represented as a relationship between two segments of string. Viewed from the level of mathematics, the instrument is a kind of extension of the diagrams used in Euclidean arithmetic, and its use in mathematical harmonics is consequently limited in ways that also limit the arithmetical use of geometrical diagrams.

The fourth level is Greek science more broadly. Here I have had to be selective. The third level raises questions about how instruments are incorporated into the methods of sciences other than harmonics, and what such a comparison can tell us about the role of the monochord as a scientific instrument. What, if anything, is unique about the way it mediates between the sensory and intelligible realms? If there is anything unique about it, what effect does this have on the way it was used by Greek harmonicists, and on the development of their scientific methods? To frame preliminary answers to these questions (the eligible material is vast, and what I offer here is only a start), I have chosen a pair of astronomical instruments discussed by Ptolemy in his *Almagest*, and have compared his introduction of them with his introduction of the monochord in the *Harmonics*.

When a geared calendrical device was found in an ancient shipwreck off the Greek island of Antikythera in 1901, it was a scientific instrument without a literature. The complexity of the Antikythera mechanism came as a surprise to students of ancient science and technology, because such a level of mechanical sophistication could not have been inferred from surviving Greek literature.[22] The monochord, by contrast, is an instrument without an archaeology. Because its history must be written entirely from books, there are a number of things we cannot know about its earliest incarnations: their dimensions, the materials of their construction, and so on. But what we can be more certain of are the uses to which it was put, so far as these are described for us by Greek authors. Precisely how the Antikythera mechanism functioned, and what exactly it was designed to do, are matters of

[22] On the Antikythera mechanism, see especially Price 1974; Bromley 1986; Freeth, Bitsakis *et al.* 2006; Freeth, Jones *et al.* 2008.

ongoing investigation. What the students of both instruments share, however, is the frustration of a silent record. Neither instrument can be assumed in a context where it is not explicitly mentioned, and this poses particular difficulties for the dating of the monochord's invention. It remains untraceable prior to about 300 BC, and yet all the fundamental theorems of mathematical harmonics had apparently been articulated in some form by this date. A persistent tradition that Pythagoras invented it surfaces for the first time around AD 100; if true, it would place the instrument's invention no later than the beginning of the fifth century BC. Responses to this tradition have varied in modern scholarship. Some have taken it at face value; on this hypothesis the instrument would have a roughly two-century prehistory before its first literary appearance. Others have been inclined to date its invention much later, later even than Euclid (on the argument that the *Sectio canonis* is at least in part a post-Euclidean document). I shall suggest that neither of these extremes is the result of a fair appraisal of the evidence (what little there is), and in the process I shall argue first that the late tradition which makes Pythagoras the instrument's inventor is unreliable and should be disregarded altogether, and second that no surviving account of harmonics before the death of Aristoxenus (in the late fourth century BC) presupposes the monochord's existence.

The book is in six chapters, the last five of which proceed in chronological order. My goal in the first is rather different. Since the documentary evidence for the monochord in the period I have chosen is framed by two important texts, the *Sectio canonis* and Ptolemy's *Harmonics*, and because these texts (despite their many differences) share some common features in their presentation of mathematical and harmonic argument, they provide a basis on which to compare the monochord with a number of other mathematical and scientific tools. Thus I shall begin by attempting to contextualise the monochord in the third and fourth levels mentioned above: as a mathematical device, and as a scientific instrument. Specifically I shall try to show how the monochord sits, somewhat precariously, between the disciplines of arithmetic and geometry, and how its roles in harmonic argument share affinities not only with those of other scientific instruments (among which the abacus, compasses, ruler, armillary sphere and parallactic instrument), but also with those of diagrams and tables.

In chapter 2, 'Mathematical harmonics before the monochord', I shall turn to chronology, and to the more narrow concerns of the first level of context outlined above. Here the task will be to establish, as far as possible, when the instrument might first have appeared, and what kind of arguments were advanced in mathematical harmonics without it. My conclusions will be primarily negative: the pioneering work in the field appears to have been done without the instrument, and two of the authors most widely credited with monochord use, Philolaus and Archytas, probably predated the instrument. On the other hand, these negative conclusions shed light on just what proportion of a science's principal doctrines and methods can be developed without demonstrations on the instruments that lend them the greatest degree of irrefutability.

The aim of chapter 3 is to show what the introduction of the monochord, and particularly its use in the *Sectio canonis* around 300 BC, owes to fourth-century developments in harmonics, acoustics and mathematical argumentation. It is clear that instruments of various sorts had been employed in harmonic demonstrations, with varying degrees of rigour, by the middle of the fourth century; indeed the first such procedure was reportedly carried out in the early fifth century by Hippasus of Metapontum. One of the features of the monochord which gives it particular demonstrative force is that it makes its quantities visible through measurable linear distances. A problem with this presentation is that harmonicists who wish to claim that the ratios of these distances possess some abstract reality (that an octave, for example, somehow 'is' 2:1 because the instrument displays these ratios when an octave is produced on it) come up against the difficulty of explaining the causes of pitched sound in a way that will allow them to argue about the exact magnitudes of intervals in the realm of arithmetic. For the uncomfortable fact is that not all musical instruments appear to display the same ratios when the same intervals are sounded on them; some do not appear to display ratios at all. The monochord-theorist must attempt to show that the ratios his instrument displays are essential to the identity of the interval itself, and not merely an attribute which appears here and not there. This is perhaps the area in which Greek harmonicists were consistently at their weakest, but the fact that strings were a hot topic in the acoustical debates of the fourth and third centuries, and that the author of the *Sectio canonis* manages one of the best attempts at this problem in Greek antiquity, shows the extent to which it was an important aspect of the development of a scientific method for the monochord.

Chapter 4 deals with one author, Eratosthenes, who wrote about harmonics, among many other subjects, in the late third century BC. He warrants a chapter to himself simply because he is the only securely datable author between the *Sectio canonis* and Thrasyllus (d. AD 36) who is credited with publishing a 'division of the monochord'. The question I address in this short chapter is whether this required the monochord at all. The answer is not straightforward, partly because we have only two brief testimonia which shed any light on Eratosthenes' reputed 'division of the monochord', and partly because the most likely interpretation of this division (or divisions, as I shall suggest) involves mathematical absurdities which cloud the picture even further. I shall offer Eratosthenes a method, built out of ingredients he himself devised, by which he might have employed the monochord in order to produce the ratios attributed to him, but this will only serve to highlight the impossibility of proving the instrument's place in the project of an author who either did not mention it, or whose discussion of it no longer survives.

Chapter 5 covers the period between Eratosthenes and Ptolemy, during which mathematical harmonics came to be called 'monochord science', or 'canonics' (*kanonikē*). The language and practice of canonics at this time reveal several important aspects of the relationship between the instrument and its science. Several authors, beginning perhaps with Ptolemaïs of Cyrene, address the question of how to define the various approaches to harmonics taken by theorists of different persuasions. One answer was that they placed differing degrees of emphasis on either reason or perception. Some were concerned so completely with the evidence of perception that they could hardly be counted theorists at all; others treated harmonics as a purely mathematical discipline and paid no attention to the evidence of perception. Within this scheme the 'canonicists' (*kanonikoi*) emerge as a group whose particular brand of harmonics requires the active participation of both faculties, and the monochord itself is identified as the link between them. The defining activity of their science was 'canonic division' (*kanonos katatomē*), a term which appears to designate a number of closely related operations. At its simplest, it constituted the bridging of the instrument's string in predetermined locations so as to exhibit the ratios of the most elementary of the musical intervals (at minimum, the octave, fifth and fourth; sometimes also the tone). At its most complex, it involved marking the ruler (*kanōn*) with bridge positions corresponding to all the notes of a complete scale-system of two octaves' range.

The emphasis on reason and perception raises difficulties of two sorts. The first is that if reason is taken to dictate that 'concord' is in fact a mathematical category, and that it includes only multiple and epimoric ratios, then while it will accept the fifth (3:2) and the octave plus fifth (3:1) as concords, it will reject the octave plus fourth (8:3) even though it will accept the fourth (4:3). This proved to be one of the most difficult and contentious issues in mathematical harmonics, and one of the authors of this period who makes the most determined attempt to include the 8:3 interval among the concords unaccountably omits it from a canonic division which includes all the other concords within a two-octave range (this is Adrastus). The second difficulty is that the instrument can only be construed as a link between reason and perception if the quantities it represents acoustically are exactly what reason prescribes. In a badly argued objection to the possibility of dividing the tone into exactly equal semitones, Adrastus contends that since the monochord's bridge must always have some width, the two string segments it generates will not add up to the one length it was supposed to divide: the sum of the two segments will always be less than the whole by an amount equal to the width of the bridge. Within Adrastus' own argument the objection fails, but nevertheless it raises a larger problem: whether canonic division can really do what it claims to do.

This difficulty was apparently not addressed seriously until Ptolemy tackled it. Chapter 6 is about Ptolemy's canonics, which included for the first time a concerted effort to establish the reliability of the monochord and related instruments. Also for the first time, the instrument appears not only as a means of demonstrating rationally conceived intervals to the ear, but also as a means of testing whether these ratios are in fact acceptable to sense-perception. The roles of reason and perception in harmonic science are nowhere more clearly articulated in antiquity than in Ptolemy's treatise, and one of the consequences of this emphasis is Ptolemy's insistence not only that the ear must be allowed to judge the attunements of reason, but that the only way it can do so reliably is to hear them in the context of melodies, where it can recognise their true character. This is something for which the monochord is ill qualified: some advances in monochord-playing technique developed by the Neronian harmonicist Didymus mitigate some of its most obvious defects, but this does not satisfy Ptolemy. His solution is to build more complex *kanones*: first with eight strings, then with fifteen. He develops these further with ideas gleaned from an attempt to improve on an instrument called the *helikōn* inherited from 'the mathematicians'.

Ptolemy's canonics is characterised by two features which set it apart from earlier approaches to the science: one is his attention to the practical details of the instruments' construction; the other is his geometricising of the instruments, both in their presentation and in the demonstrations of their accuracy. The fundamental postulates of the science are still firmly governed by arithmetic, but the instruments themselves are in Ptolemy's hands more clearly geometrical than they had ever been.

This study owes much to the work of others. Andrew Barker's annotated translations of the most important texts (*Greek Musical Writings*, vol. II, 1989) are in many ways a foundation for much of what I have attempted, and his detailed discussions of the instruments in *Scientific Method in Ptolemy's 'Harmonics'* (2000) have been a point of departure for many of my arguments about Ptolemy.[23] André Barbera's work on the *Sectio canonis*, particularly his edition of the three branches of the textual tradition (*The Euclidean Division of the Canon*, 1991), has been indispensable, and although I have disagreed with some of his conclusions,

[23] Equally important to this book are his discussions of Philolaus, Archytas, Aristoxenus and the *Sectio canonis* in *The Science of Harmonics in Classical Greece* (2007).

I have benefited greatly from his work on the document's transmission and from the reshaping of the debate about its authorship and date which followed from this. Our understanding of two early Pythagoreans who worked on harmonics, Philolaus and Archytas, would be much less clear without the annotated editions and translations of Carl A. Huffman.[24] The case made by Reviel Netz for the centrality of the lettered diagram in Greek mathematical thinking (*The Shaping of Deduction in Greek Mathematics*, 1999) offered new avenues for exploring the relationship between instruments and diagrams in Greek harmonics. Two recent translations of Ptolemy's *Harmonics* by Jon Solomon (2000) and Massimo Raffa (2002) have contributed much to a growing body of literature on what is indisputably the most important text for the history of ancient canonics.[25]

Some further preliminaries

Several basic aspects of the monochord and its use have still to be addressed. One of these is the instrument's name, *kanōn*; another is what a simple canonic division looks like; a third is how the process of canonic division was applied to the explication of Greek scales; a fourth is the structure of these scales and the names of their notes. Since these elements necessarily appear from time to time in the literature of canonics discussed in this book, readers may wish to have a brief survey of them here. My account of the Greek scale-systems aims not at completeness, but at utility: it could be extended, for example, to questions about the chronological development of these structures or about how closely they represented the attunements of musical practice, but is focussed instead on providing a summary of those elements which the reader needs in order to follow the discussions of notes, intervals and patterns

[24] *Philolaus of Croton: Pythagorean and Presocratic* (1993), *Archytas of Tarentum: Pythagorean, Philosopher and Mathematician King* (2005).

[25] Mathiesen (1999) includes many relevant and helpful interpretations of the literature of the instrument; see also the briefer summaries by West (1992: 240–2) and Landels (1999: 131–6). I regret that Fabio Acerbi's annotated translations of the complete works of Euclid (*Euclide: Tutte le opere* (Milan, 2007)), which include a critical introduction to the *Sectio canonis* and a Greek text, came to my attention too late to be taken into account when the book was being written.

of attunement that arise throughout the book, with suggestions for further reading.

How did *kanōn* come to mean 'monochord'? The basic sense of the term is 'straight rod'. In Homer it is used to refer to the staves which maintained the shape of a shield (*Il.* XIII.407). Later it is applied to a variety of objects of which straightness and rigidity were defining factors: curtain-rods, bed-posts, poles, window-bars; also tools used for judging the straightness of things: a carpenter's ruler, straight-edge, measuring-rod, yardstick.[26] In Epicurean philosophy the study of the criteria by which we can discern the truth of things from our senses came to be called *kanonikon*.[27] The musicological application of this vocabulary, (*monochordos*) *kanōn*, '(one-stringed) ruler', was taken by some ancient authors to reflect its philosophical usage. Panaetius the Younger, for instance, assumes that the musical *kanōn* got this name 'because it is a *kritērion* of [i.e. a means of judging] the quantity in the concords according to the sense of hearing'.[28] Ptolemy says that it was named *kanōn* because it straightens (*kanonizein*) 'what is deficient in sense-perception with respect to the truth'.[29]

Gaudentius provides one of the simplest specimens of a division of the instrument. He places it in the midst of a fabulous account of the musical discoveries of Pythagoras:

χορδὴν γὰρ τείνας ἐπὶ κανόνος τινὸς καὶ τὸν κανόνα διελὼν εἰς μέρη ιβ′, πρῶτον μὲν πᾶσαν κρούσας, εἶτα τὸ ἥμισυ αὐτῆς τὸ τῶν ἓξ μερῶν, σύμφωνον ηὕρισκε τὴν πᾶσαν τῷ ἡμίσει κατὰ διὰ πασῶν· . . . ἔπειτα πᾶσαν καὶ τὰ τρία μέρη τῆς πάσης κρούσας τὸ διὰ τεσσάρων ἑώρα σύμφωνον. πᾶσαν δὲ καὶ τὰ δύο μέρη τῆς πάσης κτυπήσας τὴν διὰ πέντε συμφωνίαν εὑρίσκει, καὶ τὰς ἄλλας ὁμοίως.

When he had stretched a string on a *kanōn* and had divided the *kanōn* into 12 parts, he struck the whole string first, and then half of it (the segment corresponding to six parts), and discovered that the whole was concordant with the half at the octave . . . Next he struck the whole string and three-quarters of it, and perceived the concord of the fourth. He also sounded the whole string

[26] LSJ s.v. κανών.

[27] See Striker's discussion of the words *kanōn* and *kritērion* in the context of Epicurean epistemology (1996: 31–2).

[28] ἐπεὶ κριτήριόν ἐστι τοῦ κατὰ τὴν ἀκοὴν ἐν τοῖς συμφώνοις γινομένου πλήθους, Panaetius ap. Porph. *In Harm.* 66.22–3.

[29] τὰ ταῖς αἰσθήσεσιν ἐνδέοντα πρὸς τὴν ἀλήθειαν, Ptol. *Harm.* 5.12–13. See also Oppel 1937, Huffman 2005: 219–20.

and two-thirds of it and discovered the concord of the fifth, and the rest [of the concords] in similar fashion.[30]

The presentation of canonic division in Greek harmonic texts varies greatly according to the aims and assumptions of the author. Some are nearly as skeletal as this one; others are quite detailed. Among the latter are those whose focus is on the location of specific notes, as opposed to intervals, on the *kanōn*, so as to map out a musical scale as a series of numerically designated bridge positions.

Greek harmonic theorists called their scales *systēmata*, 'systems'. The smallest *systēma* was the tetrachord, an attunement of four notes, the highest and lowest of which were always a fourth apart. All the larger systems were assembled out of tetrachords, and so the internal arrangement of these four notes was considered to be of prime importance to all harmonic structures.

The outer notes of a tetrachord were unalterable, but the two inner notes could be tuned in a number of different ways. The relative magnitudes of the three intervals internal to a given tetrachord thus came to be seen as the source of its particular harmonic identity. Three broad categories (called *genē*, 'genera') had been identified by the fourth century BC: enharmonic, chromatic and diatonic. There was no standard attunement to which each of these genera conformed; in fact, a significant amount of Greek harmonic literature is devoted to debates about how the genera should be tuned, and by what criteria each genus should be defined in the first place. These debates are further complicated by the fact that while mathematical harmonicists expressed a tetrachordal attunement (or 'division') as a series of three ratios, Aristoxenus and those who followed his approach expressed it as a series of three intervals quantified in tones and fractions of a tone, two forms of expression not easily comparable with one another.

Among the genera, the chromatic and diatonic admitted the widest variety; there were named sub-categories such as the 'soft chromatic' and the 'tense diatonic', and rival attunements of each. The number of possible tetrachords, therefore, was quite large. A rough idea of the difference between the three genera can be given

[30] Gaud. 11 (341.13–17, 19–22). Translations, unless otherwise noted, are my own.

in the following scheme, which is derived from Aristoxenus. The enharmonic is characterised by two very small intervals, which Aristoxenus treated as quarter-tones, at the bottom of the tetrachord; above it lies a much larger interval, often identified as a ditone (i.e. an interval equal to two tones). In the chromatic, the two lowest intervals were roughly a semitone each, leaving a larger interval of about a tone and a half above it. The lowest interval in the diatonic was approximately a semitone, with two larger intervals above it, each roughly a tone. In both the enharmonic and chromatic the two lowest intervals of the tetrachord could be described jointly as the *pyknon*, so called because they sound 'compressed' (*pyknos*).[31]

One of the consequences of expressing intervals as ratios is that it is impossible to express exact semitones and quarter-tones: there is no ratio of numbers which will divide the ratio of the tone equally.[32] This in itself was sometimes taken to indicate that a true 'semitone' was not a musical interval at all, or even that such an interval could not exist. Mathematical harmonicists were therefore forced to adopt rational approximations of this interval, and one of the consequences of this necessity was that a tetrachord which would have been expressed (in descending order) as 'tone + tone + semitone' by a follower of Aristoxenus was represented instead by the ratios 9:8 × 9:8 × 256:243. (This particular example was sometimes called the 'ditonic' diatonic because of the combined magnitude of its two upper intervals.)

A tetrachordal division, then, meant something slightly different to a canonicist than it did to an Aristoxenian theorist, in that the ratios by which the tetrachord was defined provided a direct connection to the monochord. A scale built out of tetrachords expressed as ratios could be constructed with precision on the instrument simply by calculating bridge positions as units of string length standing to one another in the prescribed ratios. The

[31] Aristox. *El. harm.* 48.29–31. In order to sound 'compressed', these two intervals evidently had to be small enough in combination that they took up less than half of the tetrachord; Aristoxenus insists on this both times he defines the *pyknon* (*El. harm.* 24.11–14, 50.15–19).

[32] That is, $\sqrt{9:8}$ and $\sqrt[4]{9:8}$ are not ratios of integers. (On what Greek authors meant by 'number', see n. 5 above).

following is an example of one of the competing sets of tetrachordal divisions recorded in this way:

enharmonic	*chromatic*	*diatonic*
5:4	6:5	9:8
31:30	25:24	10:9
32:31	16:15	16:15

These are the tetrachords Ptolemy attributes to Didymus (*Harm.* II.14). Ptolemy offers several others, both his own and those of his predecessors.

Tetrachords can be combined in several ways. Most simply two tetrachords can be put together with a tone to form an octave; the tone will either be above them, below them or between them. When two tetrachords are separated by a tone, they are said to be 'disjunct'; when the highest note of one tetrachord is also the lowest note of the one above it, they are said to be 'conjunct'. A two-octave *systēma* will suffice to contain all the possible permutations of a single-octave range, and so the theorists adopted a structure of this size which they called the Greater Perfect System. It could be schematised as follows:

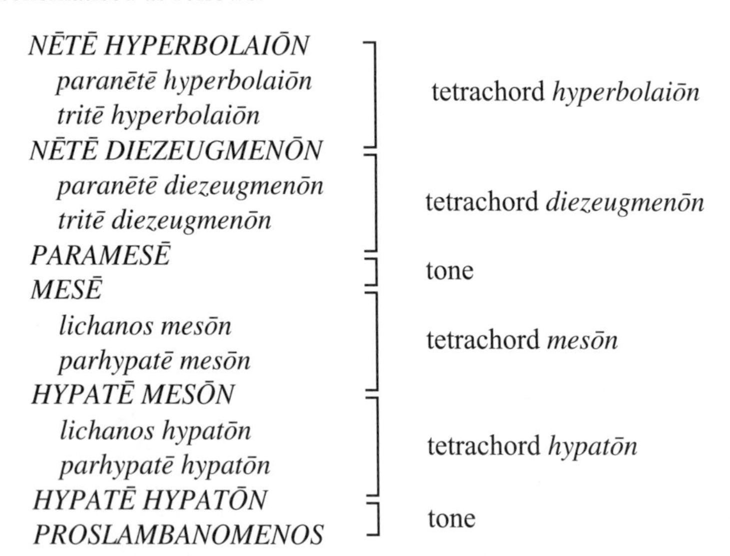

In the scheme the note-names in upper-case letters indicate the 'fixed' notes of the scale (in Greek *hestōtes*, 'standing ⟨notes⟩'), while lower-case letters indicate the so-called 'movable' notes (*kinoumenoi*). The outer notes of each tetrachord, since they were always tuned a fourth apart, were consequently fixed; the inner notes were movable. Many different octave-scales were possible depending on which eight-note segment of the Greater Perfect System one selected, and the genus in which the movable notes of its tetrachords were attuned.

The names of individual notes in this system were of some antiquity, though the system itself is a construct of the fourth century BC. Philolaus, writing in the fifth century, used the Doric forms *hypata*, *messa*, *trita* and *neata* to indicate notes standing in the same relation to one another as the Greater Perfect System's *hypatē mesōn*, *mesē*, *paramesē* and *nētē diezeugmenōn*. In discussions of the earliest theorists, I shall use the note-names of this central octave without the name of their qualifying tetrachord unless otherwise noted (i.e. I shall say '*hypatē* to *nētē*' rather than '*hypatē mesōn* to *nētē diezeugmenōn*', or '*lichanos*' rather than '*lichanos mesōn*').

Greek harmonic theorists also recognised the possibility of adding a tetrachord directly above *mesē* without an intervening (or 'disjunctive') tone. This means replacing the tetrachord 'of disjoined ⟨notes⟩' (*diezeugmenōn*) with a tetrachord 'of conjoined ⟨notes⟩' (*synēmmenōn*). When this is done a smaller *systēma*, called the Lesser Perfect System, is formed, with a range of an octave and a fourth. It could be schematised as follows:[33]

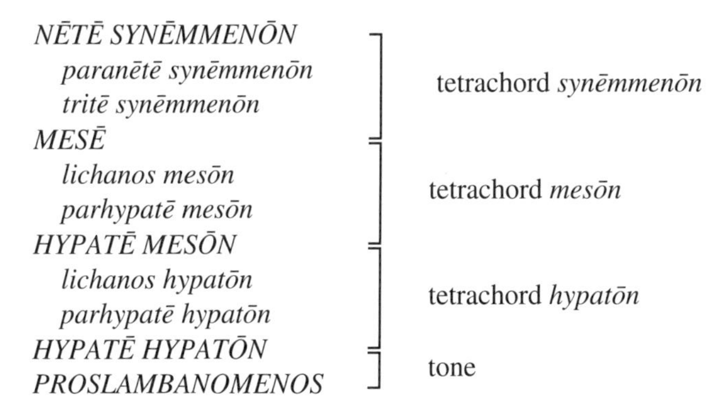

[33] For full accounts of the *systēmata* and their history, see particularly West 1992, chs. 6 and 8; and Barker 1989: 11–27.

I

HEARING NUMBERS, SEEING SOUNDS: THE ROLE OF INSTRUMENTS AND DIAGRAMS IN GREEK HARMONIC SCIENCE

Instruments belong most naturally to music, and diagrams to science. Plato's Socrates confesses to Crito that he has been taking kithara lessons; he also guides a slave-boy toward 'recollection' of mathematical knowledge with the aid of a diagram.[1] But instruments belong to science as well, and diagrams to music. Plato's Timaeus says that to attempt to describe all the motions of the heavenly bodies without looking at models (*mimēmata*) would be 'labour in vain'; Aristoxenus complains that his harmonicist predecessors made diagrams (*diagrammata*) to represent musical space without properly representing musical structures in the process.[2] The Euclidean *Sectio canonis* (*Kanonos katatomē,* 'Division of the monochord'), in presenting a diagrammatic exposition of a musical instrument, occupies a key position at the intersection of music and science. Furthermore, if (as I shall argue) it is to be dated close to the time of Euclid himself, it is one of the two earliest texts which mention the monochord.[3] It was also a central text for later authors writing on mathematical harmonics; Porphyry quotes passages from it in his commentary on Ptolemy's *Harmonics*.[4] Ptolemy, too, made much use of both instruments and diagrams, not only in his *Harmonics* but in his other treatises as well, and in the *Almagest* most of all.

The history of the monochord begins with the *Sectio canonis*, and this book attempts to trace it as far as Ptolemy. The aim

[1] Kithara lessons: *Euthd.* 272c; diagram: *Men.* 82b–85b.

[2] Pl. *Ti.* 40d, Aristox. *El. harm.* 28.1. The term *diagramma* does not always refer to a diagram, but sometimes also to a geometrical proof, as at Arist. *Eth. Nic.* 1112b21 and *An. pr.* 41b14. In the latter case the proof is expressed by reference to a diagram, now lost; the two concepts are closely related in Greek scientific discourse. As David Fowler puts it, 'the word *diagramma* seems, in Plato and Aristotle, to refer ambiguously to either a geometrical figure or a proof' (1999: 33). See LSJ s.v. διάγραμμα I.a–b.

[3] The other is Duris *FGrH* 76F23 (discussed in ch. 2 part 2). On the date and authorship of the *Sectio canonis*, see ch. 3.

[4] Porph. *In Harm.* 90.7–22, 99.1–103.25.

of this chapter is to preface the mainly chronological account of chapters 2–6 with an overview of the monochord's role in Greek scientific discourse and method in comparison with a select number of other devices with which it shares some elements of similarity: mathematical diagrams of the type found both in Euclid's *Elements* and in the *Sectio canonis*, the armillary sphere and parallactic instrument of Ptolemy's *Almagest*, the astronomical and meteorological *parapēgma*, and the numerical tables of Ptolemy's *Almagest* and *Harmonics*.

From this comparison I shall argue first that canonic division, as a scientific activity, is defined (and also limited) by mathematical as well as practical constraints, and that constructing and reading the diagram of the instrument is therefore an activity which is subject to the conventions of mathematics as well as of music. This is perhaps most apparent in the inability of mathematical harmonics ever satisfactorily to accommodate the equal division of the tone. Secondly, I shall argue that the monochord, as a kind of diagrammatic instrument, bridges the gap between the perceptible and intellectual objects of its science in ways that are both similar to and different from those by which the instruments of astronomy perform the same task. This will be evident both in the way Ptolemy introduces instruments in the *Harmonics* and the *Almagest*, and from their relationship to the diagrammatic propositions and the numerical tables in which they have an argumentative role.

I Hearing numbers: arithmetic, geometry and canonic division

Mathematical harmonicists took as one of their starting points the assertion that every musical interval corresponds to a ratio of numbers.[5] I have shown some of the consequences of the adoption of this premise already in the Introduction: irrational relationships were relegated to the world of mere 'sounds' at the earliest stage of musical thinking in the mathematical tradition, and the division

[5] On the important difference between what was meant by *arithmos* and what we now mean by 'number', see Introduction n. 5.

of the tone into two equal semitones was declared to be impossible, because such an interval would be irrational and therefore unmusical.[6] Given any three notes generated by the lengths A, B, C of the same string, where A:C = 9:8 (the ratio of the tone) and A:B::B:C, B (the geometric mean or 'mean proportional' of A and C) will not be a number.[7] In other words, $\sqrt{9:8}$ is irrational: the relationship between these quantities will not be found 'in numbers'. In fact, B will be incommensurable with A and C whenever A:C is an epimoric ratio.[8] Archytas (according to Boethius, *Mus.* III.11) had offered a proof of this in the early fourth century BC; the *Sectio canonis* gives an expanded proof (prop. 3), and then applies it specifically to the epogdoic tone (props. 13, 16):

ἐπιμορίου διαστήματος οὐδεὶς μέσος, οὔτε εἷς οὔτε πλείους, ἀνάλογον ἐμπεσεῖται ἀριθμός. ἔστω γὰρ ἐπιμόριον διάστημα τὸ ΒΓ· ἐλάχιστοι δὲ ἐν τῷ αὐτῷ λόγῳ τοῖς ΒΓ ἔστωσαν οἱ ΔΖ Θ. οὗτοι οὖν ὑπὸ μονάδος μόνης μετροῦνται κοινοῦ μέτρου. ἄφελε ἴσον τῷ Θ τὸν ΗΖ. καὶ ἐπεὶ ἐπιμόριός ἐστιν ὁ ΔΖ τοῦ Θ, ἡ ὑπεροχὴ ὁ ΔΗ κοινὸν μέτρον τοῦ τε ΔΖ καὶ τοῦ Θ ἐστί· μονὰς ἄρα ὁ ΔΗ· οὐκ ἄρα ἐμπεσεῖται εἰς τοὺς ΔΖ Θ μέσος οὐδείς. ἔσται γὰρ ὁ ἐμπίπτων τοῦ ΔΖ ἐλάττων, τοῦ δὲ Θ μείζων, ὥστε τὴν μονάδα διαιρεῖσθαι, ὅπερ ἀδύνατον. οὐκ ἄρα ἐμπεσεῖται εἰς τοὺς ΔΖ Θ τις. ὅσοι δὲ εἰς τοὺς ἐλαχίστους μέσοι ἀνάλογον ἐμπίπτουσι, τοσοῦτοι καὶ εἰς τοὺς τὸν αὐτὸν λόγον ἔχοντας ἀνάλογον ἐμπεσοῦνται. οὐδεὶς δὲ εἰς τοὺς ΔΖ Θ ἐμπεσεῖται, οὐδὲ εἰς τοὺς ΒΓ ἐμπεσεῖται. (ed. Jan with small amendments)

No mean number, neither one nor many, will fall proportionally within an epimoric interval. For let Β:Γ be an epimoric interval. And let ΔΖ and Θ be the smallest numbers in the same ratio as Β:Γ. Thus these are measured by the unit alone as a common measure. Take away ΗΖ equal to Θ. And since ΔΖ:Θ is epimoric, the excess ΔΗ is the common measure of both ΔΖ and Θ. ΔΗ is therefore the unit; thus no mean will fall between ΔΖ and Θ. For the number which falls between them will be smaller than ΔΖ, but greater than Θ, so as to divide the unit, which is impossible. No number will thus fall between ΔΖ and Θ. And as many means as fall proportionally between the smallest terms, this many also will fall between terms which have the same ratio [*El.* VIII.8]. But none will fall between ΔΖ and Θ, nor between Β and Γ. (prop. 3)

[6] See e.g. Ptolemy's summary of the Pythagorean position at *Harm.* 12.24–7. See also Barbera 1977, Barker 1991.
[7] That is, B will not be an *arithmos*.
[8] On the categories of ratio relevant to mathematical harmonics (including epimorics), see the Introduction.

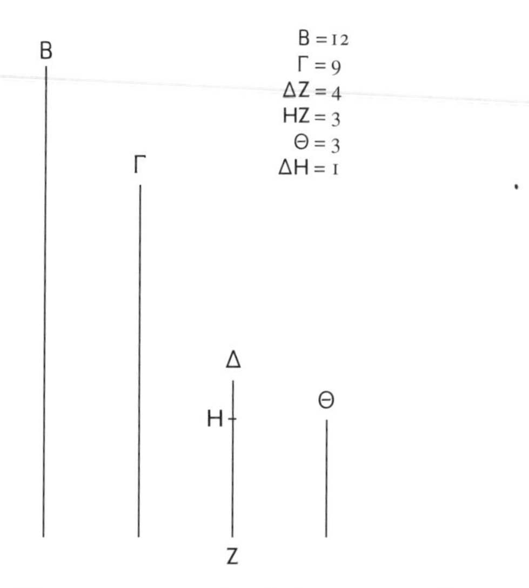

Fig. 1.1 Diagram accompanying *Sect. can.* prop. 3.*

λοιπὸν δὴ περὶ τοῦ τονιαίου διαστήματος διελθεῖν, ὅτι ἐστὶν ἐπόγδοον. ἐμάθομεν γάρ, ὅτι ἐὰν ἀπὸ ἡμιολίου διαστήματος ἐπίτριτον διάστημα ἀφαιρεθῇ, τὸ λοιπὸν καταλείπεται ἐπόγδοον. ἐὰν δὲ ἀπὸ τοῦ διὰ πέντε τὸ διὰ τεσσάρων ἀφαιρεθῇ, τὸ λοιπὸν τονιαῖόν ἐστι διάστημα· τὸ ἄρα τονιαῖον διάστημά ἐστιν ἐπόγδοον.

It remains, then, to give an account of the interval of the tone, that it is epogdoic [9:8]. For we learned that if an epitritic [4:3] interval is taken away from a hemiolic [3:2] interval, the remainder is epogdoic. And if the fourth is taken away from the fifth, the remainder is the interval of the tone; the interval of the tone is therefore epogdoic. (prop. 13)

ὁ τόνος οὐ διαιρεθήσεται εἰς δύο ἴσα οὔτε εἰς πλείω. ἐδείχθη γὰρ ὢν ἐπιμόριος· ἐπιμορίου δὲ διαστήματος μέσοι οὔτε πλείους οὔτε εἷς ἀνάλογον ἐμπίπτουσιν. οὔκ ἄρα διαιρεθήσεται ὁ τόνος εἰς ἴσα.

The tone will not be divided into two or more equal intervals. For it has been demonstrated to be epimoric [in prop. 13, since the epogdoic ratio is epimoric in form]; and neither many means nor one fall proportionally within an epimoric interval. Therefore the tone will not be divided into equal parts. (prop. 16)

* The presentation of the diagram in those MSS which contain it is varied, as is the designation of its elements in the text. The most important divergence between MS traditions is the presence or absence of the letter H. The issues raised by these differences will concern us more directly below (nn. 30 and 33); for a full account, see Barbera 1991: 260–91. The diagram above is based on that of Jan's edition (1895: 152), whose text I quote.

These three related propositions appear to constitute an attempt to challenge Aristoxenus' treatment of the tone as divisible into two, three or four equal parts:

τόνος δ' ἐστὶν ᾧ τὸ διὰ πέντε τοῦ διὰ τεσσάρων μεῖζον· τὸ δὲ διὰ τεσσάρων δύο τόνων καὶ ἡμίσεος. τῶν δὲ τοῦ τόνου μερῶν μελῳδεῖται τὸ ἥμισυ, ὃ καλεῖται ἡμιτόνιον, καὶ τὸ τρίτον μέρος, ὃ καλεῖται δίεσις χρωματικὴ ἐλαχίστη, καὶ τὸ τέταρτον, ὃ καλεῖται δίεσις ἐναρμόνιος ἐλαχίστη· τούτου δ' ἔλαττον οὐδὲν μελῳδεῖται διάστημα.

The tone is that by which the fifth is greater than the fourth; the fourth contains two and a half tones. Of the parts of the tone which are melodic,[9] there is the half, which is called the semitone, and the third, which is called the smallest chromatic *diesis*, and the quarter, which is called the smallest enharmonic *diesis*; there is no melodic interval smaller than this.[10]

These statements in book II parallel a similar set of claims in book I:

ἔστι δὴ τόνος ἡ τῶν πρώτων συμφώνων κατὰ μέγεθος διαφορά. διαιρείσθω δ' εἰς τρεῖς διαιρέσεις· μελῳδείσθω γὰρ αὐτοῦ τό τε ἥμισυ καὶ τὸ τρίτον μέρος καὶ ⟨τὸ⟩ τέταρτον· τὰ δὲ τούτων ἐλάττονα διαστήματα πάντα ἔστω ἀμελῴδητα. καλείσθω δὲ τὸ μὲν ἐλάχιστον δίεσις ἐναρμόνιος ἐλαχίστη, τὸ δ' ἐχόμενον δίεσις χρωματικὴ ἐλαχίστη, τὸ δὲ μέγιστον ἡμιτόνιον.

The tone is the difference in magnitude between the first concords [i.e. the fourth and fifth, which are 'first' in ascending order of magnitude]. It may be divided in three ways, since its half, third and quarter may be taken as melodic; but all intervals smaller than these should be considered unmelodic. Let the smallest of them be called the smallest enharmonic *diesis*, and the next the smallest chromatic *diesis*, and the largest the semitone.[11]

It was Aristoxenus' rejection of the analysis of musical intervals by ratios of numbers (a kind of analysis which, he said, falls completely outside the field of harmonics and is entirely contrary to what is perceived, 32.24–9) that enabled him to claim that the tone can be divided equally. The objection to the equal division of the tone, in the *Sectio* and elsewhere, is framed in terms of arithmetic,

[9] That is, intervals which can legitimately be used as melodic steps (Barker 1989: 140 n. 81), as opposed to merely theoretical intervals, to which there is no minimal limit (*El. harm.* 46.15–18).
[10] *El. harm.* 45.34–46.8. [11] *El. harm.* 21.21–31.

not in terms of perception ('neither many means nor one fall proportionally within an epimoric interval'); but for Aristoxenus, the existence and melodic use of equal quarter-, third- and half-tones was one of the things that are 'perceptually evident to those experienced in music' (32.31).[12] Aristoxenus' fourth will thus contain five equal semitones and ten equal quarter-tones.

Inconsistent definitions

Central to the controversy between the arithmetical approach of the *Sectio canonis* and the perception-based approach of Aristoxenus is the question of whether musical intervals must be rational (in the sense, 'expressible in ratios of numbers'), but this question masks a wider difference between the two. If both Aristoxenus and the author of the *Sectio* were using the terms 'fourth', 'fifth' and 'octave' to refer to precisely the same perceived intervals,[13] then a very real objection may be raised against Aristoxenus' definitions. For even if we reject the mathematical theorists' assumption that all musical intervals are rational, and accept with Aristoxenus that the tone may be divided equally, the two definitions of the semitone presented at 45.34–46.8 (quoted above) are at odds with one

[12] Aristoxenus treats this as axiomatic: cf., for instance, 32.31–33.1 with 43.30–44.20. For 'perceptually evident' as a translation of φαινομένας (32.31), see Barker 1989: 150 n. 12. Aristoxenus is careful to separate existence from melodic use where the smallest parts of the tone are concerned: third-tone intervals exist in music, but no melody will contain three third-tones in succession (46.9–16; cf. 28.6–17).

[13] There is no way to be certain, of course, but it is probable. Aristoxenus insists again and again on the fact that in matters of accurate tuning, trained perception can be the only reliable judge (see e.g. 43.76–9). Any kind of 'tempering' of concords (such as is used on modern keyboard instruments) to generate perfectly equal semitones requires tuning the pure concords first, and then sharpening or flattening. Aristoxenus implies the existence of a related process in his (disapproving) account of the apparently current fashion of 'sweetening' (*glukainein*) the enharmonic *lichanos* by sharpening it slightly, so that it lay less than a true ditone below *mesē* (23.11–22). The ditone *lichanos* can be found in four steps starting at *mesē* by tuning down a fifth, up a fourth, down a fifth, and up a fourth; tuning the 'sweetened' *lichanos* requires a further adjustment. But the fact that Aristoxenus draws such attention to the sharpened *lichanos*, along with the fact that the tempering itself requires an admission of pure intervals – which can be identified (whether inside or outside the proper realm of harmonics) in terms of ratios of string lengths – makes it hard to imagine that there was not fundamental agreement about what perceived magnitudes the names of the basic concords indicated. That anyone with a good musical ear can recognise these three intervals when they are properly tuned was one of the few items which were not contentious in ancient harmonics.

another. If 'the fourth contains two and a half tones', then the five equal 'semitones' which make up the fourth will each be smaller than the 'semitone' which is half of the tone which is 'that by which the fifth is greater than the fourth' (i.e. $\sqrt[5]{(4:3)} < \sqrt{(3:2 \div 4:3)}$). In this case two of the five equal 'semitones' will combine to form a 'tone' smaller than the tone which is the difference between the fifth and the fourth (i.e. $(\sqrt[5]{(4:3)})^2 < (3:2 \div 4:3)$). But if 'the difference in magnitude between the first concords' is taken as a constant measure of the tone, then the leftover 'half' designated by the statement that 'the fourth contains two and a *half* tones' will be less than half of a tone (i.e. $4:3 \div (3:2 \div 4:3)^2 < \sqrt{(3:2 \div 4:3)}$). To put it another way, if the 'semitone' which is half of 'the difference in magnitude between the first concords' is compounded fivefold, an interval greater than a pure fourth will be formed (i.e. $(\sqrt{(3:2 \div 4:3)})^5 > 4:3$).

Aristoxenus presents his definitions as though they identify a single semitone, but in fact they do not; his statements require at least two of a set of three different 'semitones', namely $\sqrt[5]{(4:3)}$, $\sqrt{(3:2 \div 4:3)}$, and $4:3 \div (3:2 \div 4:3)^2$. This last 'semitone', the interval which remains when two tones have been taken from a fourth, is used by Aristoxenus elsewhere (it is, for example, the interval between his enharmonic *lichanos* and *hypatē*, 22.26–23.11). Unlike the first two, it is rational: it is the 256:243 *leimma* of the mathematical theorists (so named because it is 'left over' when the parallel operation is performed in arithmetic: $4:3 \div (9:8)^2 = 256:243$), and it is the smallest of the three: five *leimmata* combine to form an interval smaller than a fourth (i.e. $(256:243)^5 < 4:3$). The magnitudes of the three 'semitones' admitted by Aristoxenus can be compared as follows:

$$\sqrt{(9:8)} > \sqrt[5]{(4:3)} > 256:243$$

The differences between them are so small as to be perceptible only under optimal conditions (as for example when the three intervals are played accurately one after another from the same starting pitch), and nearly imperceptible when located in different positions within the scale and heard in the context of an unaccompanied melody. In short, Aristoxenus had no reason to doubt that he had defined the semitone consistently, and without admitting irrational intervals the mathematical theorists had no means of proving otherwise in terms which Aristoxenus would have recognised.

For several reasons besides his interest in music (his Tarentine origins, his education in the Lyceum, his authorship of many works on the Pythagoreans) Aristoxenus may have been familiar with Archytas' proof that an epimoric ratio cannot be divided into equal parts.[14] If he could have been presented with *Sectio canonis* propositions 3, 13 and 16 (which, as we have seen above, jointly apply Archytas' proof to musical intervals and to the epogdoic tone specifically), Aristoxenus would certainly have agreed that an epimoric ratio cannot receive a mean proportional, but would have denied the relevance of this proposition to harmonics. In fact, the very term 'epimoric interval' (*epimorion diastēma*, *Sect. can.* 3, 16) transgresses the boundary between sciences, in attaching to something perceptible (a *diastēma*) a quality (*epimorion*) which can only apply to objects which exist outside the realm of sense-perception (ratios). The term can only be considered valid by those who hold that intervals and ratios are to be so closely associated that the limits of arithmetic constitute the limits of the melodic. Aristoxenus held no such view.

Modern equal temperament meets the need for an equal division of the tone into universally constant semitones in this way:

$$\text{octave} = 2:1 = 1200\ \text{cents}^{15}$$
$$\text{tempered fifth} = (\sqrt[12]{(2:1)})^7 = 700\ \text{cents} < 3:2$$
$$\text{tempered fourth} = (\sqrt[12]{(2:1)})^5 = 500\ \text{cents} > 4:3$$
$$\text{tempered tone} = (\sqrt[12]{(2:1)})^2 = 200\ \text{cents} < 9:8$$
$$\text{tempered semitone} = \sqrt[12]{(2:1)} = 100\ \text{cents} > 256:243$$

[14] Boeth. *Mus.* III.11 = A19 Huffman. For a discussion of the extent of Aristoxenus' awareness of Pythagorean musical theory, see Gibson 2005: 15.

[15] A cent is a logarithmic measure of pitch difference; one cent = $\frac{1}{100}$ of a tempered semitone = $\frac{1}{1200}$ of an octave = $\sqrt[1200]{(2:1)}$.

This tempered semitone is different from the three we have already encountered:

$$\sqrt{(9:8)} > \sqrt[12]{(2:1)} > \sqrt[5]{(4:3)} > 256:243$$

From the summary above, it will be observed that the intervals on a properly tuned modern piano reconcile Aristoxenus' two statements on the tone and semitone:

'The tone is that by which the fifth is greater than the fourth':	700 − 500 = 200 cents
'the fourth contains two and a half tones':	200 × 2.5 = 500 cents

No surviving ancient account of scales, however, will remedy the discrepancy contained in these definitions.[16]

Numbers and canonic division

The admission of the half, quarter and third of a tone as musical intervals in the first place is another matter; here the issue is between arithmetic and geometry.[17] Because the first apodeictic proof of the impossibility of the equal division of the tone comes to us in the text which also offers the earliest surviving canonic division, we are led to think of canonic division, both the procedure and the basic assumptions behind it, as inimical to irrational relationships between magnitudes in music.[18] The two are presented as integrated parts of the same package: canonic division requires

[16] See n. 13 above.

[17] Arithmetic is by definition the science of number (*arithmos*), and therefore cannot treat quantities that cannot be expressed as numbers, like that of the circumference of a circle in relation to its diameter. Such quantities require geometrical expression, and this regularly involved diagrams. But numbers can also be represented by lines (i.e. geometrically) even when they do not need to be: this kind of geometrical visualisation of numbers arises from mathematical convention rather than from mathematical necessity. The geometrical expression of arithmetical propositions was common at least as early as Euclid. Therefore, as we shall see, it would be naïve to assume that a proof with a diagram is by consequence a geometrical proof, or that the mere use of lines (whether actually drawn or simply indicated in the text as the distance between two lettered points) to represent numbers moves a proof from the realm of arithmetic to the realm of geometry. This issue will concern us directly below (see 'Diagrams and canonic division'.)

[18] An 'apodeictic' proof is one whose method follows the formal demonstrative logic of Aristotle's *Analytics*, in which true, necessary and universal conclusions are arrived at by syllogisms based on first principles (*archai*) which are themselves true, necessary and universal, as well as being immediate (i.e. they are plainly evident to anyone, and cannot be demonstrated; these *archai* are called *axiōmata*, 'axioms'). See Aristotle's definitions

whole-number ratios:[19]

$$\begin{aligned} \text{octave} &= 2:1 \\ \text{fifth} &= 3:2 \\ \text{fourth} &= 4:3 \\ \text{double octave} &= 4:1 \\ \text{octave + fifth} &= 3:1 \\ \text{octave + fourth} &= 8:3 \\ \text{tone} &= 9:8 \end{aligned}$$

The package itself belongs to arithmetic rather than to geometry, for only in arithmetic can the division of the unit be ruled out with the standard *reductio ad absurdum* argumentation formula ὅπερ ἀδύνατον, 'which is impossible'.[20] Because the proof of the impossibility of the equal division of epimoric intervals is predicated upon the indivisibility of the unit, the use of the terminology both of 'unit' (*monas*) and of 'interval' (*diastēma*) in the *Sectio* deserves closer attention.

The definitions of 'unit' and 'number', with which Euclid opens *Elements* VII, show the primacy of the concept of unity to arithmetic:[21]

1 μονάς ἐστιν, καθ' ἣν ἕκαστον τῶν ὄντων ἓν λέγεται.
2 ἀριθμὸς δὲ τὸ ἐκ μονάδων συγκείμενον πλῆθος.

1 A unit is that by which everything that exists is said to be one.
2 And a number is a plurality composed of units.[22]

It will also be apparent that the indivisibility of the unit is built into its very definition; it is the atom of arithmetic.[23] Furthermore,

at *An. post.* I.2 (esp. 72a17); at *Metaph.* B 2 (997a7) he distinguishes axioms in particular from first principles generally. (The word *axiōma* can also mean 'philosophical doctrine', as at Arist. *Metaph.* 1001b7, or 'logical proposition', as at Chrysipp. *SVF* 2.53: LSJ s.v. ἀξίωμα II.2.) See also Barnes 1975: 65, 79. For an account of Aristotelian *apodeixis* see Mendell 1998.

[19] This *sine qua non* of canonic division unifies the different approaches of e.g. Panaetius the Younger (ap. Porph. *In Harm.* 66.20–30), Thrasyllus (ap. Theo. Sm. 89.9–23), Nicomachus (*Harm.* 254.11–255.3), Adrastus (ap. Theo. Sm. 57.11–58.12) and Ptolemy (*Harm.* I.8, 18.22–19.15).

[20] *Sect. can.* prop. 3, quoted above; see Netz 1999a: 140, formula 61.

[21] The arithmetical books of the *Elements* are V (on proportion theory) and VII–IX (on number theory).

[22] *El.* VII def. 1–2.

[23] The point (*sēmeion*) plays a similar role in geometry; cf. *El.* I def. 1: σημεῖόν ἐστιν, οὗ μέρος οὐθέν, 'A point is that of which there is no part.'

all 265 occurrences of the word *monas* in the *Elements* are in books VII–X; the unit is an overwhelmingly arithmetical notion in Euclidean mathematical vocabulary. By contrast, the word *diastēma*, which is regularly used to mean 'interval' in musical texts, is used in the *Elements* to indicate the 'distance' between points in geometrical constructions;[24] the word occurs thirty times in the *Elements*, exclusively in the geometrical books (specifically, I–IV and XI–XII), and only in the dative singular.[25] The term *epimorios*, 'epimoric', which one might expect to encounter in book V (on proportion theory), never occurs in the *Elements*. Mathematical harmonics requires the language of both arithmetic and geometry, because it uses physical distances (e.g. of a vibrating string) as the measurable intermediary between relationships between sounds and relationships between numbers. Objects of sight are used to demonstrate the connection between objects of hearing and objects of thought. It is because mathematical harmonics attached such logical primacy to the numerical relationships that the arithmetical objection to the equal division of the tone could be so uniformly sustained.[26]

Diagrams and canonic division

The *Sectio canonis* also shares an affinity with the arithmetical books of the *Elements* in its use of diagrams.[27] In both cases,

[24] As for example the radius of a circle: κέντρῳ μὲν τῷ Α διαστήματι δὲ τῷ ΑΒ κύκλος γεγράφθω ὁ ΒΓΔ, 'Let the circle ΒΓΔ be described with centre Α and radius ΑΒ', *El.* I.1; see Netz 1999a: 110. On the meaning of the term in spherical constructions, see Sidoli 2004.

[25] Word counts are taken from the *TLG*. Fowler and Taisbak also note the exclusively dative use of *diastēma* in the *Elements* (1999: 361). Their focus is on geometry, but their broader conclusion that 'in other contexts it [sc. *diastēma*] means "distance", though geometrically, as a line, and never numerically' (363) overlooks harmonics as a relevant branch of mathematics. Their 'never numerically' is neatly disproved by the phrase *epimorion diastēma* in *Sect. can.* 3 and 16, since a non-numerical entity cannot be epimoric, and since one of the primary postulates of the *Sectio* is that *all* musical distances are numerical.

[26] 'Mathematical harmonics' was from its first definition by Aristotle arithmetical rather than geometrical, and defined in opposition to 'hearing-based harmonics' (the relevant passages are discussed in the Introduction).

[27] The only book-length study of diagrams in Greek mathematics is Netz 1999a, a work to which my argument is indebted at many points. For a treatment of the epistemology of diagrams beyond the Greek context, see Norman 2006. Barbera (1991: 260–91)

diagrams consist only of straight lines; when there are more than one, they are always parallel; they are sometimes divided at points specified in the proposition (as in *Sect. can.* prop. 3 above). No angles, arcs or polygons (geometrical entities) are ever constructed on them. In short, lines are used exclusively to represent either individual numbers or relationships between numbers. In the case of the former, a single letter will designate the line's identity (as Β and Γ do in fig. 1.1); in this case a ratio between quantities is given in the text by juxtaposition of the two corresponding letters: ἔστω γὰρ ἐπιμόριον διάστημα τὸ ΒΓ, 'for let Β:Γ be an epimoric interval'.[28] In the second case – that is, when a single line is used to indicate a relationship between numbers – the line is lettered at either end and also, often, at the point(s) where it is cut; in this case a line segment indicating a number must be expressed by juxtaposition of the two corresponding lettered *points* (rather than lettered *lines*, as in the previous case): μονὰς ἄρα ὁ ΔΗ, 'ΔΗ is therefore the unit'. A geometrical distance is identified with an arithmetical entity. Relationships between quantities are thus indicated by listing two-letter coordinates in succession: ἡ ὑπεροχὴ ὁ ΔΗ κοινὸν μέτρον τοῦ τε ΔΖ καὶ τοῦ Θ ἐστί, 'the excess ΔΗ is the common measure of both ΔΖ and Θ'. This last example illustrates the dual role of the designation-letter in arithmetical propositions; the reader needs the diagrammatic context (whether or not it is actually drawn) in order to know that while the single letter Δ is a point, the single letter Θ is a number, and that while the letter-pair ΔΖ is a number, the letter-pair ΒΓ is a ratio of numbers.[29] The letter thus stands for either an arithmetical object (e.g. Θ) or a geometrical one (e.g. Δ).

Diagrams in arithmetical proofs are therefore hermaphroditic: their lettering requires both arithmetical and geometrical

includes in his study of the *Sectio canonis* a critical edition of the diagrams in the three MS traditions (direct, indirect via Porphyry, indirect via Boethius) from a collation of 100 MSS.

[28] Or, to use Netz's formulation of this type of sentence (1999b: 284), which has the advantage of drawing attention to the shift from the general statement of the enunciation to the specific case in which it will be proven (285), 'For let there be an epimoric interval, ⟨namely⟩ Β:Γ.'

[29] This practice is common in the arithmetical books of the *Elements*: compare, for example, the diagram of *El.* IX.20, and Netz's discussion of it in answering the preliminary question 'what do letters stand for?' (1999a: 69).

interpretation. The physical diagram is rarely essential to the proof – the one which accompanies *Sectio canonis* proposition 3 could easily have been reconstructed from the text even if it had been missing in the MSS[30] – but the *idea* of the diagram is essential: the proof is diagrammatic in its exposition.[31] The visible geometry of line segments is used to prove an argument in arithmetic.[32] But a diagrammatic exposition of an arithmetical proposition does not *ipso facto* transform it into a geometrical proposition. The critical role of the indivisibility of the unit in *Sectio canonis* proposition 3 proves this: the *line segment* ΔH is divisible into equal parts (*El.* I.10); the *unit* ΔH is indivisible (*El.* VII def. 1). That the latter trumps the former in this context shows that the text of the proposition denies the reader a geometrical interpretation of the diagram:

[30] One family of MSS, in fact, contains no diagrams (Barbera's **za**, 1991: 261 n. 1 with 111 (*conspectus codicum et notarum*) and 79 (*stemma codicum*)), but as Barbera notes (40) its text also breaks off at prop. 10, and so does not contain the division of the monochord (props. 19–20), in which a line will represent a pluckable string and therefore require the designation of bridge positions by means of lettered *points*. Faint traces of original diagrams in another MS (Venetus Marcianus gr. app. cl. VI/3 = M Jan = Mm Barbera) descended from the same hyparchetype (γ) as Barbera's **za** show horizontal, rather than vertical lines (Jan 1895: 150). To this MS a later hand added marginal diagrams with vertical lines, in most cases of equal length; Barbera's edition of the diagrams relies primarily on these. Jan's diagrams are based on those of later, derivative MSS, in which lines representing different numbers differ in length; Barbera follows this practice in his 'translation' of the diagrams (261 n. 1). He also notes that in an important MS (Vaticanus gr. 2338 = Vc) derived from a different hyparchetype (δ) the diagrams are supplied by a third hand, but that 'Vc³ may be reconstructing or retracing some diagrams already extant in the manuscript' (261 n. 2). Whether the treatise in its earliest form possessed diagrams of some kind or – as Barbera concludes (15, 40) – not, its use of designation-letters nevertheless presupposes the type of diagrammatic presentation described above, unless we follow Barbera in treating props. 19–20 and the designation-letters representing points rather than numbers in propositions 3 and 6 as late additions to the treatise (on which see n. 33 below). Compare De Young's description of the very similar variations in MS diagrams in the Arabic transmission of the arithmetical books (V and VII–IX) of Euclid's *Elements* (2005: 166 (a general conspectus), 168–75 (a list of the characteristics of diagrams in fifteen MSS)). The issue of the transmission of diagrams will be discussed further in part 2 of this chapter.

[31] See n. 2 above.

[32] 'There is one exception to the use of diagrams – the *di' arithmōn*, "the method using numbers". While in general arithmetical problems are proved in Greek mathematics by geometrical means, using a diagram, sometimes arithmetical problems are tackled as arithmetical. Significantly, even this is explicitly set up as an exception to a well-defined rule, the *dia grammōn*, "the method using lines". The diagram is seen as the rule from which deviations may (very rarely) occur' (Netz 1999a: 14). For the use of διὰ τῶν γραμμῶν to mean 'by rigorous methods' from Hipparchus on, see Neugebauer 1975: 771 n. 1; cf. also ἐν ταῖς γραμμαῖς, 'a phrase that canonically denotes rigorous geometrical proofs involving diagrams' (Acerbi 2008: 122 n. 5).

arithmetic can only recognise ΔH as a numerable quantity, not as a distance. So while André Barbera was right to see the *Sectio* as an arithmetical treatise, he overinterpreted the mathematical effect of what he took to be Byzantine emendations when he claimed that 'although these emendations might initially seem to be minor, those introduced into the texts of Propositions 3 and 6 transform the entire text from arithmetic to geometry'.[33] The designation of ΔH as the unit and the statement of its indivisibility, present even in Barbera's 'geometric version' of this proposition (263), rule out his interpretation: the proposition is no more geometrical than those of the arithmetical books of the *Elements*. The reader could as easily make the numbers visible in a purely arithmetical way by following the argument on an abacus, except that the dual use of letters in the proof only makes sense if the visualisation is geometrical. Letters like Θ in *Sectio canonis* proposition 3 can stand for pebbles, but letters like H cannot.[34] This is geometricised arithmetic – but it is still arithmetic. The 'assimilation of arithmetic to geometry' (Netz 1999a: 69) is a feature of Euclidean mathematics, and thus a feature also of the presentation of canonic division in mathematical texts.[35]

[33] 1991:31. See also his discussion of 'arithmetic and geometric versions of the treatise' at 40–4. The principal difference between the two versions of the text of prop. 3 is the absence of the designation-letter H, the only geometrical entity in the proposition, from the 'arithmetic version'. The quantity ΔZ will in this case mean the sum of two integers Δ and Z, of which Δ represents the unit. The differences in prop. 6 are similar. Barbera's decision to prefer these 'arithmetic' readings and interpret the 'geometric' ones as the result of Byzantine emendations was based in part on the witness of Boethius, despite the fact that Porphyry's quotation of these two propositions includes the lettered points (Porphyry uses the designation-letter E rather than H in his quotation of prop. 3). It is essential to recognise that the lack of geometry in propositions 3 and 6 has nothing to do with which version one reads: both are proofs in arithmetic no matter how we read them.

[34] For a similar and representative example from Euclid's *Elements*, compare VII.2 (a problem on whose solution the author of the *Sectio* relies in prop. 9). The first two numbers set out are designated by two letter-pairs: ἔστωσαν οἱ δοθέντες δύο ἀριθμοὶ μὴ πρῶτοι πρὸς ἀλλήλους οἱ AB, ΓΔ, 'let AB and ΓΔ be the two given numbers not prime to one another'. Later four more numbers are introduced, designated by the letter-pairs BE and EA, ΔZ and ZΓ. Finally, we are given a number designated by a single letter, H. It will be obvious that the same possibilities for diagrammatic representation of this proposition exist as for *Sect. can.* prop. 3: no more, and no fewer.

[35] Netz's discussion of diagrams in Greek arithmetic (1999a: 267–9) is useful here; for the basic practices of diagram-lettering, see 68–71. He also points out (63) that while mathematical pebble-manipulations are attested in the fourth century BC (Arist. *Metaph.* 1092b11–12, *Ph.* 203a13–15), dot-representations based on such practices are not found

The distinctly arithmetical context of the *Sectio canonis* can be readily appreciated when we consider the combination of vocabulary and diagram presented in proposition 3. The *ekthesis* (setting out) begins with the construction of the epimoric interval in which the impossibility of inserting a mean proportional will be demonstrated: ἔστω γὰρ ἐπιμόριον διάστημα τὸ ΒΓ, 'For let Β:Γ be an epimoric interval.'[36] It is the adjective *epimorion*, the context already generated by the *protasis* (enunciation), and the accompanying diagram which enable the reader to understand ΒΓ as a ratio of numbers, not a distance between two points, as it is for example at Euclid *Optics* proposition 4:

ἔστω ἴσα διαστήματα ἐπὶ μιᾶς εὐθείας τὰ ΑΒ, ΒΓ, ΓΔ.

Let ΑΒ, ΒΓ and ΓΔ be equal distances on a single straight line.

To recapitulate: it is only in arithmetic that a Greek mathematician can speak of the impossibility of dividing the unit, and the *Sectio canonis* is an explicitly arithmetical text both in its use of vocabulary and in its use of diagrams. In geometry, by contrast, incommensurable distances and irrational proportions are at home. Consider Euclid, *Elements* VI.13, in which it is shown how to construct a mean proportional to two given straight lines. The problem's relevance to the question of the equal division of the tone can be seen more clearly when we modify the accompanying diagram so that the two particular straight lines specified in the *ekthesis* are in epogdoic (9:8), and thus epimoric, ratio, the latter being the key condition of *Sectio canonis* proposition 3.

δύο δοθεισῶν εὐθειῶν μέσην ἀνάλογον προσευρεῖν. ἔστωσαν αἱ δοθεῖσαι δύο εὐθεῖαι αἱ ΑΒ, ΒΓ· δεῖ δὴ τῶν ΑΒ, ΒΓ μέσην ἀνάλογον προσευρεῖν. κείσθωσαν ἐπ' εὐθείας, καὶ γεγράφθω ἐπὶ τῆς ΑΓ ἡμικύκλιον τὸ ΑΔΓ, καὶ ἤχθω ἀπὸ τοῦ Β

in Greek arithmetical texts before the first century AD. This, he argues (64), is a result of the fact that pebbles must be 'moved and added' if they are to be used in an arithmetical proof. The geometrical diagram thus holds an advantage over the abacus, even in arithmetic, in that as a static object it can be transferred from the oral medium to the written one and retain its argumentative force.

[36] On the parts of a mathematical proposition (*protasis*, 'enunciation'; *ekthesis*, 'setting out'; *diorismos*, 'definition of goal'; *kataskeuē*, 'construction'; *apodeixis*, 'proof'; *sumperasma*, 'conclusion'), see Proclus, *In Euc.* 203.1–207.25; Netz 1999a: 9–11 (a summary) and 1999b (a more detailed treatment). The dating of these terms is controversial.

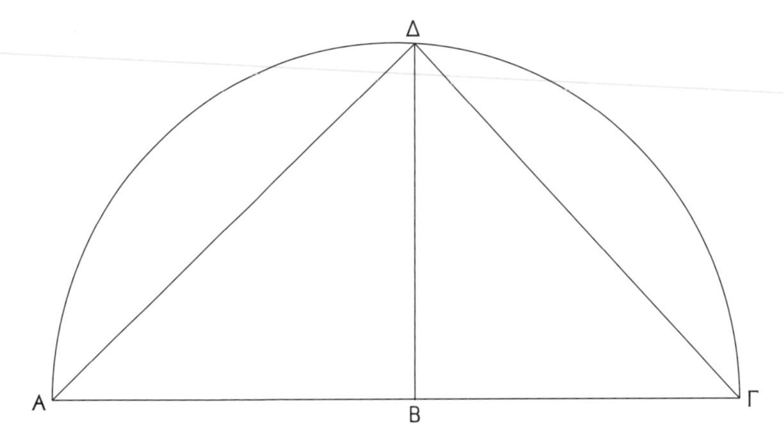

Fig. 1.2 Diagram accompanying Euc. *El.* VI.13 (after Heiberg 1883–8, vol. II).

σημείου τῇ ΑΓ εὐθείᾳ πρὸς ὀρθὰς ἡ ΒΔ, καὶ ἐπεζεύχθωσαν αἱ ΑΔ, ΔΓ. ἐπεὶ ἐν ἡμικυκλίῳ γωνία ἐστὶν ἡ ὑπὸ ΑΔΓ, ὀρθή ἐστιν. καὶ ἐπεὶ ἐν ὀρθογωνίῳ τριγώνῳ τῷ ΑΔΓ ἀπὸ τῆς ὀρθῆς γωνίας ἐπὶ τὴν βάσιν κάθετος ἦκται ἡ ΔΒ, ἡ ΔΒ ἄρα τῶν τῆς βάσεως τμημάτων τῶν ΑΒ, ΒΓ μέση ἀνάλογόν ἐστιν. δύο ἄρα δοθεισῶν εὐθειῶν τῶν ΑΒ, ΒΓ μέση ἀνάλογον προσεύρηται ἡ ΔΒ· ὅπερ ἔδει ποιῆσαι.

To find a mean proportional to two given straight lines. Let AB and BΓ be the two given straight lines; it is necessary then to find a mean proportional to AB and BΓ. Let them be placed on a straight line, and let a semicircle AΔΓ be described on AΓ, and let the ⟨straight line⟩ BΔ be drawn from point B at right angles to the straight line AΓ, and let AΔ and ΔΓ be joined. Since AΔΓ is an angle in a semicircle, it is right [*El.* III.31]. And since in the right-angled triangle AΔΓ, ΔB has been drawn from the right angle perpendicular to the base, ΔB is therefore a mean proportional to the segments AB and BΓ of the base. Thus to the two given straight lines AB and BΓ a mean proportional, ΔB, has been found; which is what it was necessary to construct [QEF].

To show how this simple proposition might have been applied to harmonics, I have placed point B such that AB:BΓ = 9:8; the interval of the irrational semitone therefore occurs twice:

$$\mathrm{AB}:\Delta\mathrm{B} = \sqrt{9:8}$$
$$\Delta\mathrm{B}:\mathrm{B}\Gamma = \sqrt{9:8}$$

It will be evident that this geometry can be applied to the diagram accompanying *Sectio canonis* proposition 3 (fig. 1.1) in order to find a mean proportional to the terms of an epimoric ratio. It is only the arithmetical context of that proposition which constrains

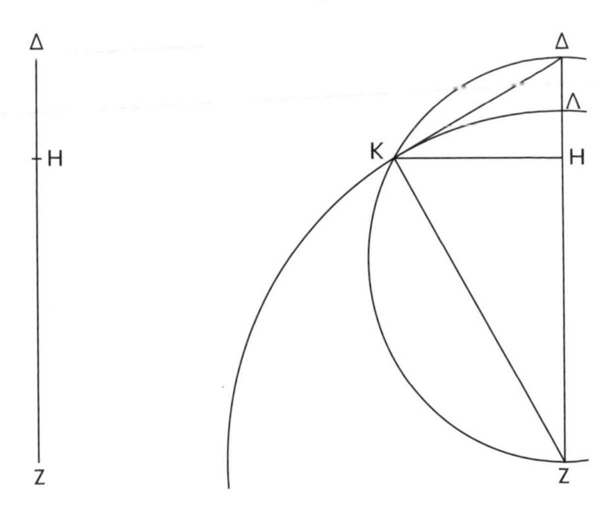

Fig. 1.3 Diagram based on that of *Sect. can.* prop. 3 (fig. 1.1).

the reader's use of its diagram: the *protasis* identifies the target as a mean proportional *number*, not simply a mean proportional. It is this which allows ΔH to be identified as a unit, and hence as indivisible. Otherwise, the reader could view ΔH as a line segment, and as such, infinitely divisible. A mean proportional could then be found through geometry. In fig. 1.3, ZΛ is the mean proportional of ΔZ, ZH. (Because the right triangles ΔKZ and KHZ are similar, ΔZ:ZK::ZK:ZH (Euc. *El.* VI.8). Therefore ZK is the mean proportional of ΔZ, ZH. And because ZK and ZΛ are radii of the same circle, they are equal. Therefore ZΛ is the mean proportional of ΔZ, ZH.) But, of course, it is not a mean *number*, since ΔZ:ZH is epimoric. Lines can always stand for *quantities*, but they cannot always stand for *numbers*. Arithmetical proofs necessarily strip their diagrams of their full geometrical potential so that only numbers are in view.

To this we could add a more complex construction recorded by Pappus (*Collectio* III.11) of the three means (arithmetic, geometric and harmonic) in a single semicircle.[37] Here, too, we can find

[37] The term b in the series a, b, c is the arithmetic mean of a, c when $a - b = b - c$ (e.g. 4, 3, 2 where $4 - 3 = 3 - 2$). The term e in the series d, e, f is the geometric mean of d, f when d:e::e:f (e.g. 24, 12, 6 where 24:12::12:6). The term y in the series x, y, z is the harmonic (or 'subcontrary') mean of x, z when $(x - y):x = (y - z):z$ (e.g. 12, 8, 6

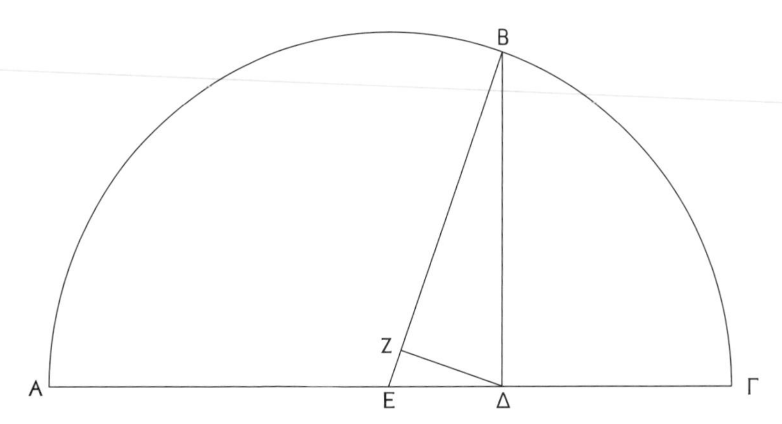

Fig. 1.4 Diagram accompanying Papp. III.11 (after Hultsch 1876–8, vol. I).

the relative string lengths needed to sound the irrational semitone (68.18–25):

ἐν ἡμικυκλίῳ τὰς τρεῖς μεσότητας λαβεῖν ἄλλος τις ἔφασκεν, καὶ ἡμικύκλιον τὸ ΑΒΓ ἐκθέμενος, οὗ κέντρον τὸ Ε, καὶ τυχὸν σημεῖον ἐπὶ τῆς ΑΓ λαβὼν τὸ Δ, καὶ ἀπ᾽ αὐτοῦ πρὸς ὀρθὰς ἀγαγὼν τῇ ΕΓ τὴν ΔΒ, καὶ ἐπιζεύξας τὴν ΕΒ, καὶ αὐτῇ κάθετον ἀγαγὼν ἀπὸ τοῦ Δ τὴν ΔΖ, τὰς τρεῖς μεσότητας ἔλεγεν ἁπλῶς ἐν τῷ ἡμικυκλίῳ ἐκτεθεῖσθαι, τὴν μὲν ΕΓ μέσην ἀριθμητικήν, τὴν δὲ ΔΒ μέσην γεωμετρικήν, τὴν δὲ ΒΖ ἁρμονικήν.

Another[38] asserted that he had constructed the three means in a semicircle: after setting out a semicircle ΑΒΓ, the centre of which was Ε, and taking a random point Δ on the ⟨straight line⟩ ΑΓ, and drawing ΔΒ away from it at right angles to ΕΓ, and joining ΕΒ, and drawing ΔΖ perpendicular to ΕΒ from Δ, he said simply that the three means had been set out in the semicircle, ΕΓ being the arithmetic mean, ΔΒ the geometric mean, and ΒΖ the harmonic mean.

Pappus goes on to find fault with this construction, but we may stop here.[39] It is clear that point Δ may be placed anywhere between Ε and Γ; in my version of the diagram I have placed it

where (12 − 8):12 = (8 − 6):6). The earliest extant definition of these first three means is in Archytas fr. 2; the passage is discussed in ch. 2 part 4. Other means were recognised later, but these constitute the canonical early triad.

38 Sc. of the three unnamed individuals claiming to have been pupils of Pandrosion, the dedicatee of *Collectio* III, each of whom had sent a problem to Pappus, one with a request for his comments (30.17–32.3). Pappus summarises and critiques each in turn. This is the second of the three problems (τὸ δὲ δεύτερον τῶν προβλημάτων ἦν τόδε, 68.17).

39 Pappus criticises the argumentation, not the result: 'the author has failed to *complete* his argument, by failing to show the fact that ΒΖ, in addition to being third proportional in relation to the other two means, also has the "harmonic" relation to the two extremes

so that AΔ:ΔΓ = 2:1. Because of this, other basic harmonic ratios will be found in the same semicircle in several ways; for example:

$$\text{A}\Delta\text{:BZ} = 3\text{:}2$$
$$\text{BZ:}\Delta\Gamma = 4\text{:}3$$
$$\text{EB:BZ} = 9\text{:}8$$

Thus ΔB:BZ = $\sqrt{(9:8)}$, the irrational semitone, because EB:ΔB::ΔB:BZ. Malcolm Brown has suggested some reasons for giving the geometrical argument underpinning this construction a date before Archytas,[40] and if he is right, it will be evident that early mathematical harmonicists were entirely capable of constructing a geometrical canonic division which divided the tone equally; Euclid certainly was, as *Elements* VI.13 demonstrates. The exclusion of irrational intervals from canonic division was therefore, I argue, a conscious choice. This choice follows naturally from a conception of the science as a species of arithmetic, in which certain geometrical constructions are thought of as 'impossible' (*adunaton*). The ramifications of this choice included the inability of the mathematical strand of Greek harmonic science ever satisfactorily to integrate the Aristoxenian insistence that the tone was equally divisible with the language and pragmatics of canonic division.

To underline the point still further, let us consider the pragmatics of marking out the *kanonion* (the ruler fitted to the string of the monochord) in order to construct intervals smaller than a tone.

directly' (Brown 1975: 175, original emphasis). This is comparatively mild criticism: Pappus is at his harshest in dealing with the first of the three problems: 'a badly flawed attempt' at a solution to the famous Delian problem, i.e. of how to double the cube (Cuomo 2000: 128).

40 Namely, (1) that although in Pappus' summary the third mean is called 'harmonic' rather than 'subcontrary' (ὑπεναντία, the older term: Huffman 2005: 178), its distinguishing feature not only in the construction Pappus critiques, but also in Apollonius, *Conics* I.5 and a construction which Brown offers, is in fact a geometrical subcontrariety in its relation to the first two means, what Brown calls 'the "inside, but backwards" relation of the smaller to the larger similar triangle', which depends upon the smaller triangle (EZΔ in fig. 1.4) having its angle 'equal, but oppositely placed, to the angle of the larger [EΔB in fig. 1.4]' (1975: 178); and (2) that of the three constructions Brown gives, only the one offered by Pappus' unnamed author, by requiring the 'oppositely placed' angle to be right, 'makes the Pythagorean theorem directly applicable to getting a needed result, in a single step' (178–9).

From this it will become apparent that there is nothing intrinsic to canonic division which would rule out the construction of the irrational semitone *in practice*, provided that the theorist first admitted the interval as musical; in this way I shall attempt to demonstrate that the objections were theoretical rather than practical, and (as a consequence of the chronological argument of chapter 2) that they predated the invention of the monochord. It will follow from this that the monochord, as an audible extension of the lettered diagram, was allowed only equally limited geometrical participation in an entirely arithmetical procedure. If the unit (*monas*) can be represented as a line segment, it can also be represented as a length of a monochord-string; but having been defined in this way (normally as the common measure of a ratio of commensurable lengths), it falls within the realm of arithmetic, and despite its representation in visible or audible geometry, its division thus becomes *adunaton*. The geometry of sound is an arithmetical geometry.

The process of marking out the *kanonion* is normally simple line-division, and our texts pay little attention to the practicalities of measurement. Consider, for example, the opening sentence of *Sectio canonis* proposition 20:

ἔτεμον τὸν EB εἰς ὀκτὼ καὶ ἑνὶ αὐτῶν ἴσον ἔθηκα τὸν EM, ὥστε τὸν MB τοῦ EB γενέσθαι ἐπόγδοον.

I cut EB into eight parts, and I constructed EM equal to one of them, so that MB:EB = 9:8.

Fig. 1.5 Part of the diagram accompanying *Sect. can.* prop. 20.*

The author does not bother to describe *how* to divide EB in such a way that EM turns out to be exactly an eighth of EB; it is taken for granted that this operation can be executed in practice. And insofar as it is merely a diagram for the reader's eye, the exact equality of its eighths is not essential; not so, however, if it is to be applied to a *kanōn*, where any inequality will be immediately audible once the division-points E and M are used as bridge positions (compare,

* I have simplified the diagram for clarity in the present context; for the complete version see Jan 1895: 166, Barker 1989: 208.

in a non-rational context, Aristoxenus, *El. harm.* 33.10–26). If the diagram is to be of any use as a canonic division, then, it must, for accuracy's sake, be generated by a geometrical procedure: EB will need to be bisected with compasses; each half of EB will then be bisected again to give quarters, and each quarter bisected to generate exact eighths (in effect, successive repetitions of Euc. *El.* I.10, which in turn relies on I.1); compasses will be needed again to find M, using E as the centre of a circle whose radius is equal to an eighth of EB. Such instructions are simply assumed in *Sectio canonis* proposition 20, as they regularly are in constructions predicated upon *Elements* I; instead, we find the ratios of whole numbers generated in the course of the operation, rather than the operation itself, at centre stage. The pragmatics of canonic division are effected through basic geometry, but right from its infancy it remains rigidly focussed on the arithmetic, not on the geometry, of its own procedures.

It is this focus, in part, which allows the mathematical harmonicist to assert the invalidity of the irrational semitone, and to prefer various rational approximations instead (e.g. 256:243, 16:15, 20:19). The most common of these, the 256:243 *leimma*, can be generated by two parallel processes, one in music, the other in arithmetic; this is the method which Aristoxenus calls *lēpsis dia symphōnias*, the 'method of concordance'.[41] The sample problem with which he introduces the method is that of finding a ditone below a given pitch (*El. harm.* 55.11–56.13); he then goes on to apply the method to the question of 'whether it was correctly

[41] *El. harm.* 55.12; see Barker 1989: 49–50. In mathematical contexts the term *lēpsis* and its cognate verb *lambanein* are used to signify the finding or obtaining of a mathematical object. Compare, for example, Eutocius' account of Eratosthenes' solution to the Delian problem: ἐπινενόηται δέ τις ὑφ᾽ ἡμῶν ὀργανικὴ λῆψις ῥᾳδία, δι᾽ ἧς εὑρήσομεν δύο τῶν δοθεισῶν οὐ μόνον δύο μέσας, ἀλλ᾽ ὅσας ἂν τις ἐπιτάξῃ, 'But an easy instrumental solution has been contrived by me, by which we shall find not only two means of the two given ⟨straight lines⟩, but as many as you please' (*In Sph. Cyl.* 90.11–13). In the *Sectio canonis* the terminology is applied to musical notes in a phrase which echoes Aristoxenus': αἱ παρανῆται καὶ αἱ λιχανοὶ ληφθήσονται διὰ συμφωνίας οὕτως, 'the *paranētai* and *lichanoi* will be found by means of concords in the following way' (prop. 17). The proposition which follows employs the method arithmetically, with reference to a lettered diagram, to find the required notes. Compare Aristoxenus: ἐὰν μὲν οὖν προσταχθῇ πρὸς τῷ δοθέντι φθόγγῳ λαβεῖν ἐπὶ τὸ βαρὺ τὸ διάφωνον οἷον δίτονον ἢ ἄλλο τι τῶν δυνατῶν ληφθῆναι διὰ συμφωνίας . . . , 'So, if we are required to construct a particular discord below a given note, such as a ditone or any other of those which it is possible to construct by means of concords . . .' (55.13–16).

postulated at the outset that the fourth is made up of two and a half tones',[42] an assumption which may be tested with greatest precision (ἐξετάσειεν ἄν τις ἀκριβέστατα), he says, by following a procedure which relies on the method of concordance (56.14–58.6).[43] It seems clear that Aristoxenus introduces the *lēpsis dia symphōnias* precisely in order to offer his empirical procedure for testing whether or not the fourth consists of two and a half tones, a discussion which concludes what now survives as book II of the *Elementa harmonica*.[44] Because of this, and also because Aristoxenus goes out of his way to claim that his construction will generate two exact and equal semitones,[45] we would be justified in seeing a direct methodological connection between his use of the method of concordance on the one hand, and his assertion of the equal division of the tone on the other. Perception, he argues (by which he means trained musical perception), is the direct and ultimate arbiter of correct attunement, and it is at its most confident when required to adjudicate concords (55.3–10; 56.31–57.3). The *lēpsis dia symphōnias* is thus for Aristoxenus a means of artificially extending the musical ear's confident reach, and the conclusion of the process is to validate his hypothesis about the equal division of the tone, and the commensurability of the resulting semitone with the fundamental concords, which, on his view, it measures exactly.

The musical version of the simplest form of the procedure can be done as follows (see fig. 1.6): take a note P; tune Q a fourth above it; tune R a fifth below Q; tune S a fourth above R; tune T

[42] πότερον δ᾿ ὀρθῶς ὑπόκειται τὸ διὰ τεσσάρων ἐν ἀρχῇ δύο τόνων καὶ ἡμίσεος, 56.14–15.

[43] As Barker notes (2007: 190), Aristoxenus does not represent the construction as a proof; it is not, formally, a 'demonstration' (*pace* Mathiesen 1999: 329), and because its conclusion is left to the reader's sensory judgement, it is in fact 'empirical' (Gibson 2005: 68).

[44] See Introduction n. 20.

[45] 'For the lowest of the notes generated was tuned to the concord of a fourth with the note which formed the upper boundary of the lower ditone, and it turned out that the highest of the notes generated was concordant with the lowest at the fifth; consequently, since the remainder is a tone and is also divided into equal parts, each of which is both a semitone and the remainder of the fourth over the ditone, it clearly follows that the fourth is made up of five semitones' (ὁ μὲν γὰρ βαρύτατος τῶν εἰλημμένων φθόγγων διὰ τεσσάρων ἡρμόσθη σύμφωνον τῷ τὸ βαρύτερον δίτονον ἐπὶ τὸ ὀξὺ ὁρίζοντι, τὸν δ᾿ ὀξύτατον τῶν εἰλημμένων φθόγγων διὰ πέντε συμβέβηκε συμφωνεῖν τῷ βαρυτάτῳ, ὥστε τῆς ὑπεροχῆς οὔσης τονιαίας τε καὶ εἰς ἴσα διῃρημένης ὧν ἑκάτερον ἡμιτόνιόν τε καὶ ὑπεροχὴ [μὲν] τοῦ διὰ τεσσάρων ἐστὶν ὑπὲρ τὸ δίτονον, δῆλον ὅτι πέντε ἡμιτονίων συμβαίνει τὸ διὰ τεσσάρων εἶναι, 57.4–15).

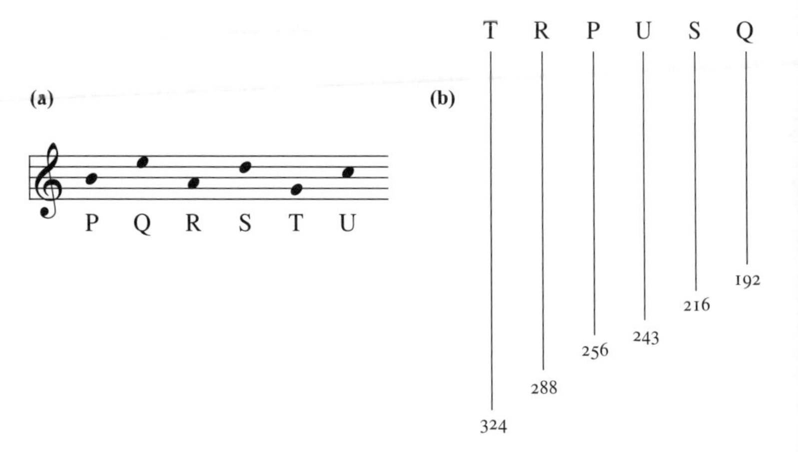

Fig. 1.6 (**a**) The *lēpsis dia symphōnias* in its simplest form (= Aristox. *El. harm.* 55.13–23, extended by one interval, T:U). The numbers in the diagram (**b**) show the relative lengths of the same string needed to sound all six notes, in lowest terms.

a fifth below S; finally, tune U a fourth above T. The final note U will lie a 'semitone' above the starting note P.

The arithmetical version of the same procedure, the method *di' arithmōn*,[46] demonstrates that the interval P:U, while it is what Aristoxenus calls 'the remainder of the fourth [P:Q] over the ditone [U:Q]',[47] is not in fact an exact semitone, but something smaller:

$$\begin{aligned}
&\text{Let } P = 256 \\
&256 \div 4{:}3 = 192 = Q \\
&192 \times 3{:}2 = 288 = R \\
&288 \div 4{:}3 = 216 = S \\
&216 \times 3{:}2 = 324 = T \\
&324 \div 4{:}3 = 243 = U \\
\therefore\ &P{:}U = 256{:}243 \\
&9{:}8 \div 256{:}243 = 2187{:}2048 \\
\text{and}\ &256{:}243 < 2187{:}2048 \\
\therefore\ &256{:}243 < \sqrt{(9{:}8)}
\end{aligned}$$

I have used this simpler procedure because it is sufficient to demonstrate that 'semitones' generated by the method of concordance will be less than half a tone when measured arithmetically,

[46] See n. 32 above. [47] 57.12–13; for the context, see n. 45 above.

and because the lowest terms of the corresponding string lengths exhibit the familiar ratio of the *leimma* (256:243). The construction which Aristoxenus advances to test his assumption that the fourth is made up of two and a half tones (56.14–58.6) is somewhat more complex, but can be refuted equally easily in arithmetic (an important qualification). Start with the interval of a fourth, he says, and use the method of concordance to construct a ditone above the lower note, and a ditone below the upper note. Then take a fourth upwards from the lower limit of the upper ditone, and a fourth downwards from the upper limit of the lower ditone, and assess the final pair of notes thus generated by ear. If they are discordant, the initial postulate is incorrect: the fourth is not equal to two and a half tones; but if they form a perfectly tuned fifth, then the postulate is correct. Perception (*aisthēsis*) has the final judgement, but Aristoxenus clearly expects that the verdict will come down in favour of concordance. If his procedure is followed exactly, and all the fifths and fourths are precisely tuned, the final interval, made up as it is of a perfect fourth plus two *leimmata*, will not be a perfect fifth: $4:3 \times (256:243)^2 = 262{,}144:177{,}147$. This interval is smaller than a hemiolic fifth by a *komma*: $3:2 \div 262{,}144:177{,}147 = 531{,}441:524{,}288$.[48]

The question of whether the difference between a perfect fifth and the flattened fifth generated by Aristoxenus' procedure ought to have been audible to him is difficult to answer. First, the difference is small: scarcely more than a ninth of a tone, just under half the magnitude of Aristoxenus' smallest melodic interval. Second, much depends on the capabilities of the instrument used. Aristoxenus does not recommend an instrument on which to perform the operation, but it cannot be done without one. For stability of pitch and ease of tuning, he will have needed a stringed instrument; the aulos is ruled out by its notorious unreliability on the first of these criteria (43.19–24); the *syrinx* (panpipe) is probably to be ruled out on the second. The fact that the procedure requires the exact tuning of twelve pitches in sequence means that either a lyre or kithara with as many strings,[49] or two with fewer strings, or a

[48] On the *komma*, see ch. 2 part 3.

[49] Probably not unavailable to Aristoxenus: witness e.g. the twelve strings of the late fifth-century kitharode Timotheus, made infamous by the comic poet Pherecrates (fr. 155.25);

harp, will have been the best possible candidates; if fewer strings are available, they must be retuned in the midst of the procedure, and then one does not have the entire array of pitches available, to check their tuning again at the end. Even on an instrument with twelve strings, if the final fifth sounded flat to him, Aristoxenus might have attributed the imperfection to the compounding of small inaccuracies in the course of the eleven successive tuning operations between the starting pitch and the end result.

It is curious, in fact, that Aristoxenus entrusts this test to an instrumental procedure. One might object that, having asserted scarcely more than a dozen pages earlier that 'neither auloi nor any other instrument will ever establish the nature of attunement',[50] he is now entrusting an argument about the nature of attunement to an instrument. To this Aristoxenus would probably reply that since 'none of the instruments tunes itself', perception retains its authority even here:[51] if a reader's ear is competent to assess the final fifth, it will be equally competent to tune the procedure's seven fourths and four fifths with precision as well.[52]

Ptolemy, who offers a refutation of Aristoxenus' argument (*Harm.* I.10), does not comment directly on the question of whether perception is adequate to judge whether the final interval is a perfectly tuned fifth; instead, he offers a simpler arithmetical construction (the method *di' arithmōn*) to show that the remainder of the fourth over the ditone is less than a semitone (*Harm.* 23.3–18), and asserts the ear's ability to recognise beyond doubt an accurately tuned hemiolic fifth, as set out on the monochord (23.21–24.1).

But in order for the method *di' arithmōn* to be used to demonstrate anything about the results of the *lēpsis dia symphōnias*, there must be a logical intermediary: the method *dia grammōn*. Because the use of diagrams, the method 'using lines', is the standard way

the twelve-stringed lyre depicted on a bronze mirror dated to c. 350 BC (Berlin Antiqu. 8519: West 1992: 63).

[50] ἀλλὰ γὰρ οὔτ' αὐλοὶ οὔτε τῶν ἄλλων οὐθὲν ὀργάνων ποτὲ βεβαιώσει τὴν τοῦ ἡρμοσμένου φύσιν, 42.22–4.

[51] ὅτι δ' οὐδὲν τῶν ὀργάνων αὐτὸ ἁρμόττεται ἀλλὰ ἡ αἴσθησίς ἐστιν ἡ τούτου κυρία, δῆλον ὅτι οὐδὲ λόγου δεῖται, φανερὸν γάρ, 43.6–9.

[52] Gibson (2005: 212 n. 16) suggests that 56.14–58.6, which now stands at the end of *El. harm.* II, may have been written in response to the *Sect. can.*, the latter being in her view a reply to *El. harm.* I.

of proving an arithmetical argument in Greek mathematics, the diagrammatic exposition of the *lēpsis dia symphōnias* is essential, and almost always included wherever it appears in Greek harmonic texts.[53] In order to complete the proof, however, the diagram must be audible; if lines are merely lines, the argument stops short of proving that a particular relationship of numbers is analogous to a particular relationship of sounds. The monochord is thus the mathematical harmonicist's method *dia grammōn akoustōn*, as it were; it is in this sense that the instrument might be said to be the metonym of mathematical harmonics, and hence a kind of *meta*-metonym of mathematics.[54] The harmonicist who performs the *lēpsis dia symphōnias* on a monochord is simultaneously drawing a geometrical diagram to prove an arithmetical argument, and tuning musical intervals by ear. The instrument is thus an audible diagram: the musician hears numbers; the mathematician sees sounds.

It will be immediately apparent from the foregoing argument, however, that the mathematician sees sounds strictly as ratios of numbers only if he first recognises the mathematical process as one belonging exclusively to the realm of arithmetic. For, of course, two given line segments may be incommensurable; the longest book of the *Elements* (X) is predicated upon this assumption (see *El.* X def. I.3). A Greek mathematician could easily have generated a division of the monochord by entirely geometrical steps, to locate not only the usual concords and the tone, but the irrational semitone as well.

Let us imagine a *kanonion* ΑΓ (see fig. 1.7), which is also the diameter of a circle whose centre is Ε. ΑΓ:ΑΕ is therefore the octave ratio, 2:1. If an equilateral triangle ΑΖΗ is then inscribed in the circle, ΖΗ will cut ΑΓ at point Θ, and ΑΓ:ΑΘ will be the ratio of the fourth, 4:3. Next, let a square ΑΒΓΔ be inscribed in the circle, let ΔΒ be joined, and let a second square ΕΒΠΓ be constructed on the side ΕΒ; the diagonal ΕΠ will intersect ΘΖ at Κ. If ΚΔ is

[53] See for example the second diagram of Ptol. *Harm.* I.10; see also *Sect. can.* prop. 17, where a diagram similar to that of fig. 1.6b is used to show how the *paranētai* and *lichanoi* 'will be found by means of concords' (ληφθήσονται διὰ συμφωνίας).

[54] My adaptation of Reviel Netz's very useful term for the lettered diagram (1999a: 12, 66).

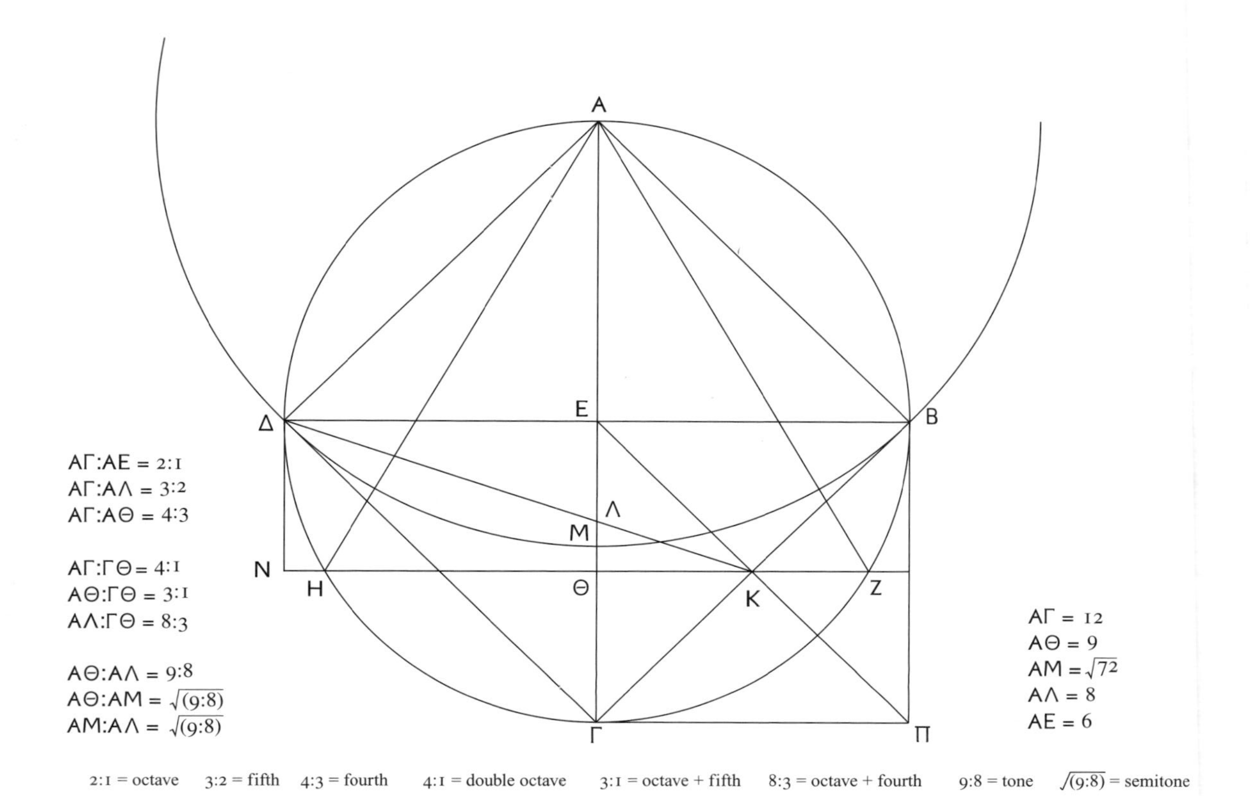

Fig. 1.7 Geometrical canonic division of the epogdoic tone (AΘ:AΛ) into two equal semitones (AΘ:AM and AM:AΛ).

joined, it will intersect ΑΓ at Λ, and ΑΓ:ΑΛ will be the ratio of the fifth, 3:2, since the right triangles ΔΚΝ and ΛΚΘ are proportional: hence ΝΚ:ΘΚ::ΝΔ:ΘΛ (Euc. *El.* VI.4). Thus ΝΔ:ΘΛ = 3:1, so ΑΓ:ΑΛ = 12:8 = 3:2. Therefore ΑΘ:ΑΛ will be the ratio of the tone, 9:8. If a second circle is drawn with centre Α and radius ΑΒ, its circumference will cut ΑΓ at Μ. ΑΘ:ΑΜ and ΑΜ:ΑΛ will be the interval of the irrational semitone, since ΑΒ is the geometric mean of both ΑΓ:ΑΕ and ΑΘ:ΑΛ, and ΑΜ = ΑΒ.

The fact that no such construction occurs anywhere in what survives of Greek mathematical harmonics, despite the fact that every step was available to mathematicians by the time of Euclid, and despite the particular attention paid to incommensurable magnitudes by more than one mathematician who also wrote about harmonics (e.g. Archytas, Eratosthenes), illustrates the extent to which the practice of canonic division was governed by the rules of arithmetic. The *kanonion* was not a geometrical line, but an arithmetical one; all its proportions were to be found in numbers.

This is easy enough to say. But how, practically, did the harmonicist divide ΑΓ so as to generate the 256:243 *leimma*? The simplest process is much more laborious, in fact, than that used to construct the irrational semitone above;[55] to do the job exactly, ΑΘ will have to be divided into 256 equal linear distances. Each of these units will be the result of eight successive bisections of ΑΘ; unless the *kanonion* is very long, the units will be quite small. But if the harmonicist got this far, he would locate his arithmetical Μ at 243 units from Α and 13 units from Θ. But then he would have to place Λ somewhere between 227 and 228; in fact, larger terms are needed to accommodate the desired three notes in whole numbers: 2304, 2187, 2048 are the smallest. Despite the prevalence of very large division numbers in later harmonic texts,[56] it is likely that monochords were marked with fewer units, and that the bridge positions for intervals like the *leimma* were approximated.

[55] Besides the method of successive bisections described here, there is also the geometrical equivalent of the *lēpsis dia symphōnias* of fig. 1.6 – also very laborious, as it requires the construction of six circles and six inscribed equilateral triangles. Neither method is anywhere attested in extant Greek literature.

[56] 10,368 units are required at *Ti. Locr.* 22; Thrasyllus, who begins at 10,368, requires 41,472 in total (Theo. Sm. 93.2–4); see ch. 5 part 2.

Ptolemy uses sexagesimal fractions in order to list most of his divisions within a sixty-unit octave (120:60 = 2:1).[57]

The irrational semitone, then, can be marked on the *kanonion* by means of a small number of simple steps, to a high degree of mathematical exactitude. The marking of the 256:243 *leimma*, by contrast, was so laborious in practice that approximations had to be accepted. The rigidly arithmetical approach to canonic division seems all the more pointed when we consider that several ancient authors who worked on mathematical harmonics also offered solutions to the problem of how to double the cube, which requires finding two means in continued proportion. This is a problem of geometry, since like the irrational semitone, it involves incommensurable magnitudes: given two cubes A and B where the volume of B is double that of A, the side of B will be incommensurable with that of A. Eutocius records geometrical solutions by, among others, Archytas, Plato and Eratosthenes, all of whom also retained a strictly rational approach to harmonics. Plato and Eratosthenes adopted the Pythagorean 256:243 *leimma*; Archytas' closest approximation to the semitone was 243:224 (the middle interval in his chromatic tetrachord).[58] Not even Eratosthenes, whose partiality toward Aristoxenus' tetrachords is clear, appears to have entertained the idea of presenting those irrational intervals on a monochord geometrically, even though the mathematical tools for the job were at his disposal; instead, he attempted a conversion into ratios. We shall consider some of the reasons for the persistence of this rigidly arithmetical approach to canonic division in chapter 3 part 2.

2 Seeing sounds: instruments, diagrams and tables

The two great advantages of the mathematical diagram are that it is static and two dimensional: unlike an instrument, it can be rolled up with the text in a book-roll (in the MSS it usually follows

[57] See for example the division-tables in *Harm.* II.14. Jones (1999: 305) points out that sexagesimal notation, though frequent in Greek astronomical writings (including papyri), occurs nowhere else in Greek except Ptol. *Harm.*

[58] Pl. *Ti.* 36b; Eratosth. ap. Ptol. *Harm.* II.14; Archyt. ap. Ptol. *Harm.* I.13. Archytas' tetrachords will be examined in ch. 2 part 4; the tetrachord implicit in the *Timaeus* will be considered in ch. 3 part 2; Eratosthenes' arithmetical conversions of Aristoxenian tetrachords will be discussed in detail in ch. 4.

its proof).[59] It requires no physical assembly; and being already inscribed with reference-letters, it meets the reader ready both to interpret and to be interpreted by the proof it accompanies. The great advantage of the instrument, by contrast, is that unlike the diagram it is physically manipulable. The monochord, as a diagrammatic instrument, holds both advantages. Furthermore, it possesses a demonstrative capability superior to that of the abacus, since it replicates the type of geometrical diagram which dominates the written form of Greek mathematics. But because it is the instrument of an arithmetical science, it does nothing more, mathematically, than the abacus – it represents discrete quantities visibly by means of physically manipulable components – except insofar as it represents these quantities as *distances*. As distances of a uniform string, quantities become audible, and the connection between the intellectual objects of arithmetic and the perceptible objects of harmonics is completed in a single instrument.

This connection is made through several different 'modes' (as I would like to call them) in which the monochord functions, some of which it shares with other instruments.[60] The four modes which I shall propose here are not an ancient way of thinking about scientific instruments; they are a modern analysis presented merely as a way of defining more precisely how these instruments do what scientists require them to do, and how this differs from the way diagrams fulfil their scientific function. Nor is this group of modes adequate to encompass the workings of modern scientific instruments, as for example those whose purpose is to improve human perception, and thus increase the quantity and detail of observable data (e.g. the telescope, the microscope and many related devices).

What is an ancient scientific instrument?

In employing the term 'ancient scientific instruments' I am imposing a modern category on an assemblage of ancient devices and the disciplines in which they were used. Deborah Jean Warner traces

[59] Netz 1999a: 35 n. 55.

[60] I use 'modes' for want of a better term, even though 'mode' (*modus*) is sometimes still used in a very different sense (and not entirely without anachronism) as a translation of the term *harmonia* in early Greek musical discourse (on which see ch. 2 part 3 n. 99, ch. 3 part 1 n. 20). I avoid the term altogether in this sense.

the origin of the term 'scientific instruments' to late eighteenth-century France ('instruments . . . à l'usage des sciences', 1787); its German ('Wissenschaftliche Instrumente') and English counterparts first appeared in the nineteenth century.[61] The earliest attempts to limit its usage in English are instructive for present purposes: in 1876 the British Committee of Council on Education, advised by James Clerk Maxwell, adopted the following definition: 'A piece of apparatus constructed specially for the performance of experiments is called an Instrument.'[62] In 1884 the National Academy of Sciences in the USA qualified Maxwell's definition: 'an instrument is philosophical [which adjective, they noted, was equivalent to Maxwell's 'scientific'], not in consequence of its special construction or function, but in consequence of the uses to which it is to be put, and many instruments may be put both to uses which are philosophical and to uses which are purely industrial or commercial'.[63] Some further distinguished between 'scientific' and 'philosophical' instruments, the former being those used for 'original investigation and professional training', and the latter for 'demonstrating the laws, principles and facts of physical science'.[64] I do not see this last pair of definitions as necessarily exclusive of one another: the 'laws, principles and facts of physical science' may be demonstrated to students engaged in professional training, or by an investigator in the process of proving a new theorem. I do, however, admit the important distinction between what an instrument was designed to do and what it is used for, but I would prefer to use it inclusively rather than exclusively: an everyday utensil may be called a scientific instrument when it is being used as one (a metal spoon, for instance, when used to demonstrate the conductive properties of its material), but the cathode ray tube may still be called a scientific instrument because J. J. Thompson discovered electrons with it in 1897, despite the fact that its most common use over a century later was in televisions.[65] Thus (to anticipate one of the points which will emerge from the discussion below of the modes in which instruments function) my 'ancient

[61] Warner 1990: 86–7. [62] Warner 1990: 88.

[63] 'Report on Customs Duty on Philosophical and Scientific Apparatus', *Report, National Academy of Sciences* (1884): 65–7, quoted by Warner (1990: 89).

[64] Warner 1990: 89 n. 38. [65] Baird and Faust 1990: 149.

scientific instruments' include any apparatus used for the finding or construction of an object (whether a perceptible object like a star or a musical note, or an abstract object like an angle of arc or a ratio), or the discovery or proof of a truth which is universal rather than merely particular. In fact the examples which will concern us most directly were all, as Maxwell insisted, purpose-built rather than adopted from common use. The monochord owes its existence to the insufficiencies of performance instruments (lyres, lutes, harps, panpipes) in the context of investigating or demonstrating the mathematical properties of musical intervals. But others, particularly those used for measurement (e.g. the ruler, the balance) could perform scientific as well as non-scientific functions; astronomical instruments could be used and adapted either for practical purposes like navigation, or for purely investigative purposes, or again for didactic purposes (whether the students were future mariners or future astronomers).

But as Warner points out, 'the task facing historians of scientific instruments is comprehension of the past in its own terms'.[66] So, rather than simply proposing an (anachronistic) category of ancient scientific instruments and proceeding to list the ways in which its members function, let us pause briefly here to note some of the terms in which authors of Greek scientific texts indicate their own use of instruments. The most common word is *organon*, which in its most general sense can signify almost any kind of instrument or tool; more specifically it can be used to indicate, among other things, a musical or surgical instrument.[67] In scientific contexts its definition was evidently broad: in the opening chapter of his *Harmonics*, in which he lays out 'the criteria in harmonics' (περὶ τῶν ἐν ἁρμονικῇ κριτηρίων, I.1), Ptolemy introduces instruments as an essential component of his scientific method. 'Since similar things occur in the case of sounds and hearing, we require, just as with the objects of sight, some rational criterion for them, with the assistance of appropriate instruments (*organa*), as for instance the ruler (*stathmē*) for straightness, or compasses for the circle and for the measurement of its parts.'[68] Nicomachus gives a list

[66] Warner 1990: 92. [67] LSJ s.v. ὄργανον I.3–4.
[68] τῶν ὁμοίων οὖν καὶ περὶ τοὺς ψόφους καὶ τὴν ἀκοὴν συμβεβηκότων καθάπερ ταῖς ὄψεσι δεῖ τινος πρὸς ἐκεῖνα κριτηρίου λογικοῦ διὰ τῶν οἰκείων ὀργάνων, οἷον πρὸς

of the types of 'instrumental aid' (*boētheia organikē*) available to investigators in a chapter of his handbook on elementary harmonics devoted to the discovery of 'the arithmetical ratios of the notes': sight is assisted by compasses, ruler (*kanōn*), and dioptra; touch by the balance and by the invention of measures.[69] Eratosthenes was said to have proposed an 'instrumental solution' (*organikē lēpsis*) to the Delian problem, in contrast to the geometrical solutions of his predecessors.[70] These few examples, to which more could be added,[71] show that instruments were conceived of in a variety of important roles in relation to the activities of reason and the senses in scientific investigation. The preliminary definition of an ancient scientific instrument offered above may thus be allowed to stand for the time being; we shall have more to add to it below.

Four instrumental modes

In general, then, ancient scientific instruments work in some or all of the following four ways:[72]

(a) by generation: the instrument generates the phaenomena which its science investigates;[73]

μὲν αὐτὸ τὸ εὐθὺ τῆς στάθμης φέρε εἰπεῖν, πρὸς δὲ τὸν κύκλον καὶ τὰς τῶν μερῶν καταμετρήσεις τοῦ καρκίνου, 5.2–6.

69 *Harm.* 6, 246.1–5. On the MS tradition of this text and the variations in its title see Mathiesen 1999: 392–3. Nicomachus' biography is obscure, but he evidently wrote around the beginning of the second century AD; one of his other works was translated into Latin by Apuleius, who was born around AD 125.

70 See n. 41 above. The words are from a letter quoted by Eutocius which is now agreed to be spurious (Thomas 1939: 256); the expression therefore cannot be taken as that of Eratosthenes himself, but it evidently predates Eutocius (who wrote in the early sixth century AD).

71 Compare e.g. the *Suda*'s definition of the *parapēgma* as a kind of 'astronomical instrument' (*organon astronomikon*): n. 131 below.

72 For another way of looking at scientific instruments, see Baird 2004; my modes (b) and (c), for example, correspond roughly to Baird's ch. 2, 'Models: Representing things'. On astronomical instruments in particular, see Price 1957, Taub 2002. Evans' classification of 'the entire material culture of Greek astronomy' is also relevant (1999: 238), though his concern is broader than mine insofar as it includes objects not used for scientific discovery or proof, and narrower than mine insofar as it excludes objects which had no direct connection to astronomy.

73 Here and subsequently I shall use the term 'phaenomena' to designate any perceptible objects of scientific enquiry, not merely visible ones. For reasons which will become clear below, I would like to avoid making a distinction between the objects of sight and the objects of hearing (τὰ φαινόμενα and τὰ ἀκουόμενα in their literal senses) as far

(b) by representation: the instrument represents an abstract concept visibly or tangibly,[74] or allows objects of one sense to represent the objects of another;[75]
(c) by analogy: this (object or phaenomenon) works like that (object or phaenomenon);[76]
(d) by direct physical manipulation.[77]

Note that I am not attempting a fully coordinate system, merely a list of attributes of use: (d) is not of the same order as (a)–(c), but its relevance here is that it distinguishes scientific instruments from mathematical diagrams. While (d) and (a) belong to the physical operation of some instruments, (b) and (c) are secondary, interpretative processes. What unifies the four as a set is simply that they are all characteristics of the scientific use of instruments in antiquity. But these modes are by no means exclusive to science: they are the common property of the instruments and tools of both scientific and non-scientific activity: instrumental music-making, for example, relies on (a) and (d); writing and reading on (b) and (d); coinage on (b), (c) and (d).[78]

as the methodology of scientific instruments is concerned. As we shall see, Ptolemy himself groups sight (*opsis*) and hearing (*akoē*) together as 'the more rational of the senses' (αἱ τῶν λογικωτέρων αἰσθήσεων, *Harm.* I.2, 5.23–4).

74 As on an abacus, for example, where pebbles represent units.

75 To use Nicomachus' example of the balance (*Harm.* 246.5), the objects of touch (weights) may be represented by the objects of sight.

76 I owe this presentation of the idea to Sylvia Berryman.

77 This central advantage of the instrument over the diagram is at the heart of the difficulty of communicating scientific ideas in writing. Baird introduces this difficulty with the example of Faraday's electric motor (1821): 'This point is more persuasive when one is confronted with the actual device. Unfortunately, I cannot build a Faraday motor into this book; the reader's imagination will have to suffice. But it is significant that Faraday did not depend on the imaginations of his readers. He made and shipped "pocket editions" of his newly created phenomenon to his colleagues. He knew from his own experience how difficult it is to interpret descriptions of experimental discoveries. He also knew how difficult it is to fashion even a simple device like his motor and have it work reliably' (2004: 3). For a modern pedagogical approach to the same difficulty, see Triadafillidis 1995.

78 'This drachma represents six obols'; 'you owe me twelve obols'; 'I give you two drachmas'; 'I have received correct payment'; 'those two drachmas represent the goods I have just bought from you'. Coinage may also go beyond representation and become an item of commercial value itself; but this could be said equally of a lyre, or an armillary sphere, or even (just possibly) a monochord, and is not an attribute specific to the nature of the instrument *per se*. The (b) + (c) + (d) formula also plays a central role in Greek magic: see e.g. Theocritus, *Idyll* 2, where physical, ritual objects (e.g. barley, bay, bran, wax) represent parts of the beloved's anatomy and are melted or burnt in fire by analogy for him, in order to draw him back to his lover.

The (b) + (c) formula is basic to demonstrative argument: *this* represents *that*; *this* also functions like *that*; condition *x* applied to *this* generates result *y*; therefore condition *x* applied to *that* will also generate result *y*. Using an instrument in this way allows a scientific argument to be predictive through an appeal to universality: all situations which conform to *this* pattern are reducible to *this* procedure carried out on *this* instrument, and will yield the same result.[79] The lettered diagram works in precisely this way: a single right triangle can be used to prove that in *all* right triangles the square on the hypotenuse is equal to the sum of the squares on the other two sides.[80]

The abacus combines the basic demonstrative (b) + (c) formula with mode (d): 'each of the pebbles in these two rows represents a unit' (b); 'pebbles can be added and subtracted like units' (c); 'I move this pebble from one row to the other' (d); 'I have just performed an arithmetical operation as well as a physical one.' The applied geometrical sciences use instruments similarly. A wall and the two distances from the observer to its summit and its base can be reduced to a triangle; an instrument (e.g. the dioptra) can be used to construct a triangle of the same proportions.[81] The height of the wall can then be calculated by similar triangles.[82] Armillary spheres and other planetaria also work in the (b) + (c) + (d) modes, as do all instruments which can be conceived of as models of the phaenomena under investigation.[83] Take, for instance, Ptolemy's introduction, in *Almagest* V.1, of his version of the armillary sphere (*astrolabon organon*, not to be confused with the plane astrolabe well known from mediaeval Arabic astronomy), which is to function as an observational instrument:

[79] There are limitations to such arguments in the mathematical sciences, however; see e.g. Mueller 1981: 13, Netz 1999a: 267–9, Norman 2006: 109–10.

[80] Euc. *El.* I.47. On the problem of 'single-diagram induction', however, see Norman 2006: 55–6.

[81] The smaller-scale triangle is constructed by the arrangement of the parts of the dioptra itself; the instrument thus represents the phaenomena. See Lewis 2001: 41–3.

[82] The same mathematics informs the use of Heron's version of the instrument; see Heron, *Dioptra* 12 (for vertical distances); 8, 10 (horizontal distances). The geometry is based ultimately on that of Euc. *El.* VI. It is worth pointing out here that measurement, as a scientific activity (a heuristic one), often relies on instruments which function in the (b) + (c) + (d) mode formula.

[83] Netz 1999a: 65–6.

δύο γὰρ κύκλους λαβόντες ἀκριβῶς τετορνευμένους τετραγώνους ταῖς ἐπιφανείαις καὶ συμμέτρους μὲν τῷ μεγέθει, πανταχόθεν δὲ ἴσους καὶ ὁμοίους ἀλλήλοις, συνηρμόσαμεν κατὰ διάμετρον πρὸς ὀρθὰς γωνίας ἐπὶ τῶν αὐτῶν ἐπιφανειῶν, ὥστε τὸν μὲν ἕτερον αὐτῶν νοεῖσθαι τὸν διὰ μέσων τῶν ζῳδίων, τὸν δ᾽ ἕτερον τὸν διὰ τῶν πόλων αὐτοῦ τε καὶ τοῦ ἰσημερινοῦ γινόμενον μεσημβρινόν·

For taking two circles, accurately turned on the lathe, square with respect to their surfaces [i.e. each being circular in plan and square in section], symmetrical in their size, and equal and similar to one another on all sides, we fitted them together with their diameters at right angles on the same surfaces, so that one of them is conceived to be the ecliptic, and the other becomes the meridian [passing] through the poles of both the ecliptic and the celestial equator.[84]

The vocabulary of representation (ὥστε . . . νοεῖσθαι) here is that regularly used in mathematical contexts when the instrument (or the diagram) needs to be conceived of as something different.[85] Ptolemy can now continue his description of the instrument's construction, applying the names of the celestial constructs (poles of the ecliptic, celestial equator) directly to the parts of the instrument itself. Representation allows the scientist to transfer his attention from the phaenomena to the instrument.

When other circles have been added and the instrument is fully assembled, it is set up so that:

τὴν τῶν ἐντὸς κύκλων περιαγωγὴν ἀποτελεῖσθαι περὶ τοὺς τοῦ ἰσημερινοῦ πόλους ἀπ᾽ ἀνατολῶν ἐπὶ δυσμὰς ἀκολούθως τῇ τῶν ὅλων πρώτῃ φορᾷ.

the inner circles revolve around the poles of the equator from east to west, in accordance with the primary motion of the universe.[86]

The analogy function of the armillary sphere is clear: the instrument's motion is analogous to that of the heavens because it models that motion. Instructions for a basic procedure are given next,

[84] Ptol. *Alm.* 351.12–19.

[85] On the use and comparative infrequency of the verb *noein* ('conceive', 'imagine') in Greek mathematics see Netz 1999a: 52–3; passive imperative forms (νοείσθω, νενοήσθω) are particularly common in the *ekthesis* (setting out) of propositions where there is 'under-representation by the diagram' (53). For this reason the verb is most frequent in the setting out of figures which represent three-dimensional objects: we need to *imagine* the third-dimension information because the figure rarely includes it (see also Sidoli 2004: 6 n. 18). The difference between tenses of the passive imperative will be discussed below.

[86] Ptol. *Alm.* V.1, 353.8–10.

and the introduction concludes with a summary of what has been achieved:

οὕτως γάρ, ποῖόν τε κατὰ μῆκος ἐπέχει τοῦ διὰ μέσων τῶν ζῳδίων τμῆμα, ἐπιγιγνώσκομεν ἐκ τῆς κατὰ τὴν τοῦ ἰσοδυναμοῦντος αὐτῷ κύκλου διαίρεσιν γινομένης τοῦ ἐντὸς κύκλου τομῆς, καὶ πόσας αὐτοῦ μοίρας ἀφέστηκεν ἤτοι πρὸς ἄρκτους ἢ πρὸς μεσημβρίαν ὡς ἐπὶ τοῦ διὰ τῶν πόλων αὐτοῦ κύκλου, διά τε τῆς αὐτοῦ τοῦ ἐντὸς ἀστρολάβου διαιρέσεως καὶ τῆς εὑρισκομένης διαστάσεως ἀπὸ μέσης τῆς ὑπὲρ γῆν ὀπῆς τοῦ ὑπ' αὐτὸν παραγομένου κυκλίσκου ἐπὶ τὴν μέσην γραμμὴν τοῦ διὰ μέσων τῶν ζῳδίων κύκλου.

In this way we read off the position [of the moon or any other desired object] in longitude on the ecliptic, from the graduation occupied by the inner [astrolabe] ring on the ring representing the ecliptic, and its deviation to north or south [of the ecliptic] along the circle through the poles of the ecliptic, from the graduations of the inner astrolabe ring; the latter is given by the distance between the mid-point of the upper sighting-hole on the inmost rotating ring and the line drawn through the centre of the ecliptic ring.[87]

The (b) + (c) + (d) formula allows the scientist to gain knowledge by means of an instrument (οὕτως γάρ . . . ἐπιγιγνώσκομεν).

The monochord works in all four modes: notes and intervals are generated (a) when the harmonicist positions the bridge and plucks the string (d); distances between points on the *kanōn* represent notes, and relative distances represent intervals (b);[88] notes work like numbers – that is, they have arithmetical relationships, as numbers do (c). And from part I of this chapter, we may now add to (c): the instrument works like a diagram.

Thus the key difference between harmonic and astronomical instruments is in their relationship to the phaenomena they elucidate, particularly in the role of generation (a) and analogy (c). The armillary sphere provides a *model* of celestial phaenomena; when it measures, it does so by analogy, in a way methodologically equivalent to that by which the dioptra measures the height of a distant wall by the principle of similar triangles. Harmonic instruments, by contrast, do not *model* the musical intervals; they

[87] Ptol. *Alm.* V.1, 354.8–17, trans. Toomer 1984: 219. Toomer adds a figure to illustrate Ptolemy's description (218 fig. F); the relevance of Ptolemy's decision not to present the instrument diagrammatically will be discussed below.

[88] The representation-mode (b) is emphasised by the language of canonic division, in which physical distances, musical intervals and their ratios are called *diastēmata*.

generate them. The monochord does not *work like* an octave, it produces one, when the bridge is correctly placed. But, on the other hand, it does *work like* a geometrical diagram in an arithmetical proof, where numerical ratios are presented visually through physical distances (*diastēmata*).

The scientific uses of instruments

Which of these four modes dominate in the use of an instrument for scientific purposes depends not only on the science and the nature of the instrument, but also on the particular purpose to which it is put. The definition of 'scientific purposes' is of course problematic. For the present argument, I shall privilege scientific enquiry for its own sake, and exclude mere practical application of scientific discovery. Thus the use of the dioptra in a written proposition whose objective is to show how to find an unknown distance will count as 'scientific', whereas the use of the dioptra by an ancient engineer in the construction of an aqueduct will not. But the ancient engineer's training may well have included the reading of texts which employed the instrument in a mathematical proof. As we shall see, some of the earliest written evidence for the use of the monochord comes from texts which are are either probably or certainly pedagogical in design (though less obviously practical in application: ancient musicians could tune a lyre without using a monochord or understanding the mathematics of the procedure). For this reason I shall avoid defining the 'scientific' use of instruments too precisely; often the most relevant issue will be the ways in which the ancient author defines and delineates the role of the instrument in his project, something which varies greatly through the texts we shall have to consider.[89]

Aristotle distinguished the type of harmonic theorists he called 'mathematical harmonicists' from those who based their investigations on hearing by saying that while the latter knew 'the fact that', the former knew 'the reason why', for it was the mathematical scientists, he said, who were 'in possession of

[89] For another approach to a similar set of difficulties in classification, compare Netz's attempt to capture a 'group picture' of Greek mathematicians (2002); he casts his net wide and draws a small catch.

the demonstrations of the causes'.[90] This provides a useful starting point for our delineation of the scientific uses of instruments, since we are concerned here primarily with the mathematical sciences. Those which involved the investigation of perceptible objects (like astronomy and harmonics) sought 'the reason why', not merely 'the fact that'; their ultimate goal, on Aristotle's rigorous criterion, was to explain (i.e. give causes for) the phaenomena, the truth of which would be shown by means of demonstrations (*apodeixeis*).[91]

Demonstration is thus the culmination of a process of investigation which may include other stages, some of which might require the use of instruments. The surviving remains of Greek harmonics, optics, astronomy and other mathematical sciences do not always rise to Aristotle's challenge, but they do very frequently employ logical structures which derive ultimately from Aristotle's *Analytics*.[92] Because these sciences came to be written about in the formulaic language of arithmetic and geometry (the abstract sciences), their uses of instruments can, in general, be corralled into the three logical categories we find prioritised there: the constructive, the heuristic and the demonstrative. An ancient Greek mathematical scientist uses an instrument either to construct something (a triangle, an arc, a cylinder, a cube), or to find something (a quantity, a distance, an angle, a ratio), or to show that something is so (often a general principle which, it is argued, holds in all similar cases). This distinction can be seen in the Euclidean conclusion-formulae 'which it was necessary to construct' (ὅπερ ἔδει ποιῆσαι, *quod erat faciendum*, QEF), 'which it was necessary to find' (ὅπερ ἔδει εὑρεῖν, *quod erat inueniendum,* QEI) and 'which it was necessary to prove' (ὅπερ ἔδει δεῖξαι, *quod erat demonstrandum*, QED).

While construction (QEF) is an important scientific purpose to which instruments are put, in mathematical contexts it is very often (like finding, QEI) a preliminary or intermediary step in a wider demonstrative project, rather than the ultimate purpose. Euclid's *Elements* provides a simple illustration of this. It is necessary to

90 *An. post.* 79a1–4, quoted in the Introduction.

91 On explanation as critical to the possession of scientific knowledge according to Aristotle, see McKirahan 1992: 209.

92 The relationship of the *Sectio canonis* to this tradition will be discussed in ch. 3 part 1.

construct an equilateral triangle (I.1), for example, before the properties of equilateral triangles inscribed in circles may be treated (XIII.12). Construction is preliminary to demonstration, just as definition (I def. 20) is to construction (I.1). This is equally true in harmonics. Ptolemy uses an eight-stringed *kanōn* in *Harmonics* I.11 to construct six tones in succession;[93] the construction is not an end in itself, but rather serves the broader purpose of demonstrating that the octave is less than six tones. In the exact sciences especially, theorems frequently rely on constructions, and constructions frequently rely on instruments, though these are not often mentioned.[94]

There is frequent overlap between the categories of use: measurement, for example, can be merely heuristic ('what is the height of that wall?'), or it can fulfil the *demonstrandum* of a proof. But even when the ultimate object is explanation, the instruments are deployed in processes which correspond to the three privileged activities in Greek mathematics: constructing, finding and proving.[95]

The lettered diagrams of Greek mathematics, functioning by representation (b) and analogy (c), are ideally suited to demonstrative argument. A word needs to be said here about the textual history of the diagrams, since my argument assumes not only their presence in the earliest copies of my primary texts (Ptolemy's *Almagest* and *Harmonics*), but also the close resemblance of these lost originals to the diagrams printed in the modern editions. Until very recently, works by authors in the Greek mathematical tradition have suffered from an editorial bias which has favoured the text; MS collation has rarely included the diagrams. Toomer's approach to the diagrams of the *Almagest* was a start (1984: 24–5),

93 The passage will be discussed in ch. 6.

94 As Fowler and Taisbak point out (1999: 361), 'the words for straightedge (κανών) and compass (διαβήτης) do not occur at all in the *Elements* nor, for that matter, in what has been transmitted from Archimedes and Apollonius', despite the assumption commonly made by modern readers of Greek mathematics, 'that Euclidean constructions are based on straightedge and compass'.

95 It is important to recognise that Greek science was extremely pluralistic and heterogeneous, as G. E. R. Lloyd has shown (1987, 1992). So, while it could be argued that proving was seen as the most important activity of Greek science, it must also be stressed that what counted as 'proving' was especially controversial. (On this last point, in the context of Greek mathematics specifically, see Lloyd 1979: 103–5.)

but despite recent advances on this front (e.g. Decorps-Foulquier 1999 for Apollonius of Perga, Netz 2004b for Archimedes), we are still in want of a critical edition of the diagrams of Ptolemy's *Harmonics*. I have not examined the MS diagrams, and must therefore base my argument on Düring's edition.[96] Düring's text is based on a collation of 84 of the 92 extant MSS of the *Harmonics*, but he does not describe the editorial basis for his reproduction of the diagrams, some of the most important of which are hand-drawn in his edition (e.g. that of I.8, a diagram whose importance not only to this chapter, but to the book as a whole, will soon become apparent).[97] For this reason my argument must be provisional; but in the absence of a critical edition of the diagrams of the *Almagest* and *Harmonics*, two points of limited reassurance may be raised. The first is that the convention and practices of the lettered diagram in mathematical texts were well established by Ptolemy's time, and are well attested in the surviving papyri.[98] The second is that the diagrams preserved in these same papyri confirm, in their particular cases, a satisfactory level of accuracy of transmission to our modern editions.[99] Diagrams are, of course, subject to the

[96] Düring 1930. Barker, in his translation (1989: 270–391), corrects some errors in Düring's printing of the diagrams (e.g. in *Harm.* I.5). The diagrams in Solomon's translation (2000) are often unreliable; for an extreme example, see the one accompanying Ptol. *Harm.* I.11, 26.3–14 Düring = Solomon 2000: 37 (representing the eight-stringed *kanōn*), where seven of the eight line segments designated by letter-references in Ptolemy's text cannot be located in the diagram.

[97] The diagram of I.8 is the only one for which Düring gives detailed editorial information (for which see the note on fig. 1.9).

[98] See, for example, *PHerc.* 1061, columns 9–10: a text of Demetrius of Laconia quoting Euc. *El.* I.9 with lettered diagram (Demetrius c. 100 BC; papyrus copied before AD 79), Fowler 1999: 210 with pl. 1; *POxy.* I.29: Euc. *El.* II.5 with unlettered diagram (papyrus copied c. AD 75–125), Fowler 1999: 210–13 with pl. 2 (Eric Turner's case for redating the papyrus from Grenfell and Hunt's 'third or fourth century' is quoted at 211); *PBerol.* 17469: Euc. *El.* I.9 with fragmentary, unlettered diagram (papyrus copied in the second century AD), Fowler 1999: 215–16 with pl. 4. Fowler argues that *POxy.* I.29 and *PBerol.* 17469 came from documents (students' notes?) containing merely the enunciation (*protasis*) of each proposition along with an unlettered diagram (213, 216), not from complete copies of the *Elements*. Also worth comparing are the (coloured) astronomical diagrams of Paris papyrus 1 (the earliest writing on which has been dated to c. 165 BC), because they provide evidence about the culture of visualisation in Greek mathematical sciences to within a century and a half of Euclid's lifetime (see Neugebauer 1975: 686–9 and 1453 (pl. VII)).

[99] Compare, for instance, the diagram of Euc. *El.* I.9 in *PHerc.* 1061 (first published in 1900), in *PBerol.* 17469 (first published in 1994), and in Heiberg's edition of Euclid (1883: 31); compare also the diagram of *El.* II.5 in the 'student version' of *POxy.* I.29

same forces of corruption as texts: diagrams may be added, elaborated, distorted and lost in the MS tradition; but isolated examples of the same diagrams in papyri and early minuscule codices suggest at the very least that the transmission of diagrams was no more fragile than the transmission of text, and in fact in some instances perhaps less so.[100] Thus when Ptolemy employs the formulaic language of the mathematical proposition which usually directs our attention to a diagram, and our text contains one, we may at least claim some justification for trusting what we see on the page.[101] Until the diagrams of Greek mathematical texts have been systematically edited, this assurance will have to suffice.

In astronomy, the role of the instrument is to reduce the phaenomena to a physically manipulable (d) model (c) whose parts represent (b) the celestial objects under investigation. From measurements taken with the armillary sphere (a heuristic activity), Ptolemy attempts to square the phaenomena with the *hypothesis* advanced to 'save' them.[102] This can now be expressed in the

(first published in 1898) and in Heiberg's edition (1883: 131; cf. Netz's version of the diagram (1999a: 10–11)).

100 Compare, for example, the diagram of *Elements* II.5 in the earliest extant minuscule MS, copied in AD 888 (Bodleian D'Orville 301, folio 35^v, Fowler 1999: 218 and pl. 5), to which extra annotations and diagrams have been added. Netz notes the similarity of diagrams in the MSS of Archimedes, *De sphaera et cylindro*: 'It should be stressed that the diagrams across the manuscript tradition are strikingly similar to each other, often in quite trivial detail, so that there is hardly a question that they derive from a common archetype' (2004b: 8). A comparison of the Archimedes Palimpsest (tenth century) with the other surviving MSS leads him to suggest a late ancient archetype (9), but he does not attempt to close the gap between Archimedes and this archetype: 'I shall not dwell on the question, how closely the diagrams of late antiquity resembled those of Archimedes himself: to a certain extent, however, the same question can be asked, with equal futility, for the text itself. Clearly, however, the diagrams reconstructed are genuinely "ancient," and provide us with important information on visual practices in ancient mathematics' (9).

101 Take, for instance, the first diagram of the *Almagest* (I.10), which is introduced by one of the most frequent construction-formulae in Greek mathematical language: ἔστω δὴ πρῶτον ἡμικύκλιον τὸ ΑΒΓ (32.10). It is no accident that this is also the first occurrence of ἔστω in the treatise, or that of the 117 occurrences of ἔστω in the *Almagest*, the 115 which govern designation-letters all refer to an extant diagram. In the *Harmonics*, likewise, the designation-letters of the first extant diagram are governed by the first occurrence of ἔστω (I.5, 12.8), and of the five occurrences of the word in this treatise, the three which govern designation-letters all refer to an extant diagram. (On construction-formulae in Greek mathematics, see Netz 1999a: 136–7, formulae 21–39; ἔστω is his formula 23).

102 On Ptolemy's use of the term *hypothesis* in his astronomical writings to mean 'geometrical model' as opposed to 'postulate', see Toomer 1984: 23–4. Bowen (2007: 354),

conventional form of the mathematical proposition, complete with lettered diagram (fig. 1.8):

ἵνα δὲ μᾶλλον ἡμῖν ὑπ' ὄψιν γένηται τὰ τῆς ὑποθέσεως, νοείσθω πάλιν ὁ ἐν τῷ λοξῷ τῆς σελήνης ἐπιπέδῳ τῷ διὰ μέσων τῶν ζῳδίων ὁμόκεντρος κύκλος ὁ ΑΒΓΔ περὶ κέντρον τὸ Ε καὶ διάμετρον τὴν ΑΕΓ, ὑποκείσθω δὲ ἅμα κατὰ τὸ Α σημεῖον τό τε ἀπόγειον τοῦ ἐκκέντρου καὶ τὸ κέντρον τοῦ ἐπικύκλου καὶ τὸ βόρειον πέρας καὶ ἡ ἀρχὴ τοῦ Κριοῦ καὶ ὁ μέσος ἥλιος.

In order to illustrate the details of the hypothesis, imagine the circle in the moon's inclined plane concentric with the ecliptic as ΑΒΓΔ on centre Ε and diameter ΑΕΓ. Let the apogee of the eccentre, the centre of the epicycle, the northern limit, the beginning of Aries and the mean sun [all] be located at point Α at the same moment.[103]

We may note first of all that we are given a diagram of the phaenomena, not of the instrument: an exactly opposite presentation awaits us in the *Harmonics*. Secondly, the visibility of the diagram is emphasised: its function is to make the details of the *hypothesis* more visible (μᾶλλον ἡμῖν ὑπ' ὄψιν); this is done through representation (νοείσθω), as always in the *ekthesis* of a proof. The other ingredients of the proof follow, with their usual sign-posts: 'I say (that)' (φημί, 358.8, 16), 'in this way, therefore' (οὕτως οὖν, 359.9), 'it is clear that' (δῆλον ὅτι, 359.12; φανερὸν δὲ ὅτι, 359.19).

The pattern of argumentation is repeated in subsequent chapters dealing with lunar problems (V.3–6): observation data obtained by instrument inform a diagrammatic exposition used to illustrate the *hypotheseis* which save the phaenomena. In V.7 Ptolemy explains the construction of a table of the complete lunar anomaly, and the table itself (*kanonion*) is given in V.8.[104] The lettered diagram was used demonstratively, but the table is a heuristic tool for the

commenting on the opening lines of Ptolemy's *Hypotheses planetarum*, puts it this way: what Ptolemy here calls a *hypothesis* 'is not a proposition but a quantified geometrical model which is used mathematically to yield results that are consistent with observations'. (I am grateful to Alan Bowen for allowing me to read a copy of this article in advance of its publication.) Ptolemy uses the term eleven times in the *Harmonics*, but here the context demands the sense of a 'postulate' rather than a 'geometrical model'; a more ambiguous usage in the *Harmonics* (5.16, with which cf. 100.25) will be discussed below. On 'saving' the phaenomena, see Lloyd 1978 and 1991: 248–77.

[103] Ptol. *Alm.* V.2, 357.21–358.6, trans. Toomer 1984: 221.

[104] There are several different types of table in the *Almagest* (calculation, observation, interpolation). As an interpolation table, this one contains items which are mathematically

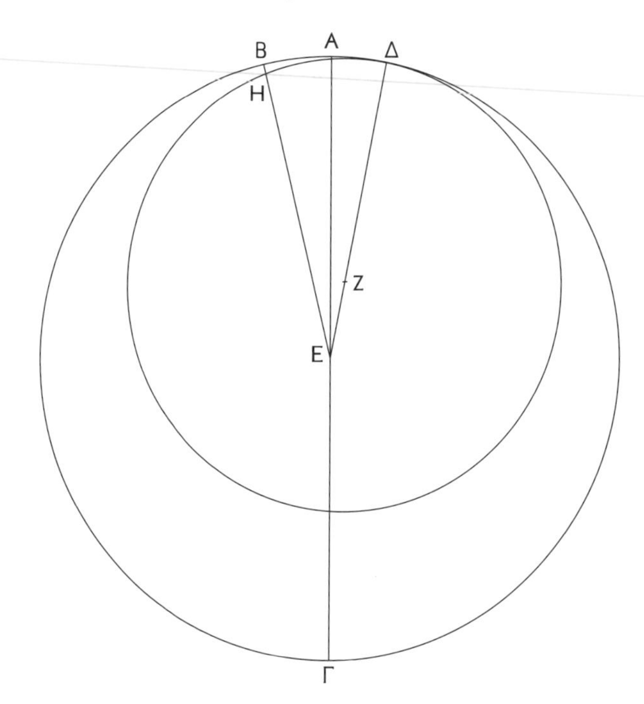

Fig. 1.8 First diagram accompanying Ptol. *Alm.* V.2 (after Heiberg 1898–1903, vol. I).

reader: the results of a demonstrative argument are laid out so as to be readily found and compared. Tables are often intended for use in conjunction with an instrument, as for example the table of northern-hemisphere constellations in VII.5. The instrument has then performed two heuristic roles, one at either end of the argument: it has supplied the initial observation data on the basis of which the *hypothesis* has been quantified, and it allows the reader finally to locate the phaenomena where they are predicted to be.[105] This latter task is demonstrative as well as heuristic, since

derived from other items by means of something more or less equivalent to what is now called an interpolation function (see Van Brummelen 1994: 299, who while acknowledging that 'the concept of a function is not explicit in ancient mathematics', nonetheless permits himself the generalisation that 'almost every *Almagest* table may be considered to represent a function of one or more arguments without a significant loss of historicity').

[105] Tables were used to present predictions from at least the first century AD: e.g. *POxy.* LXI. 4137, which contains a table listing eclipse forecasts for AD 56 and 57 (Jones 1999:

the successful location of a celestial body in a determined part of the sky at a particular time will both confirm the initial observations and illustrate that the *hypotheseis* save the phaenomena.

The fact that few of Ptolemy's observation data were his own need not be seen as detrimental to the demonstrative force of this type of instrument use.[106] My concern is with the method by which instrument, proof and table are introduced as parts of a persuasive argument in favour of certain *hypotheseis* advanced to save certain phaenomena.[107] A scientific method may still provide a logical framework for persuasion even if parts of the argument itself are flawed, or its data borrowed, plagiarised, selective, fudged, distorted or erroneous. The issue at *Almagest* V.3–8 is about Ptolemy's method of presentation, not his method of *heuresis*. Mark T. Riley, in discussing Ptolemy's tabular presentation of data, rightly emphasises the 'pedagogical' character of the *Almagest*: 'it must be remembered that the *Almagest* describes the construction of the tables in such a way that the student can see the logical progression; it does not give the process by which these tables were actually derived' (1995: 236).

As a pedagogical tool, the table does something else as well: it functions as a sort of map, whereby the reader can locate the

333). Jones stresses the importance of Ptolemy's tabularising tendencies on this front: 'One significant event that was certainly felt in Greco-Egyptian astronomy was the advent of Ptolemy's tables after AD 150, which apparently for the first time popularized kinematic methods of astronomical prediction' (336).

106 The question of whether Ptolemy obtained any of his observation data by his own observations is a subject which has generated significant discussion, much of which is summarised by Riley (1995). S. J. Goldstein (1982: 201–3) concludes that Ptolemy made no lunar observations himself, noting that he 'failed to mention the scale divisions of his astrolabe [= armillary sphere], or the procedure for orienting or checking it', and that 'the description of the astrolabe in Book V, Chapter 1 is best described as obfuscatory'; but when he asks, 'could it be that the concept of checking had not yet been invented?' (201), Goldstein overlooks Ptolemy's *Harmonics*, where checking is an essential part of the initial set-up of the instruments (I.8, 18.9–19.1; I.11, 26.15–27.13). Goldstein and Bowen (1999: 342–3) take Ptolemy at his word when he claims that several sets of lunar data employed in *Alm.* IV–V are the product of his own observations (e.g. IV.6, V.3), adding that 'it is hardly surprising that Ptolemy relies on his own observations'; the real questions are why he relied on those of his predecessors (e.g. those reported by Hipparchus), and on what basis he made his selections from among the data available to him from either source.

107 Similarly, Goldstein and Bowen (1999: 343) focus their attention not on 'the geometric proofs Ptolemy invokes in order to derive parameters from specified observations', but rather on 'the observations themselves and their cognitive force in his argument'.

particular in relation to the whole (the hour within the day, the day within the year), or one phaenomenon in relation to another (the planet in relation to the constellations of the zodiac, the sun in relation to the moon) or in relation to a fixed numerical system (e.g. ecliptic coordinates in degrees). Harmonic theorists had evidently employed diagrams for similar purposes from at least as early as the mid-fourth century BC,[108] since Aristoxenus mentions (repeatedly and disapprovingly) his predecessors' practice of mapping out musical space on diagrams (*diagrammata*) in which each note was placed in succession according to pitch, rather than melodic function, combining or compressing several different pitch-series into a single diagram.[109] No such diagrams survive, but from Aristoxenus' vague description they appear to have consisted of a line representing pitch-continuum, on which notes were marked;[110] if any attempt was made to illustrate the magnitudes of the intervals between notes (a point on which we cannot be certain), equal spaces would certainly have been used to represent intervals of equal magnitude.[111]

The *Sectio canonis* engages with this tradition by combining diagrammatic exposition of scale-structures with the mathematical diagram through the instrument itself. The canonic division of proposition 20 is the culmination of a demonstrative argument whose proofs have made use of Euclidean geometrical diagrams, but it is also a map of musical space in which all the pitches from *proslambanomenos* to *nētē hyperbolaiōn* are listed in order: a visible order of bridge positions which corresponds to an audible order of pitches. And because musical space is set out proportionally, the diagram is itself a *kanonion*, which can be fixed under the string of a monochord and used to locate all the notes inscribed on it. It is an instrumental diagram, just as the monochord is a diagrammatic instrument. Like the *kanonia* of Ptolemy's *Almagest*, it has both a

108 Possibly earlier: see West 1992: 218, Gibson 2005: 181 n. 55. On the diagrams see Barker 1978a: 5–6, 16; Gibson 2005: 16–20, 47; Barker 2007: 39, 41–3.

109 *El. harm.* 2.15, 7.32, 28.1. The term *katapyknōsis*, 'compression', is used in discussions of these diagrams (*El. harm.* 28.1; see also 7.32, 38.3, 53.4–5). On the meaning of the term, see Barker 1978a and 2007: 42.

110 Barker 1989: 127 n. 6.

111 For a later and different type of diagram used to map out harmonic space, compare Aristid. Quint. I.11.

heuristic and a demonstrative function, for hearing (like seeing) is believing.

From this it is evident that although the process of argumentation in the *Sectio canonis* is apodeictic, it is aimed at the construction of a diagram which outlives its apodeictic context (unlike those of Euc. *El.* VII–IX).[112] Ptolemy's handling of the relationship between instrument, diagram, proof and table in the *Harmonics* is slightly different both from that of the *Sectio canonis* and from his own in the *Almagest*, and consequently it is in looking at these differences that we can see more clearly not only the monochord's place in the wider context of Greek science, but also the very heterogeneity of Greek science itself.

Introducing instruments

Ptolemy gives detailed instructions for the building of two instruments in *Almagest* V: the armillary sphere in V.1, and the parallactic instrument in V.12. In both cases the descriptions are practical, and based explicitly on his own experience, but in neither is a diagram included to elucidate the description.[113] 'Taking two circles . . . we fitted them together' (δύο γὰρ κύκλους λαβόντες . . . συνηρμόσαμεν, V.1, 351.12–15); 'We constructed an instrument' (κατεσκευάσαμεν ὄργανον); 'We made two four-sided rods' (ἐποιήσαμεν γὰρ κανόνας δύο τετραπλεύρους, V.12, 403.2–3, 9). In subsequent chapters, as we have seen, mathematical arguments are presented with diagrams; these are all of the same type (that is, representing the phaenomena, not the instrument: cf. V.2–7 with V.13–17), and the summaries are presented in tabular form (cf. V.8 with V.18). In the *Harmonics*, by contrast, the monochord is introduced diagrammatically:

νοείσθω δὴ κανὼν ὁ κατὰ τὴν ΑΒΓΔ εὐθεῖαν καὶ μαγάδες πρὸς τοῖς πέρασιν αὐτοῦ πανταχόθεν ἴσαι τε καὶ ὅμοιαι σφαιρικάς, ὡς ἔνι μάλιστα, ποιοῦσαι τὰς ὑπὸ τὴν χορδὴν ἐπιφανείας, ἥ τε ΒΕ περὶ κέντρον τῆς εἰρημένης ἐπιφανείας τὸ Ζ, καὶ ἡ ΓΗ περὶ κέντρον ὁμοίως τὸ Θ, ληφθέντων τε τῶν Ε καὶ Η σημείων κατὰ

[112] On *apodeixis*, see n. 18 above.
[113] Toomer, very helpfully, adds one for V.1 in his translation (1984: 218 fig. F, based on a drawing illustrating Rome 1927).

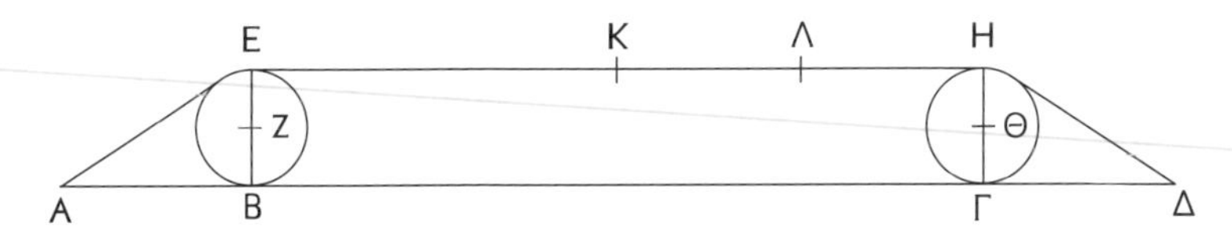

Fig. 1.9 Diagram accompanying Ptol. *Harm.* I.8.*

τὰς διχοτομίας τῶν κυρτῶν ἐπιφανειῶν. θέσιν ἐχέτωσαν τοιαύτην αἱ μαγάδες, ὥστε τὰς διὰ τῶν Ε καὶ Η διχοτομιῶν καὶ τῶν Ζ Θ κέντρων ἐκβαλλομένας, τουτέστι τὴν ΕΖΒ καὶ τὴν ΗΘΓ καθέτους εἶναι πρὸς τὴν ΑΒΓΔ.

Let us think of a *kanōn* on the straight line ΑΒΓΔ, and at its limits, bridges that are in all respects equal and similar, with the surfaces that lie under the string[114] spherical, as nearly as is possible. Let one bridge, ΒΕ, have Ζ as the centre of the surface mentioned, and let the other, ΓΗ, have Θ, similarly, as the centre, where points Ε and Η are found by bisection of the convex surfaces. Let the bridges be so placed that the lines drawn through the points of bisection Ε and Η and through the centres Ζ and Θ are perpendicular to ΑΒΓΔ.[115]

We are not presented with a report of the author's construction procedure; rather, from the first word of Ptolemy's description, we are asked to *imagine* (νοείσθω) an instrument, using the conventional *ekthesis*-language and the lettered diagram of mathematical discourse. In place of the aorist indicatives and participles of *Almagest* V.1 and 12 (κατεσκευάσαμεν, ἐποιήσαμεν, παραγράψαντες, προσεθήκαμεν, διατρήσαντες, ἐναρμόσαντες, ἐλάβομεν), here imperatives and future constructions dominate (νοείσθω; ἐχέτωσαν; ἐὰν . . . διατείνωμεν . . . ἔσται . . . λήψεται, ποιήσεται). The present tense of the passive imperative νοείσθω is an important aspect of the presentation of the instrument. As Fowler and Taisbak have pointed out (1999: 362), the perfect passive imperative form of the verb 'to draw', γεγράφθω ('let [some mathematical object] have been drawn'), constitutes the 'verbal mask' of what they call 'the Helping Hand', by which operations

* Düring (1930: cii) reported that the MS tradition of Ptolemy's *Harmonics* was quite uniform in its transmission of the diagrams with the exception of this one, which was apparently variously understood by copyists; his drawing of it (which I have followed here) was based on that of Vaticanus Graecus 186, copied in the thirteenth century (Düring's E).

114 On Barker's reading (τὴν χορδήν) versus Düring's (τὰς χορδάς), see ch. 6.

115 *Harm.* I.8, 17.27–18.4, trans. Barker 1989: 292. For the rest of the passage and fuller discussion, see ch. 6.

such as drawing lines, describing circles and the like are carried out. 'Its main effect and interest is to keep us mortals out of the play. Greek geometry is not about what we can do, but about what can be done.' What is curious, and relevant here, is that while there is indeed a Helping Hand in Greek mathematics, there seems not to be a Helping Imagination. A search of 133 mathematical authors in the *TLG* corpus reveals that while present passive imperative forms of *graphein* are as good as non-existent (I can find, in fact, only one instance of γραφέσθω in a Greek mathematical text[116]), the tendency is toward the opposite in the case of *noein*, where present passive forms are over three and a half times as common as their perfect counterparts. Two-dimensional objects may be drawn for us, but three-dimensional objects, more often than not, are thought of as requiring the active participation of our imagination.[117]

And yet, as much as Ptolemy's language and use of the diagram suggest the *ekthesis* of a mathematical proof, at the same time he asks his reader to imagine a concrete object, not merely a perfectly formed intellectual object. 'The geometer makes no use of the faculty of perception', said Aristoxenus, 'but for the student of music accuracy of perception stands just about first in order of importance.'[118] The geometry of sound requires of its practitioners both the contemplation of the perfect instrument, and the construction of a necessarily imperfect one. Ptolemy presents his reader with the conflicting demands of intellectual and concrete construction in the first sentence of the passage from I.8 quoted above: we must imagine (νοείσθω) a *kanōn* with bridges altogether equal, similar, and spherical in surface, 'as nearly as is possible' (ὡς ἔνι μάλιστα).[119] This parenthetical remark would seem unnecessary if the instrument were to be merely an object of thought. There is no 'nearly possible' for the imagination; the *kanōn* of our mind's eye can be built without imperfections of any sort. The collision of the mathematical and physical construction of the instrument

[116] Procl. *Hypotyp.* 4.33.

[117] On the association between νοείσθω/νενοήσθω and three-dimensionality in the *ekthesis*, see n. 85 above.

[118] *El. harm.* 33.14–15, 21–4, trans. Barker 1989: 150. For the entire passage and fuller discussion, see ch. 5 part 2. Compare Arist. *An. post.* 76b39–77a3.

[119] The sphericity of the bridges' surfaces will be discussed in ch. 6.

results directly from Ptolemy's decision to handle the introduction of the *kanōn* differently from that of the armillary sphere and the parallactic instrument. Unlike these astronomical instruments, the diagrammatic nature of the *kanōn* is what defines its demonstrative capability. The armillary sphere and the parallactic instrument will allow diagrammatic expositions of the phaenomena, but the *kanōn* is itself a diagram of the arithmetical relationships which underpin the phaenomena. The diagram of *Harmonics* I.8 is the diagram of an instrument of a diagram.

A contrast to Ptolemy's method here can be seen in Diocles' way of introducing the parabolic mirror (*On Burning Mirrors* = Περὶ πυρείων, prop. 1).[120] The *ekthesis* requires the construction of a two-dimensional parabola, with diagram, on which the remainder of the proof is based: 'Let there be a parabola KBM, with axis AZ...' (38). Only after the mathematical proof is complete does Diocles use the two-dimensional figure to construct the three-dimensional instrument: 'Hence, if AZ is kept stationary, and KBM revolved (about it) until it returns to its original position, and a concave surface of brass is constructed on the surface described by KBM, and placed facing the sun, so that the sun's rays meet the concave surface, they will be reflected to point Δ, since they are parallel to each other' (51–2). Like Ptolemy's monochord, Diocles' parabolic mirror is set out in a mathematical *ekthesis* with a lettered diagram (now lost; see Toomer 1976: 27, 32); in both cases the diagram 'flattens' the three-dimensional instrument into a two-dimensional representation, and plane geometry is then employed to demonstrate its mathematical properties. But whereas we are clearly meant to imagine the fully formed instrument as we read Ptolemy, *Harmonics* I.8, and see the diagram as a side-view of the monochord, Diocles makes his reader focus on a two-dimensional parabola first, and only afterward revolves it to generate the three-dimensionality of the finished instrument: Ptolemy wrote νοείσθω; Diocles almost certainly wrote ἔστω. Thus we have a different version of the (a) + (b) + (c) + (d) formula: the diagram represents an axial section of the instrument (b), whose plane geometry works

[120] Arabic text and translation in Toomer 1976; no Greek MSS survive. The translation quoted here is Toomer's.

like the solid geometry of the instrument (c); the instrument, when it has been correctly arranged (d), generates the phaenomenon under investigation (a), namely focal reflection of solar rays.

Compare, on the other hand, Heron's introduction of the dioptra at *Dioptra* 3 (lacunose but useful nonetheless): the instrument is described first in practical detail; then the description is applied to a lettered diagram (now lost) with the words ἔστω οὖν followed by the designation-letters (AB, ΓΔ, EZ and so on). This is the language of representation, but not of imagination (contrast Ptolemy's νοείσθω δή), and the instrument is already fully formed by description before it is represented by diagram.[121]

Instruments between diagrams and tables

Once Ptolemy has introduced the *kanōn* in *Harmonics* I.8, an argumentative pattern broadly similar to that of *Almagest* V can be adopted: proofs can be made, based either on principles which govern the instrument's operation, or on observations taken from its use, and these often use diagrams; the summary is then presented in tables. In *Harmonics* I.9, for example, Ptolemy gives a brief proof that intervals of the same magnitude will correspond to different lengths of the same string depending on their pitch:

ὑποτεθείσης γὰρ τῆς AB διαστάσεως τοῦ διὰ πασῶν, τοῦ A νοουμένου κατὰ τὸ ὀξύτερον πέρας, καὶ ληφθεισῶν δύο τοῦ διὰ πέντε, μιᾶς μὲν ἀπὸ τοῦ A ἐπὶ τὸ βαρύ, ὡς τῆς AΓ, ἑτέρας δ' ἀπὸ τοῦ B ἐπὶ τὸ ὀξύ, ὡς τῆς BΔ, ἐλάττων μὲν ἔσται ἡ AΓ διάστασις τῆς BΔ διὰ τὸ κατ' ὀξυτέρων πίπτειν τάσεων, μείζων δὲ ἡ BΓ ὑπεροχὴ τῆς AΔ.

For if we assume that the distance AB is an octave, A being thought of as the higher limit, and take two fifths, one downwards from A – call it AΓ – and the other upwards from B – call it BΔ – the distance AΓ will be smaller than BΔ because it falls between higher pitches, and the difference BΓ will be greater than AΔ.[122]

[121] See Lewis 2001: 53–6, 260–1; Coulton 2002.

[122] *Harm.* 21.3–8, trans. Barker 1989: 294. For another instance of ὑποτίθημι governing designation-letters in a genitive absolute construction where the diagram is even more important, cf. Ptol. *Alm.* X.4: καὶ ὑποτεθέντος τοῦ ἀστέρος κατὰ τὸ K σημεῖον (308.4–5).

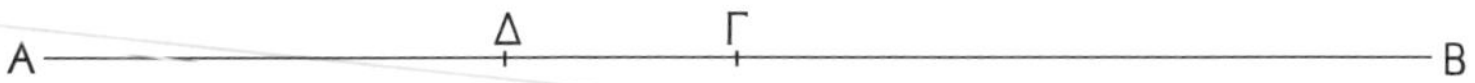

Fig. 1.10 Diagram accompanying Ptol. *Harm.* I.9 (after Düring 1930).

To understand the diagram, we must see it as representing the right-hand section of a monochord string which extends to the left past A by a distance equal to AB. B thus functions as a fixed bridge position; a movable bridge will be placed as needed at A, Δ or Γ, and the string will always be plucked to the left of the movable bridge. If the entire string were, for example, 120 units long, B would be the 120-unit mark, A would be placed at 60, Γ at 90 and Δ at 80. Thus AΓ = 30 and BΔ = 40, but both AΓ and BΔ = a fifth. I shall refer to this type of diagram as 'canonic', since it represents intervals proportionally, as the *kanōn* does. The proof, though internally sound, is not well employed in the argument of the chapter, and Ptolemy has been charged with either deliberate obfuscation or confusion, but for present purposes that is irrelevant.[123] What counts here is that a diagram representing a string can be used in conjunction with a simple proof. This is only the third diagram in the treatise as we now have it: the second was that of I.8 introducing the *kanōn*, and the first was a simple line diagram in which parallel lines of equal length represented notes of different pitches (I.5). The diagrammatic introduction of the *kanōn* has permitted the use of canonic diagrams in subsequent chapters. But there are remarkably few diagrams generally in the *Harmonics* compared with the *Almagest*; some diagrams of the canonic type accompany the explication of instrumental properties or procedures (e.g. I.11 on the eight-stringed *kanōn*; II.2 on the *helikōn* and Ptolemy's adaptation of it; III.2 on curved bridges), though less elaborate diagrams which merely presuppose an understanding of the monochord's basic principles also occur (I.11, III.1, III.2). Since the instrument is the diagram, there is no separate role for the apodeictic diagram, as there is in the *Almagest*;[124]

[123] Barker 1989: 294 n. 85. The problem is not with this proof, considered in isolation, but with the way Ptolemy attempts to use it in arguing 'that the Aristoxenians are wrong in measuring the concords by the intervals and not by the notes' (I.9 title, trans. Barker 1989: 293).

[124] By 'apodeictic diagram' I mean a lettered diagram whose parts are identified by and used in a mathematical proof in the accompanying text. On the term 'apodeictic', see n. 18 above.

most of the apodeictic diagrams of the *Harmonics* are sections of the *kanōn*-diagram, redrawn to fit the context, like that of I.9 above.[125]

As in the *Almagest*, mathematical arguments (e.g. theorem, problem, analysis, computation) are followed by tables.[126] Ptolemy sets out fifteen tables in *Harmonics* II.15, the first fourteen of which list one octave's worth of bridge positions for the attunements of his seven keys (*tonoi*) in different mixtures of genera; the last is a summary of bridge positions for each note of the octave.[127] The columns of bridge positions are given according to a standard measure, and all fractional quantities are given in sexagesimals, as in the *Almagest*.[128] The table is therefore a list of numerical coordinates which allow the reader to locate or generate the phaenomena using the instrument, as at *Almagest* VII.5.

The difference between location (in astronomy) and generation (in harmonics) is intrinsic to the nature of the phaenomena, the instruments and the sciences. Both activities are, on their own, heuristic; both, however, become part of a wider demonstrative project when the phaenomena turn out to be located or generated as expected. Here we see the difference between the primary QEI (ὅπερ ἔδει εὑρεῖν) and QEF (ὅπερ ἔδει ποιῆσαι) purposes of astronomical and harmonic tables respectively, and the secondary QED (ὅπερ ἔδει δεῖξαι) purpose which they both serve within their treatises. QEF is never applied to the phaenomena of astronomy because, by their nature, stars and planets cannot be 'constructed'

[125] There are other types of diagram in the *Harmonics*. Most work in the same way as that of I.5 (with different pitches represented by parallel lines of equal length), but many are made more complex by the inclusion of additional material. Those of II.1, for example, include not only the designation-letters used in the accompanying proofs, but also (as we receive them, at least) the note-names, string lengths on Ptolemy's standard *kanonion*, and the ratios of these lengths to one another. These diagrams, though not proportional like that of I.9, rely equally firmly on the *kanōn* for their interpretation.

[126] On the different types of table in the *Almagest*, see n. 104 above. Riley (1995) emphasises tabular presentation as a consistent feature of Ptolemy's method in optics, geography and astronomy.

[127] There are fourteen tables for seven *tonoi* because the *tonoi* are two-octave scales; Ptolemy separates them according to their upper (tables 1–7) and lower (tables 8–14) octaves.

[128] On the uniqueness of this application, see n. 57 above.

by the astronomer; hence astronomical instruments do not function in mode (a). Geometry, by contrast to astronomy, never requires the finding of a mathematical object which cannot be constructed; this is because it is a science without phaenomena. In a science with only intelligible rather than sensible objects, QEF must (at least in assumptions) precede QEI. Harmonics, like astronomy, has sensible objects, but 'finding' them means necessarily 'constructing' them too, and 'constructing' them is for the harmonicist part of proving that he really has 'found' them. Thus in harmonics, QEF is often the QED of QEI.

Ptolemy calls the first fourteen tables *kanones*, and the fifteenth a *kanonion*, the usual vocabulary familiar from the *Almagest*. By a fortuitous overlap of vocabulary, the connection between the harmonic instruments and the tables of attunements is emphasised: the marked monochord-rule (*kanonion*) is also a proportional version of the table of bridge positions (*kanonion*); the *kanōn* is now the instrument and the diagram and the table.[129] It makes the necessary connection between the objects of perception (audible pitches and intervals; visible distances) and the objects of thought (ratios of numbers) through both diagrammatic and tabular presentation.

The table is the scientist's textual equivalent of the *parapēgma*, an astronomical and meteorological calendar usually inscribed on stone, which contained holes bored beside each item into which a peg could be inserted to mark the current date.[130] The *Suda* defines it as 'a table (*kanōn*); also a kind of astronomical instrument'.[131] It is a heuristic device which allows the user to see the particular (today) in relation to the whole (the year), to understand a large structure by means of a single synoptic view, and to predict what

[129] It is unlikely that Ptolemy is either innovating or playing with the senses of *kanōn* and *kanonion*; the use of *kanōn* to mean a numerical table appears from Plutarch's use of it (κανόνες χρονικοί, *Solon* 27) to have been current in second-century AD literary Greek, and *kanōn* had also meant 'monochord' (among many other things) as early as Duris of Samos, c. 300 BC (*FGrH* 76F23 = Porph. *VP* 3). See LSJ s.v. κανών I.10, II.4; κανόνιον IV, VI.

[130] LSJ s.v. παράπηγμα; see Geminus, *Intro. astr.* 17 and the *parapēgma* appended to that treatise (text and translation in Lehoux 2007: 226–39; translation with notes in Evans and Berggren 2006). Ptolemy included a *parapēgma* in his *Phaseis* (Lehoux 2007: 261–309). See also Evans 1999: 256–9, Hannah 2002, Lehoux 2005.

[131] παράπηγμα: κανών. καὶ εἶδός τι ὀργάνου ἀστρονομικοῦ, π 410.

will happen next based on a summary of prior observation.[132] Several *parapēgmata* are extant, including one recorded on papyrus.[133]

The instrument was used to record the cycles of astronomical, meteorological and associated phaenomena, but in one instance its name was applied to a harmonic context. Ptolemaïs of Cyrene claimed that through the science of 'canonics' (ἡ κανονικὴ πραγματεία), reason discovers both what is correct (τὸ ὀρθόν) and 'the *parapēgmata* of what is well attuned' (τὰ τοῦ ἡρμοσμένου παραπήγματα).[134] In this same passage Ptolemaïs counters the apparently common assumption that the science called *kanonikē* was named after the instrument called *kanōn*. Her claim is incredible: just as Heron's term 'dioptrics' (ἡ διοπτρικὴ πραγματεία) is certainly derived directly from the name of the instrument on which this science was practised (the dioptra, ἡ διόπτρα), 'canonics' must likewise be derived from *kanōn*.[135] The primary postulates of the musical and mathematical constituents of the science are brought together through demonstrations on the instrument; the conclusion of this process is the canonic division, in which the ruler applied to the string is marked with bridge positions. The marked *kanonion* functions as a *parapēgma*, which works like a table. Tables function in mode (b): numbers represent bridge positions or celestial coordinates; the *parapēgma* adds mode (d), requiring the physical movement of pegs to indicate one's current position within the wider scheme. It is through the physical manipulation of the instrument that reason 'discovers the *parapēgmata* of what is well attuned'.

Ptolemy's use of the instruments of harmonics and astronomy shows the extent to which the polyvalent function of the monochord (its (a) + (b) + (c) + (d) formula) on the one hand, and its

[132] On the predictive functions and uses of *parapēgmata*, see Lehoux 2007: 19, 55–69.

[133] *PHib*. I.27, col. 4. Text and translation in Lehoux 2007: 217–23. See also Fowler 1999: 229–30, Evans 1999: 256–7, Evans and Berggren 2006: 58–63.

[134] Ptolemaïs ap. Porph. *In Harm*. 22.29–30, trans. Barker 1989: 239. The passage is quoted more fully in ch. 5 part I. Ptolemaïs' dates are unknown; she may perhaps have lived as late as the first century AD.

[135] Ptolemaïs makes the claim at Porph. *In Harm*. 22.27–8; it will be discussed in ch. 5 part I. For 'dioptrics', see Heron, *Dioptra* 1.1, 2.19, 34.1 and Lewis 2001: 259–60, 283.

multiple identities within a single argument (instrument-diagram-table) on the other, contribute to making it what I have called (with apologies to Reviel Netz) the 'metonym of mathematical harmonics'.[136] It will hardly be surprising, then, that Aristotle's term 'mathematical harmonics' (*An. post.* 79a1) is often displaced in later authors by Ptolemaïs' term 'canonics'. Practitioners of this science were called 'canonicists' (*kanonikoi*): 'A *kanonikos*, in general, is a harmonic theorist who constructs ratios in connection with attunement.'[137] Ptolemy, while he does not use the term *kanonikē*, does refer to 'the *kanonikoi*' (*Harm.* 66.18), and shares the fundamentals of Ptolemaïs' view of their science:

ἁρμονικοῦ δ' ἂν εἴη πρόθεσις τὸ διασῶσαι πανταχῇ τὰς λογικὰς ὑποθέσεις τοῦ κανόνος μηδαμῇ μηδαμῶς ταῖς αἰσθήσεσι μαχομένας κατὰ τὴν τῶν πλείστων ὑπόληψιν, ὡς ἀστρολόγου τὸ διασῶσαι τὰς τῶν οὐρανίων κινήσεων ὑποθέσεις συμφώνους ταῖς τηρουμέναις παρόδοις, εἰλημμένας μὲν καὶ αὐτὰς ἀπὸ τῶν ἐναργῶν καὶ ὁλοσχερέστερον φαινομένων, εὑρούσας δὲ τῷ λόγῳ τὰ κατὰ μέρος ἐφ' ὅσον δυνατὸν ἀκριβῶς. ἐν ἅπασι γὰρ ἴδιόν ἐστι τοῦ θεωρητικοῦ καὶ ἐπιστήμονος τὸ δεικνύναι τὰ τῆς φύσεως ἔργα μετὰ λόγου τινὸς καὶ τεταγμένης αἰτίας δημιουργούμενα καὶ μηδὲν εἰκῇ, μηδὲ ὡς ἔτυχεν ἀποτελούμενον ὑπ' αὐτῆς καὶ μάλιστα ἐν ταῖς οὕτω καλλίσταις κατασκευαῖς, ὁποῖαι τυγχάνουσιν αἱ τῶν λογικωτέρων αἰσθήσεων, ὄψεως καὶ ἀκοῆς.

The aim of the student of Harmonics must be to preserve in all respects the rational postulates (*hypotheseis*) of the *kanōn*, as never in any way conflicting with the perceptions that correspond to most people's estimation, just as the astronomer's aim is to preserve the postulates (*hypotheseis*) concerning the movements of the heavenly bodies in concord with their carefully observed courses, these postulates having been taken from the obvious and rough and ready phenomena, but finding the points of detail as accurately as is possible through reason. For in everything it is the proper task of the theoretical scientist to show that the works of nature are crafted with reason and with an orderly cause, and that nothing is produced by nature at random or just anyhow, especially in its most beautiful constructions, the kinds that belong to the more rational of the senses, sight and hearing.[138]

Ptolemy emphasises the important similarities between astronomy and harmonics, some of the consequences of which we have seen in his use of instruments, diagrams and tables in the *Harmonics* and

136 See n. 54 above.

137 κανονικὸς δ' ἐστὶ καθόλου ὁ ἁρμονικὸς ὁ περὶ τοῦ ἡρμοσμένου ποιούμενος τοὺς λόγους, Ptolemaïs ap. Porph. *In Harm.* 23.5–6, trans. Barker 1989: 239.

138 *Harm.* I.2, 5.13–24, trans. Barker 1989: 278–9.

in *Almagest* V. At the same time, however, we are confronted both here and in Ptolemaïs' briefer summary with the key difference between instrument use in harmonics and astronomy. The 'rational postulates of the *kanōn*' (τὰς λογικὰς ὑποθέσεις τοῦ κανόνος)[139] are compared with those 'of the movements of the heavenly bodies' (τὰς τῶν οὐρανίων κινήσεων ὑποθέσεις): the *hypotheseis* of astronomy are about motion (*kinēsis*), while those of harmonics are about ratio (*logos*).[140] The mathematical harmonic theorist, as Ptolemaïs says, uses the monochord to construct ratios in connection with attunement (περὶ τοῦ ἡρμοσμένου ποιούμενος τοὺς λόγους, 23.5–6); her vocabulary (the verb ποιεῖν) designates ratios as the QEF of canonics.

The difference, then, is that which separates arithmetical and geometrical sciences. The ultimate object of an arithmetical science is *arithmos*, number. Arithmetic is, by definition, *about* numbers and their relationships. Thus in harmonics the ratios corresponding to the musical intervals are objects of scientific interest in their own right. But in the geometrical sciences, arithmetic enters the picture only as an intermediary – a reckoning stage in the method – and not as an aim. Ratios are involved in the process of determining the height of a distant wall without approaching it, but only for thinking through the similar triangles which allow a solution. After that, the ratios which obtain in the given situation

139 Ptolemaïs uses exactly the same phrase, αἱ λογικαὶ ὑποθέσεις τοῦ κανόνος, at Porph. *In Harm.* 24.5 – unless Porphyry, our only source for Ptolemaïs, has paraphrased, using Ptolemy's language: the quotation occurs in Porphyry's commentary on the very passage quoted above (Ptol. *Harm.* I.2, 5.13–24). It is also possible that Porphyry's quotation from Ptolemaïs terminates at *In Harm.* 23.31 (though neither Düring nor Barker read it so), or for that matter that Ptolemaïs did use the phrase first, and that Ptolemy had direct access to her book. But it may be that Ptolemy, like Porphyry, read quotations from Ptolemaïs in Didymus' book (see Barker 1989: 230). If so, it would certainly have fallen within Porphyry's stated aims to use this quotation as a way of exposing Ptolemy as a plagiarist (Porph. *In Harm.* 5.7–15, discussed briefly in ch. 6).

140 The rational *hypotheseis* of the monochord are mathematical and logical starting points, for example that intervals are in ratios of numbers, or that there exist concordant and discordant intervals (see Ptolemaïs ap. Porph. *In Harm.* 23.10–22, of which Düring prints only the first three lines as a quotation from Ptolemaïs; Barker extends the quotation to 23.22); the *hypotheseis* of the movements of the heavenly bodies might be understood to mean the quantified geometrical models advanced to 'save' the phaenomena (an astronomical usage in Ptolemy: see n. 102 above), but this would disrupt the parallelism between the two uses of *hypothesis* in the same sentence. More probably we are meant to understand the similar mathematical and logical starting points of the investigation of the heavenly movements.

may be forgotten: they are particular, not universal. But for the harmonicist, the ratios are always universal.

A consequence of this is that in the geometrical sciences ratios occupy a middle position between the starting data (Ptolemy's 'obvious and rough and ready phenomena') and the *quaerendum*. But in harmonics, the ratios are at one end of the apodeictic process, and the sensible data are at the other: depending on the demonstration, the *quaerendum* will be either a note (or interval), or a ratio.

For this reason ratio form is important to the harmonicist and not to the astronomer (or not directly as a concern of astronomy *per se*). Hence a harmonicist like Adrastus (see ch. 5 part 2) or Ptolemy (ch. 6) will be concerned to show that reason and perception agree about the identification of analogous ratios and intervals, or even that the best formed of each will be apparent to each faculty independently. These concerns are part of the scientific project which Ptolemy calls 'showing that the works of nature are crafted with reason and with an orderly cause, and that nothing is produced by nature at random or just anyhow'.[141] The predilection for multiple and epimoric ratios, and also for concords and simple melodic intervals, follows from this.[142]

In astronomy, as a geometrical science, it is types of motion rather than types of ratio which occupy this place. As simplicity of ratio form dominates mathematical harmonic thinking, simplicity of motion dominates astronomical thinking.[143] Epicycles and eccentrics, as ways of combining more than one simple motion in order to account for the phaenomena, are the astronomer's equivalent of the harmonicist's compound ratios. The ancient assumptions not only of the central position of the earth, but also of a celestial model containing only perfectly circular components (conceived geometrically as spherical sections), rather than elliptical ones (oblique conic sections), are akin to those made about multiple and epimoric ratios. Both

[141] τὸ δεικνύναι τὰ τῆς φύσεως ἔργα μετὰ λόγου τινὸς καὶ τεταγμένης αἰτίας δημιουργούμενα καὶ μηδὲν εἰκῇ, μηδὲ ὡς ἔτυχεν ἀποτελούμενον ὑπ᾽ αὐτῆς, 5.20–2, trans. Barker 1989: 278–9.

[142] On multiple and epimoric ratios, see the Introduction.

[143] See Bowen 2007, and cf. Ptol. *Harm.* 100.32–101.1.

systems of thought are predicated upon a view of nature which is rooted in the idea that simple, elegant, beautiful structures underlie the entire universe in all its manifestations. These structures are both arithmetical (e.g. the 2:1 octave) and geometrical (e.g. the circular orbit). They are also, for the mathematical scientist, unquestionable.[144]

The following chapters attempt a more or less chronological presentation of the development of the scientific method which underpins Ptolemy's use of instruments in the *Harmonics*. Many of the basic assumptions about ratio form, I shall argue, predate the invention of the monochord; but once the monochord was available to harmonic scientists, the demonstrative possibilities for its use were developed in increasingly complex ways. Ptolemy stands quite alone at the culmination point of ancient Greek scientific achievement in mathematical harmonics, but we shall see that several earlier figures deserve some credit for advancing the role of the instrument within the science. The first task, however, will be to establish a *terminus ante quem* for the monochord.

[144] See e.g. Gem. *Intro. astr.* 1.19–22.

2

MATHEMATICAL HARMONICS BEFORE THE MONOCHORD

In the first place, it may be taken as certain that Pythagoras himself discovered the numerical ratios which determine the concordant intervals of the scale.[1]

The invention of the canon is traditionally attributed to Pythagoras (and whatever one may think of the half legendary figure of Pythagoras, this dating is not to be doubted).[2]

Those who attempt to explore the historical edifice of Greek mathematical enquiry from its foundations inevitably find the darkest recesses of the cellar haunted by the spectre of Pythagoras. It is self-assured, authoritative, larger than life; it repeats to all who will hear, 'I thought of that first.' And it has convinced many. But it does not speak for itself; it merely parrots lines written for it by others. Some of its lines were composed as early as the fourth century BC, but its script was assembled and improved by Nicomachus, Diogenes Laertius, Porphyry, Iamblichus, Proclus and others, those necromancers of late antiquity for whom it has done long service reshaping the origins of the mathematical sciences (and much else besides) in ancient Greece.[3] In harmonics, the cellar echoes with the sound of the monochord. Our first task in this chapter will be to separate the ghost stories from the reliable accounts, to shut the cellar door, and to begin our investigation from the more brightly lit corridors of the ground floor.[4]

1 Burnet 1920: I.45. 2 Szabó 1978: 138.

3 The reshaping was varied, even within single disciplines. For an example, see Cuomo's summary (2000: 81) of the way Iamblichus and Proclus 're-write or opportunely construct previous mathematical traditions' for their own Pythagoreanising purposes; she contrasts 'Proclus' attempt at "annexing" Euclid to his own tradition' with 'Iamblichus' attempt at pushing Euclid aside to make space for his own people'.

4 Kahn (2001) offers a helpfully balanced general account of Pythagoras and Pythagoreanism.

Nicomachus of Gerasa (c. AD 100) is our earliest extant witness to what is surely among the most memorable and widely related anecdotes about Pythagoras: the one in which a chance encounter at a blacksmith's shop leads to the discovery of the ratios of the concords.[5] It was retold, with variations, by many authors, including Boethius.[6] The ringing of hammer-blows on the anvil attracted Pythagoras' attention, the story goes, because the pitches he heard happened to be in perfectly tuned musical intervals. There were four hammers, and the intervals between their pitches were the octave, fifth, fourth and tone. Drawn to investigate this prodigy, Pythagoras discovered through many clever experiments (*peirai*) that it was the weight of the hammers, and no other factor, which governed their pitches. He found, furthermore, that the weights of the hammers which had sounded the concords were in the ratios 2:1, 3:2 and 4:3; the weights of the two that had emitted the tone between the fifth and the fourth were in 9:8 ratio. Next, the story continues, he went home (to the relief, no doubt, of the inconvenienced blacksmiths), suspended weights of the same proportions from identical strings, and discovered that the strings, when plucked, gave the same intervals. He went on to prove the correspondence between these intervals and the same ratios in a wide variety of instruments and objects capable of emitting pitched sounds, including bowls, pipes and stringed instruments of several varieties. In Boethius' account, his examination of string length and thickness as causes of pitch led him to innovate still further: and 'in this way he invented the monochord' (*itaque inuenit regulam*, 198.23–4).

Many have pointed out the various ways in which the story is impossible. In the first place, when hammers of different weights are used in smithing there is next to no variation in the pitch of the different ringing sounds, since it is the anvil, and not the hammers, whose ringing is heard. Secondly, even if the procedure

[5] Nicom. *Harm.* 6, on which see Burkert 1972: 376; Levin 1994: 86; annotated translation in Barker 1989: 256–8.

[6] Boeth. *Mus.* 1.11; also Adrastus, Gaudentius, Censorinus, Iamblichus, Macrobius, Fulgentius, Calcidius and Isidore of Seville: see Levin 1975: 74 and 1994: 86, 95 n. 2; Stahl 1952: 186 n. 4, 187 n. 6; Lloyd 1979: 144 n. 95 and 1987: 295 n. 39; Barker 1991: 52 n. 9.

were modified so that, for example, four metal objects of 12, 9, 8 and 6 units of weight were struck in pairs,[7] the intervals heard would not necessarily be those of the fundamental concords and the tone: the mass of a metal body is not what determines its pitch when struck. Likewise, many of the procedures by which Pythagoras is said to have tested his discovery will fail to yield the desired ratios: when strings of the same thickness and length are tuned in concords, the ratios of their differing tension are not the same as the ratios of the differing lengths of strings of the same thickness and equal tension, similarly tuned. The ratios of string tension are the squares of the ratios of string length (octave $= (2:1)^2 = 4:1$, fifth $= (3:2)^2 = 9:4$, fourth $= (4:3)^2 = 16:9$), and their application to pitch is inversely proportional: four times the tension raises the pitch an octave, whereas double the length lowers the pitch an octave. Unlike those of length, the tension-ratios do not seem to have been known in antiquity.[8] Furthermore, they are not easily obtained from the procedure Nicomachus describes; if it was actually carried out before the time of Ptolemy, no honest report of its results survives.[9] And even if the ratios of string tension could have been extracted from such a procedure, they would have allowed no ready connection between music and, for example, the *tetraktys*.[10] Given the many subsequent philosophical applications of the harmonic ratios by the Pythagoreans and Platonists, the disjunction between the true results of this procedure and those claimed for it is not insignificant. The Pythagoras

[7] Nicom. *Harm.* 6, 247.10. The series 12, 9, 8, 6 gives the ratios in lowest terms (12:6 = 2:1, the octave; 12:8 and 9:6 = 3:2, the fifth; 12:9 and 8:6 = 4:3, the fourth; 9:8 is the tone).

[8] Burkert 1972: 376 n. 24.

[9] Ptolemy rejects it as a 'source of controversy' (*Harm.* I.8, 17.2; Barker 1989: 291). He was evidently aware that it would not work (17.7–12), but his explanation (the impossibility of acquiring truly identical strings, and the greater distortion of the strings' length and density by the heavier weights than by the lighter ones, increasing their inequality) suggests that if he tested it himself, he was unable to find the principal cause of its failure to generate the expected ratios. Ptolemy's account is uniquely honest, however, and he rightly recognises the impossiblity of making any scientific use of the procedure, given the primary postulates of his harmonics.

[10] The *tetraktys* was a representation of the ratios 4:3:2:1 set out as an equilateral triangle composed of ten dots. Because it contained the simplest and most musically significant ratios within a simple regular geometric figure, it became an important symbol of Pythagorean musical and mathematical doctrine.

of the discovery narrative possesses a sort of scientific Midas touch.[11]

Discovery, invention and experiment

The 'harmonious blacksmith' tradition, taken as a whole, rests on two important assumptions: first, that for every significant discovery or invention there must be a *prōtos heuretēs*, a 'first discoverer/inventor';[12] and second, that the discovery of the harmonic ratios and the invention of the monochord (as the instrument on which these were most easily demonstrated) must have occurred more or less at the same time. The assessments of John Burnet and Árpád Szabó (quoted at the head of this chapter) rest, in different ways, on these assumptions.[13] To these we could add a third: that the initial role for which the monochord was devised was empirical, rather than (merely) demonstrative; that it was invented not simply for scientific proof (*apodeixis*), but for scientific experimentation (*peira*).[14]

Not even in antiquity was the first assumption uncontroversial. While many writers of Nicomachus' age clearly saw it as one of the first duties of the historian of a discipline to name the *prōtos heuretēs* of its every aspect,[15] others did not. Theon of Smyrna, for example, repeats Plato's anecdote about tragic portrayals of Palamedes inventing number in order to muster the troops and

[11] As Barbera remarks, 'that the acoustical experiments of the myth did not work out must have only enhanced the perception of Pythagoras as magical. After all, the experiments did work when he performed them' (1984a: 200). To these procedures we may compare the similarly incredible proofs of the kinship between beans and humans given by Porphyry (*VP* 44), which help to contextualise the explanation of key Pythagorean doctrines (e.g. the prohibition against eating beans) by means of misleading demonstrations in the Pythagoras narratives of late antiquity. (Riedweg (2005: 69–70), remarkably, defends one of the bean demonstrations.)

[12] On this *topos* in Greek literature, see Kleingünther 1933; on its importance in the historiography of science in antiquity, see Zhmud 2006, ch. 1.

[13] Burnet was hardly alone in being uncritical of the tradition that Pythagoras discovered the concord ratios: compare the reactions of Heath 1921: I.85; Cornford 1922: 144–5; Taylor 1928: 164. Others thereafter (e.g. Ross 1924 and Brunet and Mieli 1935) show increasing caution; see Guthrie's summary, 1962: 221.

[14] It is important to note that experiment and demonstration are not mutually exclusive: experiments may be presented as demonstrations, or as components of demonstrations.

[15] [Plut.] *De musica* is an extreme example in the field of music. The heurematography of earlier centuries is discussed by Zhmud (2006: 12–14, 23–44).

count the ships at Troy, 'as though before that time, they were uncounted and Agamemnon apparently did not even know how many feet he had, if in fact he did not know how to count'.[16] My approach to the Pythagoras narratives will be to regard them as a product of this first assumption among authors for whom Pythagoras was already a figure of great stature in the history of science and philosophy. In Burkert's terms, it is the lore, rather than the science, which dominates the writing of such accounts, whether they are of Pythagoras and the forge, or of Archimedes and the bath.[17] The naming of a *prōtos heuretēs* amounts to the setting of a chronological boundary point for a narrative, and I too must establish my *termini* in this chapter. But rather than asking 'who first?', I shall ask 'by when?'; the paltry state of our evidence for early Greek investigations in harmonics and acoustics permits little more. On the question of who invented the monochord I shall go only so far as to argue that the ancient evidence which names Pythagoras is not compelling, that an alternative recorded by Duris of Samos is unverifiable, and that further speculation would be unwise.

The second assumption, I shall propose, is unfounded, firstly because despite reliable evidence for a demonstration of the harmonic ratios in the early fifth century BC, there is no evidence of the monochord's existence before 300 BC; and secondly because none of the harmonic achievements of those who worked before 300 BC rely on the instrument.

The third assumption was made by later authors labouring under the first assumption, for whom the retelling of a discovery necessitated an account of how it was made and by what means it was confirmed. The discovery narratives, not unreasonably, present instruments as components of an apparatus of discovery, in which the method of *heuresis* often includes tests (*peirai*). In contrast to this feature of the 'harmonious blacksmith' tradition is the absence of the language of experiment from the presentation of instrumental

[16] ὡς πρὸ τοῦ ἀναριθμήτων ὄντων καὶ τοῦ Ἀγαμέμνονος ὡς ἔοικεν οὐδὲ ὅσους εἶχε πόδας εἰδότος, εἴγε μὴ ἠπίστατο ἀριθμεῖν, 4.6–8; cf. Pl. *Resp.* 522d. Later, however, Theon has no qualms about quoting Adrastus on what Pythagoras was reportedly the first to discover (πρῶτος ἀνευρηκέναι, 56.10) in the field of music.

[17] Burkert 1972 (title); Archimedes: Plut. *Non posse* 1094b–c.

procedures in other discussions of mathematical harmonics until Ptolemy. This does not mean that early investigations in acoustics and harmonics did not include 'rudimentary empirical tests', as G. E. R. Lloyd has put it (1991: 83); but it does force us to distinguish between methods of investigation and methods of presentation. The presentation of the earliest instrumental procedures in harmonics and acoustics, I shall argue, is not clearly empirical (in comparison, at least, with the presentation of instrumental procedures in the 'harmonious blacksmith' tradition), though the researches which preceded them may well have been. I shall argue, further, that in one case (Archytas' presentation of instrumental examples) the method of presentation is much more similar to that in which the monochord first appears: the demonstrative context of the *Sectio canonis*. This distinction is important not only because it will allow us to appreciate more fully the difference between Ptolemy's argumentative presentation of the monochord and those of his predecessors, but also because the assumption of experimentation which is so unequivocal in the discovery narratives of Nicomachus and others has often been applied, sometimes uncritically, to important figures such as Hippasus, Philolaus and Archytas.

The chapter is divided into four parts. Parts 1 and 2 seek to establish the two *termini* already mentioned: (1) the date by which a demonstration of the harmonic ratios had been carried out, and (2) the date by which the monochord had come into use. This chronology will place Philolaus and Archytas, both early Pythagoreans often credited with monochord use, between the two *termini*. Parts 3 and 4 accordingly attempt to show that there is nothing in the surviving work of either author which presupposes the monochord's existence. We shall see, in chapter 3, that this chronological argument helps to suggest an answer to the question which arises from chapter 1 about why mathematical harmonicists consciously chose to limit the scope of canonic division to rational intervals.

Closing the cellar door

Before launching into part 1, however, we must deal with the spectre of Pythagoras. We have seen already that the discovery narrative which first appears in Nicomachus, some six centuries after the discovery was said to have occurred, gives methodological prominence to scientifically impossible observations, which are said (in later versions) to have led to the invention of a scientifically significant instrument. I shall focus on Nicomachus' version of the story here because I take it either to be or to be based on the common source of those which include the forge and prioritise weights as the key to the discovery.[18] The fact that the list of subsequent tests by which Pythagoras supposedly confirmed his discovery contains two which would under the right circumstances generate the expected ratios has been taken as an indication that these procedures, at least, may date from a period close to the time of Pythagoras himself: these are the tests which involve lengths of pipe and lengths of string on the monochord.[19] But there are several reasons why these two ingredients of the story should not be counted as useful evidence for harmonic research in the sixth or fifth centuries, either by Pythagoras or by anyone else. The most important of these is that Nicomachus presents the weight experiments first, giving them heuristic priority; the arithmetical principle supposedly discovered through them is then said to form a stepping-stone or foundation (ἐπιβάθρα) and an 'infallible indicator' (ἀνεξαπάτητος γνώμων) for the instrumental tests listed subsequently, in which the same proportions are claimed to be manifest (*Harm.* 248.13–18). Nicomachus presents his list as a mere sampling of the 'great variety of instruments' (ποικίλα ὄργανα) on which Pythagoras worked; to the five he specifies he adds, 'and the like' (καὶ τὰ παραπλήσια). These are then merely the tip of Pythagoras' experimental iceberg; he went on to find the ratios of the *tetraktys* (4:3:2:1) in some unnumbered quantity of objects that can be made to emit pitched sounds. The five Nicomachus cites (beaten vessels, auloi, panpipes, 'monochords', triangular harps)

[18] Barker makes this assumption about the source of Aristid. Quint. III.1 (1989: 495 n. 4).

[19] The question of whether Nicomachus meant 'monochords' by *monochorda* will be addressed below. Lloyd (1979: 144–5), though he is doubtful of Pythagoras' personal involvement, does include 'accounts relating to the measurement of lengths of pipe corresponding to different notes or to investigations of the monochord' as 'evidence that tends to confirm that empirical investigations were carried out in this area [harmonics or acoustics] at least by the early fourth century' (144; he goes on to cite Archyt. fr. 1 and Pl. *Resp.* 530d–531c as further support).

are given as representative, but they are not in fact 'like' in the only way that matters: not all of them (e.g. beaten vessels) will exhibit the interval-ratios under discussion. True, monochords and panpipes will work under the right conditions; but the manner of their presentation in this late narrative undermines their value as evidence for the work of either Pythagoras or his earliest followers. Nicomachus' point is that everywhere Pythagoras looked, he found the same ratios: his very exhaustiveness makes the list an all-or-nothing proposition.

It is also important to note the role of Pythagoras in Nicomachus' *Handbook* more generally. The work is dedicated to an unnamed lady who has requested a simple account of the main points of the science of harmonics 'without complicated (mathematical) constructions or proofs' (χωρὶς κατασκευῆς καὶ ποικίλης ἀποδείξεως, 238.1–2). Nicomachus promises to write a more detailed and lengthy treatment of the subject at a later date, but limits himself for the time being to the basics, starting at the beginning, with the explicit aim of being 'easy to follow' (ῥᾴονος ἕνεκα παρακολουθήσεως, 238.13). Since his starting points are those of 'the Pythagorean school' (238.18–19), it is reasonable to see his introduction of Pythagoras in chapter 5 as a natural combination of his basic assumptions and his stated aims. Stories make elementary instruction easier to understand, and discovery stories all the more so, since they lead their readers and their protagonists together from ignorance to knowledge. Pythagoras thus takes over the narrative for two chapters, like some scientific Anchises, and guides the reader through the notes of the scale (ch. 5) and the ratios of the concords (ch. 6). In chapter 5 he is credited, quite improbably, with the introduction of the eighth string to the originally seven-stringed lyre 'for purely intellectual purposes', as Barker has noted (1989: 255 n. 38), thereby creating the first octave scale. In chapter 9 'Pythagoras' successor' Philolaus takes over briefly.[20] Plato has a moment of glory in chapter 8. The succession of guides seems calculated for didactic effect; the science

[20] Φιλόλαος ὁ Πυθαγόρου διάδοχος, 252.13; Nicomachus quotes from the first book of his *On Nature* (= Stob. *Flor.* I.21, 7d = DK 44 B 6 = fr. 6a, discussed in part 3 of this chapter).

is taught (pseudo-)historically, and discovery narrative is part of a pedagogical strategy aimed at assisting the reader's comprehension of the discipline's simplest principles. In such a project, it is easy to imagine how even a deliberate fiction could be countenanced if it made what was often regarded as a dry and difficult subject more interesting.[21]

This is not to claim that experimentation with panpipes (for instance) was not a part of sixth- or fifth-century Pythagorean investigations in harmonics – merely that the accounts which date from Nicomachus' time are useless as evidence for those early centuries. In short, the entire narrative should be discounted, pipes and all. The remainder of the tradition stands on equally tenuous foundations. The earliest attribution of the discovery of the harmonic ratios to Pythagoras is in a fragment of Xenocrates, quoted at second-hand by Porphyry in his commentary on Ptolemy's *Harmonics*: 'Pythagoras, as Xenocrates says, discovered that the intervals in music, too, do not have their *genesis* without number; for they are a comparison of quantity to quantity'.[22] Porphyry is quoting from an *Introduction to Music* by a writer named Heraclides, possibly not the Academic Heraclides Ponticus (*fl.* c. 360–320 BC), but a first-century AD scholar of the same name.[23] If the fragment is authentic to Xenocrates (head of the Academy, 339–314 BC), it presents a late fourth-century Platonist view of Pythagoras' contribution to the philosophy of number-speculation. Xenocrates followed Pythagorean tradition in defining the substance of the soul as 'number moving itself',[24] and this on its own suggests that the fragment may be authentic to him: Pythagorean number theory, and its origins to some extent, were concerns of

[21] See for example Varro's assessment of the difficulty of harmonics, quoted by Aulus Gellius, *NA* XVI.18.6.

[22] Πυθαγόρας, ὥς φησι Ξενοκράτης, εὕρισκε καὶ τὰ ἐν μουσικῇ διαστήματα οὐ χωρὶς ἀριθμοῦ τὴν γένεσιν ἔχοντα· ἔστι γὰρ σύγκρισις ποσοῦ πρὸς ποσόν, Xenocrates fr. 9 Heinze (= Porph. *In Harm.* 30.2–3); cf. Sext. Emp. *Math.* VIII.94–5. The lines which follow this sentence in Porphyry's *Commentary* provide further elaboration of the connection and Pythagoras' role in investigating it; it is unclear from the context whether they derive from Xenocrates or from Heraclides.

[23] In favour of the fourth-century Heraclides, see Jan 1895: 135, Schönberger 1914, Düring 1934: 154, Guthrie 1962: 222. Against this identification, see Heinze 1892: 6 n. 2, Wehrli 1953: 113, Burkert 1972: 381, Gottschalk 1980: 157, Barker 1989: 230, Creese 2008b. The question is still a matter of uncertainty; see Barker 2007: 373 n. 17.

[24] Plut. *De an. procr.* 1012d; see Burkert 1972: 65.

the late fourth-century Platonists. But even if Xenocrates really said or wrote something of the kind, we must not forget the first of the assumptions mentioned above. The attribution to Pythagoras of discoveries and innovations to do with number was common in philosophical authors of Xenocrates' generation. One need only look, for example, at the fragments of Aristoxenus, so many of which were quoted by later writers to flesh out the legend of Pythagoras: he was claimed to be, among other things, 'the first to introduce weights and measures among the Greeks'.[25] If it was credible to some late fourth-century writers that before Pythagoras there had been no weights or measures among the Greeks, we may not unreasonably be circumspect about their belief that before him the harmonic ratios were unknown as well.[26] Even Iamblichus, although he maintains a central role for Pythagoras, allows that the discovery may have predated him (*In Nic.* 118.23–4).

The invention of the monochord is not attributed to Pythagoras in any extant text before the third century AD. It is not even clear whether the *monochorda* in Nicomachus' list of Pythagoras' instrumental tests should be understood to mean the laboratory instrument: it is quite possible that Nicomachus means 'lutes'.[27] In any case, he does not say that Pythagoras invented the instrument; it is merely one of a great variety on which he tested the universality of the principle he had discovered. Diogenes Laertius and Gaudentius are the first to claim that Pythagoras devised the *kanōn*; Aristides Quintilianus merely makes it a part of Pythagoras' teaching.[28] Many modern scholars have accepted the dating, if not

[25] Aristox. fr. 24 = Diog. Laert. VIII.14; Burkert 1972: 415.

[26] Burkert saw the Xenocrates fragment as part of 'a falsified tradition about Pythagoras emerging from the Old Academy' (1972: 376). Kirk, Raven and Schofield took it to be the ultimate source for the Pythagoras ratio-discovery narrative, but did not allow it any historical value (1983: 234; see also Neugebauer 1957: 149; Levin 1975: 74 n. 77).

[27] With Nicom. *Harm.* 248.16 (the list of Pythagoras' instrumental tests) cf. 243.14–15: 'and the single-stringed instruments (*monochorda*) . . . which most people call lutes (*phandouroi*), but which the Pythagoreans call *kanones*' (τά τε μονόχορδα . . . ἃ δὴ καὶ φανδούρους καλοῦσιν οἱ πολλοί, κανόνας δ' οἱ Πυθαγορικοί). If Nicomachus means 'lutes' by *monochorda* at 248.16 (as Pollux probably does at *Onom.* IV.60), this is an anachronism: there is no evidence to suggest that lutes had reached Greece before the second half of the fourth century BC, when the first artistic representations of the instrument begin to appear (Higgins and Winnington-Ingram 1965: 62, 68; see also Paquette 1984: 192–3, 200–1).

[28] Diog. Laert. VIII.12, Gaud. *Harm.* 11 (341.13), Aristid. Quint. 97.3. With Diogenes Laertius and Gaudentius cf. Iambl. *VP* 119 and Boeth. *Mus.* I.11 (198.23–4, cited

the substance, of this late tradition. John Burnet, who believed that Pythagoras discovered the concord ratios, thought that the discovery was made with the monochord (1920: I.46). W. K. C. Guthrie, though he did not take a position on the monochord's invention, did conclude that Pythagoras must have used it if the discovery of the concord ratios was indeed his own (1962: 220–4). Kathleen Schlesinger assumed that the monochord 'was used in the schools, as enjoined by Pythagoras' by the time of Aristoxenus (1933: 96). Others have followed suit.[29] Although more cautious voices have been raised in recent decades,[30] the late-antique tradition remains persuasive: Leonid Zhmud has maintained that Pythagoras used the monochord in a scientific investigation of harmonics (1997: 187–201). But the lack of any evidence for this position datable to within seven centuries of Pythagoras' lifetime, and the fact that the testimonies we do possess seem to be accretions on a narrative which we have already had cause to discount *in toto*, render Pythagoras' personal involvement in the history of the monochord nothing more than a mirage. It is not impossible that the monochord had been invented by the late sixth or early fifth century BC, but there is no credible evidence to suggest it.

The tenacity of the mirage appears to stem mainly from the second of the assumptions cited above. Since the monochord had little use outside the role of demonstrating the ratios of musical intervals,[31] authors from late antiquity on have been tempted to date the instrument concurrently with their earliest accounts of investigations of the concord ratios. Some even went so far as to date the invention of the monochord *before* the discovery of the concord ratios. A ready example can be found in Porphyry. In commenting on Ptolemy's discussion of why the *kanōn* is the best instrument for displaying the concord ratios to perception (*Harm.* I.8), he lists various instruments said to have been used

above). It is possible that Diogenes Laertius and Gaudentius read Nicomachus and interpreted his *monochorda* as *kanones*.

29 For examples, see Guthrie 1988: 47.

30 E.g. Burkert 1972: 373–5; Barker 1989: 497 n. 14; Huffman 2005: 135.

31 The monochord may also have been used to play melodies, but probably as part of a harmonicist's *epideixis* (display); see Ptol. *Harm.* II.13, and Barker's conclusions on the passage (1994a: 71).

for the procedure.[32] Ptolemy's topic was the available methods of offering proof of the ratios' correspondence to heard intervals, but Porphyry goes out of his way to link that discussion with their discovery, rather than merely their demonstration:

ἄλλοι δὲ τούτων δοκοῦντες ἔτι ἄμεινον φρονεῖν ἔλεγον, ὅτι ἐκ τῆς τοῦ κανόνος κατατομῆς εὑρέθησαν οἱ λόγοι, καὶ δοκεῖ μοι καλῶς λέγεσθαι. διότι καὶ ὁ Πτολεμαῖος πάντα τὰ προειρημένα παραιτησάμενος, δι' ἃς εἴρηκεν αὐτὸς αἰτίας, ἐπὶ τὴν τοῦ κανόνος κατατομὴν ἦλθεν.

But others considered still wiser than these said that the ratios were discovered from the division of the *kanōn*, and I think they are right. Therefore Ptolemy, too, rejected the methods discussed above, for the reasons he himself gives, and turned his attention to the division of the *kanōn*.[33]

Following this brief parenthesis Porphyry returns to an elaboration of Ptolemy, *Harmonics* I.8, which deals only with why 'the string stretched over what is called the *kanōn* will show us the ratios of the concords more accurately and readily' than other devices.[34] Porphyry's assertion that the discovery of the ratios depended on the use of the monochord is therefore tangential rather than integral to the discussion at hand, and it is the intrusiveness of this statement as much as its content that helps to illustrate the extent to which the harmonic ratios and the monochord were associated by the age of Porphyry: doing mathematical harmonics, however basic, without a monochord was nearly unimaginable.

The argument of the remainder of this chapter is that doing mathematical harmonics without the monochord was not only possible, but that there is no credible ancient evidence to suggest that there was any other way to study the subject before the late fourth century; furthermore, that results of some sophistication could be achieved without the instrument; and most importantly that it was precisely during this period, probably before the monochord was introduced, that the science came into existence as a separate branch of learning (a *mathēma*) and that both its conception as a species of arithmetic and its basic postulates came

[32] Porph. *In Harm.* 119.12–121.14. [33] Porph. *In Harm.* 120.17–20.

[34] ἡ δὲ ἐπὶ τοῦ καλουμένου κανόνος διατεινομένη χορδὴ δείξει μὲν ἡμῖν τοὺς λόγους τῶν συμφωνίων ἀκριβέστερόν τε καὶ προχειρότερον, Ptol. *Harm.* I.8, 17.20–2, trans. Barker 1989: 291.

to be established with such rigidity that not even the introduction of a geometrical instrument could challenge the way in which its practitioners defined musical intervals.

1 The demonstration of the concord ratios

With the cellar door now safely shut, we can proceed to the first of our *termini*: the date by which a demonstration of the harmonic ratios had been carried out. By 'demonstration' I mean in this context no more than a procedure in which the correspondence was exhibited, without implying the existence of formal notions like axioms, *hypotheseis* or the type of deductive logic that underpins the apodeictic demonstrations of the *Sectio canonis*.

The account of the earliest such demonstration comes down to us independently of the traditional stories of Pythagoras. It is a procedure reportedly carried out by 'a certain Hippasus', the Hippasus of Metapontum known to Aristotle and Theophrastus[35] and claimed in late sources as a Pythagorean,[36] who lived in the early part of the fifth century.[37] The source of the account is a fragment of Aristoxenus quoted by the scholiast to Plato's *Phaedo* as a gloss on the phrase 'skill of Glaucus', which was said of things either difficult to accomplish or 'made with great care and skill':

Ἵππασος γάρ τις κατεσκεύασε χαλκοῦς τέτταρας δίσκους οὕτως, ὥστε τὰς μὲν διαμέτρους αὐτῶν ἴσας ὑπάρχειν, τὸ δὲ τοῦ πρώτου δίσκου πάχος ἐπίτριτον μὲν εἶναι τοῦ δευτέρου, ἡμιόλιον δὲ τοῦ τρίτου, διπλάσιον δὲ τοῦ τετάρτου, κρουομένους δὲ τούτους ἐπιτελεῖν συμφωνίαν τινά.

For a certain Hippasus made four bronze discs in such a way that while their diameters were equal, the thickness of the first disc was epitritic [4:3] in relation to that of the second, hemiolic [3:2] in relation to that of the third, and double [2:1] that of the fourth, and when they were struck they produced a concord.[38]

35 Arist. *Metaph.* 984a7, Theophr. fr. 225 (= Simpl. *In Phys.* I.2, 24.1).

36 Theo. Sm. 59.8 (= DK 18.13). Iamblichus reports in one passage that he was from Metapontum (*Comm. math.* 76.19), and in another that he was from Sybaris (*VP* 267).

37 See Barker 1989: 31 n. 5. Hippasus' dates are uncertain; Huffman ventures c. 530–450? (1993: 8).

38 Aristox. fr. 90 (= Schol. Pl. *Phd.* 108d4 = DK 18.12), trans. Barker 1989: 31, with my additions. Zhmud's translation (2006: 195) is erroneous: he renders ἐπίτριτον 'one-third' and ἡμιόλιον 'half', which would generate the proportions 2:6:4:1 (consonant, but not what the text says).

Glaucus of Rhegium is then claimed to have been the first to play music on such discs.[39]

Unlike any of the other procedures involving metal objects (either sounded by percussion or as weights for strings, whether attributed to Pythagoras or not), this one will work as described. If the discs have an equal diameter, varying their thicknesses in the ratios described (12:9:8:6 in lowest terms) will produce exactly the opposite result as varying the speaking length of a true string or a cylindrical pipe: the disc that is twice as thick will sound an octave above the first, and so on.[40]

The fact that the procedure will work as described does not in itself guarantee that Hippasus actually carried it out. We have seen already that not all fragments of Aristoxenus are credible. Paul Tannery, noting the lack of an earlier version of the story, and the fact that Hippasus appears to have left no writings himself, treated the account with some doubt.[41] Burkert was less dubious, and Barker treats Aristoxenus in this context as a 'reliable authority'.[42] M. L. West notes that as far back as the eighth century BC there was a tradition of disc-chime-making in southern Italy, where Hippasus lived.[43] This suggests that the materials and method of manufacturing such discs would have been available to Hippasus. The concord ratios had probably been known empirically for centuries: Greek panpipes (*syringes*), which had been around long before Hippasus, could have provided a basis for their observation. The *syrinx* of the classical period was made from hollow reeds of more or less the same diameter and equal length, pitch variation being obtained primarily by means of wax plugs which reduced the individual pipes to different speaking lengths. A *syrinx*-maker would recognise, whether or not he counted it a scientific discovery, that to tune pipe A an octave above pipe B,

[39] On what this music-making may have involved, see Barker 2007: 84–5. Glaucus was a performer and a writer about music, active in the late fifth century.

[40] Fletcher and Rossing 1998: 60–3. Ptolemy was aware of this principle (*Harm.* 8.10–12; see also Porphyry's comments, *In Harm.* 51.16–52.3).

[41] Tannery 1913: 319. Several works were later ascribed to Hippasus (Diog. Laert. VIII.7). Thesleff takes these to be forgeries of the third century BC or later (1961: 14, 113–14; 1965: 91–3).

[42] Burkert 1972: 206 n. 71, 377 with n. 33; Barker 1989: 31 n. 6 and 2007: 84 n. 32 (quoted). See also Mueller 1997: 292.

[43] West 1992: 234 and 126, citing Zancani Montuoro 1974–6: 27–42 and pll. IX–XVI.

pipe A would have to be blocked with wax to half the speaking length of pipe B.[44] The instrument is mentioned in Greek literature as early as the *Iliad* (X.13, XVIII.526); it is attested in the sixth century by a vase-painting dated c. 575 BC,[45] and in the early fifth by a reference in the *Prometheus Bound* (574–5).[46]

Further support for the fragment's credibility comes from the fact that unlike Xenocrates' or Aristoxenus' testimony about Pythagoras,[47] Aristoxenus' report here does not (at least in the brief quotation preserved) make Hippasus the *prōtos heuretēs* of anything. Rather, it is Glaucus who is credited with being first, and not as inventor, but as performer. The fragment is so short, however, that what it does not say cannot be made to carry much weight. Something more substantial, perhaps, can be drawn from the mathematical consequences of the procedure it relates. It is noteworthy that Hippasus' discs generate ratios in which higher numbers go with higher pitches, rather than the opposite. Unlike demonstrations with cylindrical pipes (such as we shall find in a fragment of Archytas a century later) and stopped strings, the discs present the concord ratios in a non-linear, non-geometric, non-canonic fashion. This presentation of the arithmetical relationships between notes has the advantage of suggesting that height and depth of pitch can be thought of as greater and lesser quantity, not lesser and greater distance: in other words, the behaviour of pitched sound becomes visibly and directly (not inversely) analogous to the behaviour of numbers. In Hippasus' procedure this connection is made through visible thicknesses of bronze; later, after Archytas' problematic thesis that force and speed are the determinants of pitch, the raising and lowering of pitch would come to be thought of as addition and subtraction of movement (*Sect. can.* 149.3–11), which brought the complication – never

[44] This procedure is detailed in the pseudo-Aristotelian *Problems*; see part 4 of this chapter and n. 131 below.

[45] The François Vase (Florence 4209), where it is played by Calliope: Landels 1999: 69, Bundrick 2005: 42 with n. 107. Roughly contemporary is the *Homeric Hymn to Hermes*, where the instrument is mentioned by name (512).

[46] Burkert was apparently unaware of this evidence when he considered the question of how the concord ratios might have been known in Pythagoras' lifetime, and his statement that 'the syrinx [panpipes] was not used in the music of the classical period' (1972: 374) continues to mislead scholars interested in the question (e.g. Mueller 1997: 291).

[47] Xenocrates fr. 9, Aristox. fr. 24 (both quoted above).

elegantly handled in antiquity – of applying ratios of temporal density of impacts inversely to lengths of string.[48] The advantage of Hippasus' procedure, then, and an important corroboration of Aristoxenus' report of it, is that such complexities are altogether absent: the arithmetic of pitch is the arithmetic of thickness. There are separate discs for separate notes, each of which possesses its own numerable quantity. The procedure as we receive it thus fits well in the chronology of developments by which physical acoustics was connected to mathematical harmonics in the fifth and fourth centuries. (This chronology will occupy our attention more directly in chapter 3.)

It has often been assumed that Hippasus' procedure amounted to an experiment.[49] Our account of it, however, is devoid of the language either of empiricism or of demonstration. Hippasus merely 'constructed' (κατεσκεύασε) four bronze discs in certain dimensions, and a certain acoustic result followed from the manner of their construction. If Aristoxenus' choice of vocabulary tells us anything about the scientific context of the procedure (and it may well not), his use of the verb *kataskeuazein* could suggest that he thought of Hippasus as engaged in a mathematical project, in which a construction (*kataskeuē*) can be the solution of a problem (*problēma*) or a stage leading toward a demonstration (*apodeixis*).[50] While it would be anachronistic to see Hippasus as a sort of harmonic Archimedes of the early fifth century, it is certainly possible that Aristoxenus thought of Hippasus' construction as a component of the sort of rigorous mathematical argument which had been developed in the fourth century. The

48 Archytas' acoustics will be discussed in part 4 of this chapter; on the complications it raised for harmonic demonstrations using strings, see ch. 3 part 3; on the notion that the quantities displayed in the division of the monochord are inversely proportional to those of the movements that are the cause of the pitches (as Nicomachus and Thrasyllus insisted), see ch. 5 part 2.

49 E.g. by Mueller 1997: 292, Solomon 2000: 25 n. 135, Kahn 2001: 35, Gibson 2005: 10, Huffman 2005: 135, Riedweg 2005: 109, Zhmud 2006: 195. Lloyd (1991: 82) merely counts Hippasus' discs as evidence for 'empirical investigations in acoustics' among the early Pythagoreans, which is not the same thing as suggesting that Hippasus actually presented the discs themselves (in whatever way he may have done so) as an experiment.

50 On the mathematical usage of these terms prior to Proclus' scheme (*protasis*, *ekthesis*, *diorismos*, *kataskeuē*, *apodeixis*, *sumperasma*), including the sense of *kataskeuazein* in Eudemus, see Netz 1999b: 300–1.

point of this speculation is to show that interpretations of Hippasus' procedure depend largely on what readers expect to see. A brief hearsay report by Theon of Smyrna about Lasus of Hermione (a poet-composer of the late sixth and early fifth centuries) and 'the followers of Hippasus of Metapontum, a Pythagorean' pursuing 'the speeds and slownesses of the movements, through which the concords arise' has been taken by some as evidence of empirical approaches to quantitative acoustics in the early fifth century.[51] Those who accept this view also tend to read Hippasus' procedure as an experiment.[52] If it was, our only source does not tell us so. All that we can be sure of is that Hippasus' discs were capable of illustrating the point that certain ratios appear to correspond to certain concords; anyone who was interested enough to strike them and observe their dimensions could see and hear the correspondence between ratio and concord which was a consequence of their design. This no doubt ensured their preservation. The story about Glaucus shows that by the end of the fifth century, the discs, or a copy of them, were still around to suggest their transformation into a performance instrument.

2 The first appearance of the monochord

Our earliest references to the monochord are in the *Sectio canonis* and in a fragment of the Samian tyrant and historian Duris,

[51] Λᾶσος δὲ ὁ Ἑρμιονεύς, ὥς φασι, καὶ οἱ περὶ τὸν Μεταποντῖνον Ἵππασον Πυθαγορικὸν ἄνδρα συνέπεσθαι τῶν κινήσεων τὰ τάχη καὶ τὰς βραδυτῆτας, δι᾽ ὧν αἱ συμφωνίαι, Theo. Sm. 59.7–10, trans. Barker 1989: 31. Hiller (Theon's editor) marks a lacuna after συμφωνίαι, and there are other textual problems on the same page. Compare the different report of Lasus at Aristox. *El. harm.* 3.21–33. As Barker points out (n. 9 ad loc.), 'there is no good evidence that the early investigators associated pitch with velocity of movement'; the first such theory appears in Archytas fr. 1 (discussed in part 4 of this chapter). See Huffman 2005: 138–9 for a brief but detailed discussion of the passage in Theon and the problems of its interpretation.

[52] Zhmud even goes so far as to suggest that Hippasus' 'acoustical experiment', whose proportions necessarily included the arithmetic and harmonic means, 'implies that the first three means were known to Pythagoras, whose experiments Hippasus followed' (2006: 174 n. 31). Evidence that Hippasus worked on means is late and by no means certain (Iambl. *In Nic.* 100.21–4, 113.16–17, 116.1–4, where he is consistently lumped in with Archytas, whose interest in means (fr. 2) is certain). Mueller (1997: 292) is willing to entertain the possibility that Hippasus worked with 'at least the first three means'. I am not convinced that Hippasus (let alone Pythagoras) studied the means at all, or that experiments were involved. In Kahn's blunt assessment, 'we know almost nothing about Hippasus' (2001: 38).

preserved by Porphyry. Neither can be dated with precision, but the *Sectio* was probably written around 300 BC, and Duris (c. 340–c. 260 BC) was active at the same time. The *Sectio*, which provided a starting point for our examination of the ways in which the monochord was made to function as a kind of audible diagram in chapter 1, will be discussed further in chapter 3, and the problems of its date and authorship will be delayed until then. Since it is Porphyry's quotation from Duris' *Annals*, a local history of Samos ([*Samiōn*] *Hōroi*), which provides an independent confirmation of the *terminus* suggested by the *Sectio*, we shall deal with it here.[53]

The date of Duris' *Annals* is unknown, but his history (possibly *Macedonian History*) included the events following the battle of Corupedium (281).[54] His passing reference to the monochord in the *Annals*, therefore, can be dated only very approximately.

Porphyry's quotation runs as follows:

Δοῦρις δ' ὁ Σάμιος ἐν δευτέρῳ τῶν Ὥρων παῖδά τ' αὐτοῦ [sc. Πυθαγόρου] ἀναγράφει Ἀρίμνηστον καὶ διδάσκαλόν φησι γενέσθαι Δημοκρίτου. τὸν δ' Ἀρίμνηστον κατελθόντ' ἀπὸ τῆς φυγῆς χαλκοῦν ἀνάθημα τῷ ἱερῷ τῆς Ἥρας ἀναθεῖναι τὴν διάμετρον ἔχον ἐγγὺς δύο πήχεων, οὗ ἐπίγραμμα ἦν ἐγγεγραμμένον τόδε·

Πυθαγόρεω φίλος υἱὸς Ἀρίμνηστός μ' ἀνέθηκε
πολλὰς ἐξευρὼν εἰνὶ λόγοις σοφίας.

τοῦτο δ' ἀνελόντα Σῖμον τὸν ἁρμονικὸν καὶ τὸν κανόνα σφετερισάμενον ἐξενεγκεῖν ὡς ἴδιον. εἶναι μὲν οὖν ἑπτὰ τὰς ἀναγεγραμμένας σοφίας, διὰ δὲ τὴν μίαν, ἣν Σῖμος ὑφείλετο, συναφανισθῆναι καὶ τὰς ἄλλας τὰς ἐν τῷ ἀναθήματι γεγραμμένας.

Duris of Samos, in the second book of his *Annals*, records that his [sc. Pythagoras'] son was Arimnestus, who (he claims) was Democritus' teacher. Arimnestus, on returning from exile, dedicated a votive offering in bronze in the temple of Hera, which had a diameter of nearly two cubits. On it was inscribed the following epigram:

'Arimnestus the dear son of Pythagoras set me up
because he discovered many *sophiai* in *logoi*.'

But Simos the harmonicist destroyed[55] the votive, appropriated the *kanōn* and proclaimed it as his own. Seven *sophiai* had been recorded, but because of the

[53] On the *Annals*, from which about twenty fragments survive, see Pédech 1989: 274–88, Landucci Gattinoni 1997: 205–58.

[54] Pédech 1989: 314. [55] Or perhaps 'carried off' (see Landucci Gattinoni 1997: 255).

one which Simos had stolen for himself, the others which had been written on the votive as well disappeared along with it.[56]

What exactly did the inscription claim Arimnestus had discovered or invented? The oblique phrase πολλὰς ἐξευρὼν εἰνὶ λόγοις σοφίας in the pentameter line of the distich is not easy to interpret.[57] It has been suggested that the *sophiai* referred to *akousmata* or bits of Pythagorean wisdom, but the *kanōn* would be difficult to describe as a 'saying' (*akousma*).[58] Outside Pythagorean contexts *sophia* could have a more general sense: 'skill', 'craft', 'learning'; it was also applied to scientific knowledge. The word *logoi* suggests reasoning, reckoning or calculation: *logos* in this sense is also the usual word for 'ratio'. Accordingly we might suggest 'having made many discoveries/inventions in the mathematics of ratio' as a possible paraphrase.[59] This may be too generously scientific, but since the invention of the *kanōn* was evidently one of the seven *sophiai* attributed to Arimnestus on the monument, it is unlikely to be far off the mark. Knowing what the other six were would help to provide context; but then Duris would have had no story to tell.

If Porphyry's faithful quotations and paraphrases of known fourth-century authors such as Aristotle and Aristoxenus are any indication,[60] we may be safe in assuming that this fragment has been faithfully transmitted, despite the fact that aside from the epigram, he reproduces it entirely in indirect discourse. Duris' reputation for accuracy was never very good,[61] but questions of accuracy in this fragment rely less on Duris for present purposes

56 Duris fr. 23 (*FGrH* 76F23 = Porph. *VP* 3).

57 Page was baffled: 'we have no idea what is meant by all this' (1981: 405). See also Diels 1922, I.343 n. 15; Jacoby, *FGrH* II C.122 (on 76F23); Michel 1950: 378.

58 Diels' interpretation of the seven *sophiai* as the seven means (*mesotētes*: see previous note) is similarly ruled out: the monochord can exhibit means, but cannot be described as one.

59 Thesleff makes a similar suggestion (1961: 11). It is possible that εἰνὶ λόγοις σοφίας means something like 'skills in reasoning' in a philosophical sense, but it is hard to imagine how a *kanōn* that would tempt a harmonicist to theft could be placed under such a heading.

60 Compare, for example, Porph. *In Harm.* 52.15–17 with Arist. *De an.* 419b25–7; Porph. *In Harm.* 49.16–21 with Arist. *De an.* 420a19–26; Porph. *In Harm.* 84.12–15 with Aristox. *El. harm.* 10.25–8. Cf. also Porphyry's quotations of the *Sect. can.* in his commentary on Ptol. *Harm.* I.5 (Porph. *In Harm.* 90.7–22, 99.1–103.25).

61 See e.g. Cic. *Att.* VI.1.18, Plut. *Alc.* 32 and *Pericl.* 28; Pédech 1989: 258, Dalby 1991: 539 n. 2.

than they do on Porphyry. Thus we need not be unnerved by Duris' claim that Democritus was a pupil of Arimnestus,[62] nor need we believe that Arimnestus was literally Pythagoras' son.[63] The fact that this fragment seems to contain the only ancient reference to Arimnestus[64] could be taken to suggest either that it is authentic (because it is not merely a repetition of a rumour that appears elsewhere) or that it is a complete fabrication (the conveniently named Samian Arimnestos, 'Most Worthy of Memory', ironically suffers a *damnatio memoriae* at the hands of a (foreign?[65]) harmonicist who plagiarises a seventh of his wisdom and ditches the rest); the word ἐξευρών, 'having discovered/invented', shows that this is another *prōtos heuretēs* narrative.[66] But neither picture affects the value of the fragment for the history of the *kanōn*, since even if the entire account (epigram and all) were Duris' invention, he still provides us with the first use of the word in an unequivocally musical sense.

There can be little doubt that the word *kanōn* in the Duris fragment means 'monochord'. The Simos accused of having stolen

[62] 'Duris was fond of inventing master-pupil relationships, . . . however improbable' (Page 1981: 406); cf. Müller's reaction: 'De Democrito Arimnesti discipulo jure dubitabis' (*FHG* II.482, Duris fr. 56 n.).

[63] If he was, then κατελθόντ᾽ ἀπὸ τῆς φυγῆς cannot refer to the return of exiles in 322 after the fourth-century Athenian cleruchy of Samos had ended. On the analogy of οἱ παῖδες αὐτοῦ [i.e. Ἀσκληπιοῦ] for 'physicians' (Pl. *Resp.* 407e), we might read παῖδά τ᾽ αὐτοῦ [sc. Πυθαγόρου] as merely 'a Pythagorean'. But the epigram's φίλος υἱός seems specific enough to rule out this reading. Fr. 23 is the only extant passage from Duris which mentions Pythagoras; Diogenes Laertius gives Pythagoras a son named Telauges (I.15 and VIII.43).

[64] So Thesleff 1961: 11, Landucci Gattinoni 1997: 255. The only other candidate is the Arimnestus whose aphorism about the greatest good for a man being a good death is quoted by Stobaeus (Ἀρίμνηστος ἐρωτηθεὶς τί μέγιστον ἀγαθὸν ἀνθρώπῳ, εἶπε 'τὸ καλῶς ἀποθανεῖν', *Flor.* IV.51, 26). The remark was probably meant to have been said by the fifth-century Spartan of the same name, who killed Mardonius at Plataea and who was himself later killed in the Messenian Helot revolt while commanding 300 men, all of whom died with him (Hdt. IX.64); I take the interpolation in the *Vatican Sayings* which specifies 'the son of Pythagoras' to have been the result of a confusion caused by Duris fr. 23 (Ἀρίμνηστος, ὁ Πυθαγόρου υἱός, ἐρωτηθεὶς τί μέγιστον ἀνθρώπῳ ἀγαθὸν ἔφη· 'τὸ καλῶς ἀποθανεῖν', *Sent. Vat.* 118).

[65] Iambl. *VP* 267; see n. 67 below.

[66] Page regarded 'the whole story told by Duris, including the epigram, as a product of his own imagination', and judged it 'a waste of time to inquire what Duris may have meant by the details of his fiction' (1981: 406). Landucci Gattinoni, on the other hand, argued that since the primary audience of the *Annales* would have been readers with local knowledge of Samos, it would have been difficult for Duris to have passed off a deliberate fabrication such as this without losing credibility (1997: 218).

credit for the invention is called a *harmonikos* – not, for instance, a *mousikos*; the term is specific (but probably does not yet carry the particular refinements of meaning later outlined by Ptolemaïs).[67] The context is also explicitly Pythagorean: this is the right place to hear about the monochord. Finally, *logoi* (ratios) are exactly, and only, what a *harmonikos* demonstrates by means of a *kanōn*.

The instrument may have been well known enough by the end of the fourth century that Duris did not need to explain what it was.[68] The most generous interpretation of the fragment would be to accept Arimnestus as a son of Pythagoras and as inventor of the monochord. This would place the instrument in the fifth century, possibly even earlier than Hippasus' disc demonstration. The possibility of a fifth-century origin for the monochord cannot be ruled out conclusively, but this dating is unlikely for at least one reason besides the suspicion that Duris may not be a reliable witness.[69] It is, in fact, a corollary of one of my arguments for accepting the account of Hippasus' disc procedure. We noted in support of the latter that the necessary materials, methods and knowledge were available in southern Italy by the beginning of the fifth century. Now while the technology of stringed instrument manufacture was well developed by this time, the instruments themselves would not have suggested the monochord; the disc-chimes of southern Italy, by contrast, would have offered a more suggestive starting point for Hippasus. In instruments of the lyre family, whose strings were either parallel or minimally radial, pitch variation was achieved primarily by varying string thickness. In harps, at least one variety of which is attested in Greek lands by the late sixth century,[70] it was obtained by a combination of string length and thickness. In neither instrument was string division the primary way of

[67] Porph. *In Harm.* 23.5–9. Duris' 'Simos the harmonicist' may be the Pythagorean Simos from Posidonia mentioned by Iamblichus (*VP* 267; see also Michel 1950: 268–9).

[68] Unless, of course, he was attempting to impress his readers with his own learning by being deliberately cryptic. Pédech (1989: 268–70) gives some reasons for thinking that Duris may have had a special interest in music, but I am not convinced by his arguments for supposing that Duris' *Peri nomōn* was a book about musical *nomoi* (despite the instruments mentioned in fr. 81).

[69] See n. 61 above; Arimnestus is also clearly starring in a *prōtos heuretēs* role here.

[70] Athenaeus (635d–e), quoting Pindar; Hdt. I.17.1; Alcaeus 36.5; Sappho 22.11, 156; Anacreon, *PMG* 373, 386.

changing pitch.[71] This is not to say that these instruments did not have the capacity to reveal the concord ratios,[72] but rather that they did not suggest string division as a simple means of demonstrating the ratios to perception. In fact, string division is suggested by no instrument available to the Greeks until the arrival of the lute in the late fourth century BC. Nicomachus' exercise in onomastics and classification in chapter 4 of his *Handbook* reveals a conceptual connection between the lute and the monochord which may go back to the Hellenistic period. He refers to one group of instruments as 'the *monochorda*' (the single-stringed instruments), 'which most people call lutes (*phandouroi*), but which the Pythagoreans call *kanones*'.[73]

All this hints at a fourth-century origin for the monochord. Beyond this we cannot say anything with certainty. Faced with such an irremediable scarcity of information, some have offered guesses for a *terminus post quem*: 300 BC (Van der Waerden), or the time of Aristotle (Burkert).[74] Van der Waerden's is probably too late; Burkert's may be too early, but it cannot be very far from the mark. None of the arguments for a pre-Aristotelian date for the monochord are convincing. One of the most problematic was put forward by Árpád Szabó, who claimed that 'the canon can be conclusively proved to have existed at least at the time of Plato' (1978: 118). The sole basis for this claim was a passage in the pseudo-Platonic *Epinomis*, where it is observed that the hemiolic and epitritic ratios, by which the arithmetic and harmonic means are found, happen to fall within the ratio 6 to 12.[75] By taking a passage of Gaudentius (*Harm.* 341.13–342.6: Pythagoras' first canonic division) as an indication that the division of the string

[71] In lyres and harps with strings at high tension, placing a finger lightly at the mid-point of a string would generate an octave above the pitch of the open string. But there is no unambiguous mention of this technique in Greek literature (see ch. 3 part 3).

[72] Triangular harps were later used in the pseudo-Aristotelian *Problems* to demonstrate the octave-ratio (XIX.23, 919b). The compilation of the *Problems* may date from the early third century BC, and the materials from which it was assembled may differ in date.

[73] The Greek is quoted in n. 27 above.

[74] Van der Waerden's conjecture follows from his rejection of *Sect. can.* props. 19–20 as post-Euclidean in date (1943: 172–7); Burkert 1972: 375 n. 22.

[75] ἐν μέσῳ δὲ τοῦ ἕξ πρὸς τὰ δώδεκα συνέβη τό τε ἡμιόλιον καὶ τὸ ἐπίτριτον, *Epin.* 991a. The discussion depends heavily on the ideas of *Republic* VII and *Timaeus* 35b–36b (on which see ch. 3 part 2).

of the monochord into twelve was more or less standard,[76] Szabó concluded that the numbers 6 and 12 for the octave ratio in the *Epinomis* passage are only explicable if one assumes the existence of the monochord:

> I believe that these questions cannot be answered properly unless one bears in mind that the measuring instrument of the Pythagoreans (the canon) which was used to illustrate the proportional numbers of the consonances was *divided into twelve parts*. In other words, Plato's remark proves convincingly that the canon existed at that time. Modern attempts to regard the canon as 'an artificial piece of apparatus which was devised later' have not been successful.[77]

Because he credited Gaudentius' account of the invention of the monochord with some historical value, accepting the date if not the attribution to Pythagoras,[78] Szabó thought that the content of Gaudentius' division could be applied directly to the harmonic landscape of the fourth century BC simply because some of the numbers were the same.[79] But Gaudentius himself evidently knew very little about the instrument's early history: his repetition of the Pythagoras-and-the-forge myth proves as much. Szabó adduced no other evidence from any of the undisputed dialogues of Plato, nor did he attempt to integrate the monochord into the harmonics of Plato's generation. He may not have been wrong about the date, but his argument fell far short of proving that he was right.

On the other hand, the fact that Aristotle does not mention the instrument (to say nothing of Aristoxenus and Theophrastus) cannot be considered sufficient basis for a *terminus post quem*, as Burkert suggested. I prefer the opposite approach: rather than attempting to establish a *terminus post quem* (which, given the state of our evidence, is impossible), we must satisfy ourselves with an approximate *terminus ante quem*, and Duris provides one

[76] Szabó 1978: 115–16. Gaudentius may have lived as late as the fourth century AD. A twelve-unit monochord may have been standard – if there was ever such a standard – by the time of Thrasyllus (see Theo. Sm. 89.10; Thrasyllus died in AD 36), but this does not help Szabó's argument. Part of the Gaudentius passage is quoted in the Introduction.

[77] Szabó 1978: 119, original emphasis. The quotation appears to be from Burkert (cf. 1972: 116).

[78] See n. 2 above.

[79] The *Epinomis* was probably written in the fourth century even though Plato was almost certainly not its author; Diogenes Laertius (III.37) attributes it to Philip of Opus (claimed to be Plato's literary executor); see Tarán 1975.

(c. 300 BC) even if the *Sectio canonis* does not.[80] The fact that we cannot trace the instrument before Duris means that at least a century and a half elapsed between the earliest date by which we can be certain that the concord ratios had been demonstrated (Hippasus) and the first surviving literary reference to the monochord. The two most prominent early Pythagoreans who wrote about harmonics, Philolaus and Archytas, lived between these two dates. It remains to be shown that what survives of their work neither requires the *kanōn* nor presupposes its existence.

3 Philolaus

Nearly all that survives of Pythagorean harmonics before the time of Aristotle is represented by a few brief fragments from the writings of Philolaus and Archytas. The remnants of Philolaus' discussions of musical intervals do not appear to have been conceived as contributions to an independent branch of scientific investigation; rather, they seem to have served as an illustration of the principle of cosmic order and unity. Until Archytas the ratios of musical intervals, and especially those used to divide the tetrachord, seem not to have been studied either with a view to explaining their concordant or melodic qualities, or for their application to contemporary musical practice. Both of these concerns were of interest to Archytas, however, the latter of which drew criticism from Plato (*Resp.* 531b–c), and from Archytas' pursuit of them many scholars have concluded that he, at least, must have used the *kanōn*. Those who assume this much have usually been generous enough to share the instrument with Philolaus as well.

Proponents of this hypothesis have included Hermann Koller, who dated the monochord to the second half of the fifth century BC, assuming that it had been instrumental in the discovery of the concord ratios (1960: 67). Those who have accepted an even earlier date for the monochord also make it available, by extension, to both Philolaus and Archytas. Among the more cautious,

[80] Some reasons for considering the *Sectio canonis* a unified document of Euclidean date (i.e. c. 300 BC) will be considered in ch. 3; if accepted they would allow the *Sectio* to support the *terminus ante quem* provided by Duris.

M. L. West entertained the possibility that Archytas used the monochord to construct and test the intervals of his tetrachordal divisions (1992: 237, 240); among the less cautious, Giovanni Comotti placed the monochord in the hands of the earliest Pythagoreans (1991: 26). Despite the doubts which had been raised for over half a century,[81] Comotti's argument was taken up soon afterward by Robert Wallace:

> As G. Comotti so ably demonstrated, since the late sixth century Pythagoreans in southern Italy and elsewhere had been analysing music experimentally, by the monochord and other devices, as a means of revealing the mathematical basis of the universe.[82]

> Philolaus describes the three main intervals of the *harmoniai* in terms of mathematical ratios obtained by means of experiments on the monochord. This type of research continued in the fourth century (and later), for example in the work of the Tarentine Archytas.[83]

We have seen already that there is no evidence to support the first set of assumptions quoted above. The goal of parts 3 and 4 of this chapter is to show that the second set is equally unfounded: there is nothing in the fragments of Philolaus to suggest that his presentation of the magnitudes of musical intervals owes anything to experiments on the monochord; Archytas, unlike Philolaus, actually mentions musical instruments by name, but they are not stringed instruments, and he introduces them only to demonstrate the truth of his acoustical theory. His instrumental procedures are presented exclusively within statements of the form 'if one does this, such-and-such will happen', and never within statements such as 'I did this, to find out if such-and-such would happen'. However experiment may have participated in the investigative method of early Pythagorean harmonicists, it was never (so far as we can tell) part of the presentation. As we saw in chapter 1, the monochord, when it was eventually introduced, performed a role broadly consistent with this picture: it appeared in the final propositions of the *Sectio canonis* as a diagrammatic instrument, where its function was demonstrative. My aim is not to prove that Philolaus and Archytas made no use of the monochord, or that they did not

[81] Van der Waerden 1943: 179, 192; Burkert 1972: 375 n. 22; Barker 1989: 50.
[82] Wallace 1995: 17. [83] Wallace 1995: 20–1.

conduct acoustical or harmonic experiments with instruments, but simply that conclusions to the contrary are not supported by any of the available evidence.

For the moment I shall ignore many aspects of the fragments of Philolaus and Archytas which tell us important things about their notions of harmonics, and focus simply on showing that there is no reason to assume their knowledge of the monochord. Indeed, we shall see that the presence of the monochord can never be inferred. Not even an author's description of an instrumental procedure can be taken as unequivocal evidence that he actually performed it, as we saw from the array of acoustic experiments retailed by late antique writers who either never attempted them or were not candid about the results. But the monochord can sometimes be presupposed even when its physical presence is not strictly required; the 'canonic' diagrams of Ptolemy's *Harmonics* are a case in point (see ch. 1 part 2). We do not need a monochord to follow the argument of *Harmonics* I.9, for example, but the demonstration nevertheless relies on the principles of its operation (intervals are conceived as ratios of linear distances) which were spelled out in detail in I.8. My approach to Philolaus and Archytas will be to show that on the most generous reading of the extant fragments, their achievements did not require the monochord, and the presentation of their conclusions does not presuppose its existence. The simplest way to demonstrate this will be to show that everything they say about musical sounds can be obtained directly or indirectly from the connections between observable phaenomena and numerical relationships already demonstrated by Hippasus, or from observations made in their own accounts.

The magnitude of harmonia

The earliest writings that contain an explicit link between musical intervals and numerical ratios are those of the fifth-century Pythagorean Philolaus, of whose works only a small number of fragments survive.[84] Roughly a contemporary of Socrates,

[84] A small number of genuine fragments, that is; many more have been discounted as spurious. The fragments and testimonia are collected, with commentary, by Huffman (1993).

Philolaus was born in about 470 BC and lived into the early fourth century.[85] He was an Italian Greek, probably from Croton,[86] who later spent time in Thebes.[87] His name is firmly attached to the Pythagorean tradition, and the genuine fragments of his work represent the earliest writings of that school.[88]

Those fragments which most scholars now accept as the authentic work of Philolaus indicate a Pythagorean concerned with the means by which the parts of the universe cohere and through which it maintains its unity.[89] Two of these are important for present purposes: number and *harmonia*. Number is of primary epistemological significance for Philolaus; in fr. 4 he presents it as the key to knowledge and understanding: 'and all things, indeed, that are known have number: for it is not possible for anything to be thought of or known without this'.[90] *Harmonia*, 'fitting together', is a unifying principle, invoked to explain the apparent unity of a world composed of things which are dissimilar.[91] 'All the things that exist must necessarily be either limiters or unlimiteds, or else both limiters and unlimiteds' (fr. 2);[92] 'it is clear that the universe and all the things in it are harmonised together (συναρμόχθη) from

[85] Huffman 1993: 5–6; he allows that Philolaus 'could have been born as early as 480 or as late as 440'.

[86] Tarentum is also a possibility; see Guthrie 1962: 329.

[87] Pl. *Phd.* 61d.

[88] Diog. Laert. VIII.84–5. 'After Pythagoras himself there are really just three prominent names in Pythagoreanism: Hippasus (c. 530–450?), Philolaus (c. 470–385), and Archytas (c. 430–350). There is no good evidence for any Pythagorean, including Pythagoras himself, writing a book before Philolaus' (Huffman 1993: 8).

[89] The authenticity of the Philolaus fragments was the subject of much disagreement in the nineteenth and twentieth centuries; for a complete account of the debate from 1819–1961 see Thesleff 1961: 41–5; see also Michel 1950: 258–61, Guthrie 1962: 329–33. Burkert (1972) and Kirk, Raven and Schofield (1983) also offer important (and divergent) discussions. Huffman (1993) assesses each of the fragments and testimonia and groups them as either 'genuine' or 'spurious or doubtful'; about half the fragments are in each category. His assessments have been widely accepted (see Kahn 2001, ch. 3), but not universally: on the fragments relevant to harmonics see especially Barker 2007, ch. 10.

[90] καὶ πάντα γα μὰν τὰ γιγνωσκόμενα ἀριθμὸν ἔχοντι. οὐ γὰρ ὁτιῶν ⟨οἷόν⟩ τε οὐδὲν οὔτε νοηθῆμεν οὔτε γνωσθῆμεν ἄνευ τούτω, Stob. *Flor.* I.21, 7b = DK 44 B 4, trans. Barker 1989: 36.

[91] Compare the use of the word in Homer to denote the physical means of joining the parts of a ship together (*Od.* V.361). In Herodotus it is used of caulking (II.96); in the Hippocratic treatise on surgery the word *harmoniai* (in the sense 'joins' or 'fittings together') apparently refers to sutures (*Off.* 25).

[92] ἀνάγκα τὰ ἐόντα εἶμεν πάντα ἢ περαίνοντα ἢ ἄπειρα ἢ περαίνοντά τε καὶ ἄπειρα, Stob. *Flor.* I.21, 7a = DK 44 B 2.

both limiters and unlimiteds' (fr. 2).[93] In fr. 6, Philolaus says that since limiters and unlimiteds were 'neither similar nor of the same kind' (οὐχ ὁμοῖαι οὐδ' ὁμόφυλοι), it required *harmonia* for them to be 'ordered with one another' (αὐταῖς κοσμηθῆναι) if they were to be 'kept together in a *kosmos*' (ἐν κόσμῳ κατέχεσθαι).[94]

Harmonia for Philolaus is thus primarily a principle of 'fitting together' at the cosmic level; our understanding of the universe and its structure will consequently rely on our understanding of *harmonia*. The naturally immiscible fundamental principles may have been understood numerically, although the link between limiter and unlimited on the one hand (frs. 1, 2, 6), and odd and even on the other (fr. 5), may have been made by Aristotle and not by Philolaus himself.[95] In fragment 6a, Philolaus employs the terminology of mathematical ratio to describe the sizes of the intervals in music; here, *harmonia* is the intervallic framework of the octave scale. Fragment 6a comes not from a work on harmonics, but from a book *On Nature* which was also the source of fragment 1.[96] *Harmonia* is a structural, epistemologically significant concept based on number, and music is of primary value to such a discussion because it can be shown to possess *harmonia*.[97] The first point, then, is a general one: the scope of Philolaus' project was much broader than merely to investigate the intervals in music. This does nothing to exclude the possibility of his having used the monochord, but it is nevertheless important to keep in mind that harmonics was not Philolaus' central concern, and *harmonia* was not for him a narrowly musical idea.

The second and more specific point is that nothing Philolaus says about musical intervals relies in any way on principles uniquely demonstrable on the monochord. There are only two fragments and one testimonium which tell us anything about how Philolaus

[93] δῆλον τἆρα ὅτι ἐκ περαινόντων τε καὶ ἀπείρων ὅ τε κόσμος καὶ τὰ ἐν αὐτῷ συναρμόχθη, Stob. *Flor.* I.21, 7a, trans. Barker 1989: 36. Compare the similar statement of fr. 1 (Diog. Laert. VIII.85 = DK 44 B 1), where the verb is the uncompounded ἁρμόχθη.

[94] Stob. *Flor.* I.21, 7d = DK 44 B 6.

[95] See Huffman's discussion (1993: 179–84).

[96] Fr. 6a (= DK 44 B 6) is quoted by Stobaeus (*Flor.* I.21, 7d) and Nicomachus (*Harm.* 252.17–253.3). Nicomachus quotes from the first book of Philolaus' φυσικός ⟨λόγος⟩ (252.14); Diogenes Laertius quotes fr. 1 as the first sentence of Philolaus' Περὶ φύσεως (VIII.85): 'Nature in the universe was harmonised (ἁρμόχθη) from unlimiteds and limiters, both the whole universe and all things in it' (trans. Barker 1989: 36).

[97] See Michel 1950: 71; Burkert 1972: 378; Huffman 1988: 19, 28.

thought about musical intervals. None of these three has been accepted unanimously as the genuine work of Philolaus, but in order to grant him the widest possible latitude, we shall consider them all. The first is fragment 6a, already mentioned; the second and third are fragment 6b and testimonium A26, preserved only in Latin by Boethius (*Mus.* III.8 and III.5).

Fragment 6a runs as follows:

ἁρμονίας δὲ μέγεθός ἐστι συλλαβὰ καὶ δι' ὀξειᾶν. τὸ δὲ δι' ὀξειᾶν μεῖζον τᾶς συλλαβᾶς ἐπογδόῳ. ἔστι γὰρ ἀπὸ ὑπάτας ἐπὶ μέσσαν συλλαβά, ἀπὸ δὲ μέσσας ἐπὶ νεάταν δι' ὀξειᾶν, ἀπὸ δὲ νεάτας ἐς τρίταν συλλαβά, ἀπὸ δὲ τρίτας ἐς ὑπάταν δι' ὀξειᾶν. τὸ δ' ἐν μέσῳ μέσσας καὶ τρίτας ἐπόγδοον, ἁ δὲ συλλαβὰ ἐπίτριτον, τὸ δὲ δι' ὀξειᾶν ἡμιόλιον, τὸ διὰ πασᾶν δὲ διπλόον. οὕτως ἁρμονία πέντε ἐπόγδοα καὶ δύο διέσιες, δι' ὀξειᾶν δὲ τρία ἐπόγδοα καὶ δίεσις, συλλαβὰ δὲ δύ' ἐπόγδοα καὶ δίεσις.

The magnitude of *harmonia* is *syllaba* and *di' oxeian*. *Di' oxeian* is greater than *syllaba* by an epogdoic. For from *hypata* to *messa* is a *syllaba*, from *messa* to *neata* is a *di' oxeian*, from *neata* to *trita* is a *syllaba*, and from *trita* to *hypata* is a *di' oxeian*. Between *messa* and *trita* is an epogdoic, the *syllaba* is epitritic, the *di' oxeian* is hemiolic, and the *dia pasan* is duple. Thus *harmonia* is five epogdoics and two *diesies*, *di' oxeian* is three epogdoics and a *diesis*, and *syllaba* is two epogdoics and a *diesis*.[98]

In this passage, *harmonia* has the specific sense 'octave scale'.[99] Its internal structure is defined in terms of named intervals (*syllaba*, 'fourth'; *di' oxeian*, 'fifth'; *dia pasan*, 'octave') which lie, as the author seems to assume his reader already knows, between certain named notes. Philolaus' *hypata*, *messa*, *trita* and *neata* are the Doric forms of the notes called *hypatē*, *mesē*, *tritē* and *nētē* which we encountered in the Introduction, and with one exception he uses them to indicate the same scale degrees: his *trita* is located where later theorists placed the *paramesē*, a fourth below *neata*/*nētē*.[100]

[98] Stob. *Flor.* I.21, 7d; Nicom. *Harm.* 252.17–253.3 (= DK 44 B 6). My translation is based on that of Barker 2007: 264, retaining the punctuation of Huffman's text (1993); I include the six words omitted from Barker's translation (συλλαβά, ἀπὸ δὲ τρίτας ἐς ὑπάταν). The fragment's authenticity has been defended at length by Burkert (1972: 387–94) and Huffman (1993: 147–56). I accept their arguments, but this has little bearing on the present discussion, since my aim here is to show that even an overly generous reading of Philolaus will not support a pro-monochord hypothesis.

[99] *Harmonia* in the sense 'fitting-together of notes within an octave' is to be distinguished from *dia pasōn*, 'octave interval'; see Huffman 1993: 161, Barker 2007: 265.

[100] This difference in usage is one of the aspects of the fragment which convinced Burkert (1972: 394) and Huffman (1993: 156) of its authenticity.

The intervals between these notes are then quantified by terms now familiar from the language of ratio, in ascending order of magnitude: epogdoic (9:8), epitritic (4:3), hemiolic (3:2), duple (2:1). Finally the three largest intervals are quantified, in descending order of magnitude, according to the number of epogdoics and *diesies* they each contain.[101]

The mere fact that Philolaus names four ratios in this fragment cannot be taken to suggest that a monochord lurks between the lines. The connection between octave, fifth and fourth and the duple, hemiolic and epitritic ratios had been demonstrated at least as early as Hippasus; Glaucus' adaptation of Hippasus' discs shows that the demonstration was still known in Magna Graecia during Philolaus' lifetime.[102] The epogdoic ratio is calculable once one knows the hemiolic and epitritic ratios,[103] and nothing in this fragment indicates that Philolaus bothered to calculate the ratio of the *diesis*.[104] An important and curious aspect of the way in which he names his ratios, however, may cast further light on how he thought of their combinations. Andrew Barker has drawn attention to the fact that the substantive adjective 'epogdoics' (*epogdoa*) in the final sentence is neuter, and therefore cannot be assumed to refer to ratios (*logoi*, masculine), but must mean 'epogdoic *intervals*' (*diastēmata*).[105] In fact, it is not only the epogdoics in the final sentence to which this feature applies: every named ratio in the fragment is in the neuter.[106] But lest we assume that this feature of Philolaus' language implies, on its own, a kind of analysis in which intervals and their combinations are not thought of primarily in terms of the mathematics of

[101] Philolaus' *diesis* (Doric plural *diesies*) is equivalent to what later harmonic writers called the *leimma*; its ratio was 256:243. (Compare Aristoxenus' use of *diesis* to indicate a smaller interval: *El. harm.* 21.21–31 and 46.2–8, quoted in ch. 1 part 1).

[102] The relative proximity of Metapontum, Rhegium and Croton may be a significant element in the transmission of harmonic ideas in the fifth century.

[103] If the operation $2:1 \div 3:2 = 4:3$ was already known, $3:2 \div 4:3 = 9:8$ was an easy step.

[104] The simplest way to demonstrate that the *diesis* is smaller than half of an epogdoic, however, involves the calculation of its ratio; in our terms it is the statement that $(4:3 \div (9:8)^2)^2 < 9:8$.

[105] Barker 2007: 270.

[106] This grammatical point is clearest in the clause ἁ δὲ συλλαβὰ ἐπίτριτον, which can only mean 'the *syllaba* is an epitritic [interval]', since the adjective does not agree with the noun.

ratio,[107] it is worth noting that the same feature reappears in the *Sectio canonis*, where intervals are both defined and limited by their rational expressions, and where the monochord has a culminating demonstrative role.[108]

The strongest case that could be made for the monochord would be to read fr. 6a as a canonic division. To do this, we would note that the way in which Philolaus leads his reader's ear up and then down the octave scale in the second sentence can be followed very economically on the monochord, where only three bridge positions would be necessary. Of the four intervals constructed in that sentence, the first pair are ascending, the second descending. On a twelve-unit monochord where *hypata* was sounded by the open string, we would place the movable bridge at the nine-unit mark for *messa*, the six-unit mark for *neata* and the eight-unit mark for *trita*. Finding *messa* at 9 and *trita* at 8 would immediately show us that the interval between them is epogdoic, and that is precisely what Philolaus tells us next; the rest of the ratios are simple arithmetic.

But the order in which Philolaus presents the intervals can be explained just as simply without recourse to the monochord, and in such a way as to account for the peculiar names he uses for the fourth and fifth. Later harmonic writers call the fourth *dia tessarōn* and the fifth *dia pente*; the terms mean 'through four (notes or strings, counting inclusively)' and 'through five'. The octave interval, by extension, was called *dia pasōn*, 'through all', and Philolaus uses this term (in the Doric form *dia pasan*) when he needs to specify the octave interval, as opposed to the arrangement of notes within it. If he knew the terms *dia tessarōn* and *dia pente*, he preferred *syllaba* and *di' oxeian* instead.[109] Andrew Barker draws attention to a passage in Porphyry which suggests

[107] Barker (2007: 269–71) offers other reasons in addition to this one for making this suggestion.

[108] Compare e.g. *Sect. can.* prop. 9: τὰ ἓξ ἐπόγδοα διαστήματα μείζονά ἐστι διαστήματος ἑνὸς διπλασίου, 'six epogdoic intervals are greater than one duple interval'. Cf. also Porphyry's discussion of the uses of *diastēma* and *logos* by his predecessors in his commentary on Ptol. *Harm.* I.5 (especially *In Harm.* 92.9–94.26).

[109] It is possible that the terms *dia tessarōn* and *dia pente* had not yet been coined; they do not occur in a musical sense in any text written in the fifth century. *Syllaba* and *di' oxeian* may have been the more usual names for the intervals in the fifth century; this is suggested not only by the remarks with which Nicomachus prefaces his quotation of

that *syllaba*, literally 'grasp', meant 'fourth' because this was the compass of the four lowest strings on the lyre which lay under the performer's left hand in its 'starting position'.[110] To say 'from *hypata* to *messa* is *syllaba*' is therefore to state the obvious to anyone familiar with the lyre, as a philosophically literate Greek of the late fifth century might be expected to be, and this, Barker points out (2007: 266–7), would explain the causal connective 'for' (γάρ) with which the sentence is introduced. *Di' oxeian*, literally 'through the higher-pitched', thus means 'fifth' because the remainder of the octave, a fifth, fell under the player's left hand in its 'upper' position.

The truth of Philolaus' next statement, 'from *messa* to *neata* is *di' oxeian*', will have been equally obvious to a reader of the sort he seems to assume. The third and fourth clauses of this sentence then appear to serve the purpose, as Barker argues, of showing the equality of magnitude between the ascending fourth and fifth and the descending fourth and fifth: the *syllaba* on the way up is the same size as the *syllaba* on the way down, even though these are located between different strings on different ends of the lyre. What must be assumed (as Barker admits, 267) is a demonstration somewhere else in Philolaus' book in which *syllaba* and *di' oxeian* were identified as (respectively) epitritic and hemiolic, for otherwise he could not simply proceed with 'the ⟨interval⟩ between *messa* and *trita* is epogdoic', as though this fact had already been established. Since we know that such a demonstration was possible without the monochord, the assumption of the instrument's absence is simpler than the one which our first hypothesis necessitates (that a monochord is presupposed and that we have a canonic division in fr. 6a), and it has the advantage of allowing us to make sense of Philolaus' terminology in a way that the canonic hypothesis does not.

fr. 6a (*Harm.* 252.4–10; cf. Porph. *In Harm.* 96.21–2 and 96.29–97.11, Aristid. Quint. 15.8–10), but also by a passage in the Hippocratic treatise *On Regimen*, which has been dated, somewhat insecurely, to about 400 BC (Joly 1967: xiv–xvi). The identification relies on an emendation of the otherwise unintelligible MSS readings συμφωνίας τρεῖς, συλλήβδην, διεξιόν (or διεξιών), διὰ πασέων (*Vict.* I.8); see Huffman 1993: 152, Barker 2007: 280. Barker considers it likely that the author knew Philolaus' book *On Nature* (281).

[110] Barker 2007: 264, interpreting Porph. *In Harm.* 97.2–8.

One further feature of Philolaus' method supports this reading: the magnitudes of the intervals are not expressed according to a consistent direction of musical travel (up and down, in our terminology). This is important: on a monochord, where ratios are always in view, we must say 'the interval from *messa* to *trita* is epogdoic', and not the other way round, because the length of string required to sound *messa* is longer than that needed to sound *trita* by 9:8. Now we might read Philolaus' phrase 'the ⟨interval⟩ between *messa* and *trita* is epogdoic' as meaning the same thing, and in a certain sense it does. But to say τὸ δ' ἐν μέσῳ ⟨διάστημα⟩ is not the same as to say, for instance, ἀπὸ ὑπάτας ἐπὶ μέσσαν, 'from *hypata* to *messa*'. It merely states position, not direction, and direction is an essential aspect of the means by which the monochord presents intervals as ratios.[111] Furthermore, while ἀπὸ ὑπάτας ἐπὶ μέσσαν presents direction, so does ἀπὸ δὲ νεάτας ἐς τρίταν, 'from *neata* to *trita*', but this time it is the opposite direction. Both phrases give us the fourth, but they present it as an interval conceived spatially as a hand-position (*syllaba*), not arithmetically as a ratio (an epitritic).

This is not so very surprising if the instrumental demonstration of the concord ratios Philolaus had in mind was that of Hippasus' discs. Philolaus' *hypata*, *messa*, *trita* and *neata* – even if they are conceived of as lyre strings – are sounded like discs, and they give out the same concords whether one strikes them in ascending or descending order. The four lyre strings which Philolaus seems to be thinking of, unlike Hippasus' discs, do not exhibit the concord ratios in their physical dimensions (late fifth-century vase-paintings show strings virtually equal in length), but they do not need to: this arithmetical reality has been established otherwise. The fact that the ratios of disc thickness must be applied inversely to string length in order to sound the same notes is thus irrelevant to the discussion in fr. 6a: ratios are here merely a measure of intervallic magnitude (μέγεθος), not of string length.

[111] Direction can be stated in purely mathematical terms, e.g. ὥστε τὸν ΜΒ τοῦ ΕΒ γενέσθαι ἐπόγδοον, 'so that the ⟨ratio⟩ of MB to EB is epogdoic' (*Sect. can.* prop. 20, 165.6).

Dividing the tone

Fragment 6b and testimonium A26 may be dealt with more briefly. The *diesis*, which appeared in the final sentence of fr. 6a, was undefined there, leaving the expectation that Philolaus had specified elsewhere what it was. What it was not, clearly, is a semitone, as his usage makes clear: *harmonia* is not six tones, but five epogdoics and two *diesies*; two *diesies* therefore do not combine to make one epogdoic. It is not even certain that the word 'tone' had come into use by Philolaus' time, although it is difficult to accept that musicians had no name for this important interval;[112] there can of course be no 'semitone' until there is a 'tone'. The expectation of a definition for the *diesis* left by fr. 6a is met in fr. 6b, complete with the word for 'tone':

diesis, inquit, est spatium quo maior est sesquitertia proportio duobus tonis. comma uero est spatium, quo maior est sesquioctaua proportio duabus diesibus, id est duobus semitoniis minoribus. schisma est dimidium commatis, diaschisma uero dimidium dieseos, id est semitonii minoris.

A *diesis*, he says, is the interval by which the epitritic ratio is greater than two tones. A *komma* is the interval by which the epogdoic ratio is greater than two *diesies* (that is, than two smaller semitones). A *schisma* is half of a *komma*, and a *diaschisma* is half of a *diesis* (that is, of a smaller semitone).[113]

In testimonium A26 the counterpart of the *diesis* is introduced: this is the *apotomē*, and together they make up a 'tone'. The magnitudes of both are calculable as ratios.[114] The *komma* is defined in two different ways. In fr. 6b above, it is 'the interval by which the epogdoic ratio is greater than two *diesies*'; in testimonium A26 it is the difference between the *diesis* and the *apotomē*. The two definitions of the *komma* are consistent, and if followed mathematically they will generate the same ratio.[115] The *schisma* and *diaschisma*

[112] Barker (2007: 267) suggests both the possibility and the objection to it, citing Rocconi 2003: 23–4 on the history of the word *tonos*.

[113] Fr. 6b = Boeth. *Mus.* III.8.

[114] Tone = *diesis* + *apotomē*: 9:8 = 256:243 × 2187:2048.

[115] The ratio is 531,441:524,288, the result of either 9:8 ÷ $(256{:}243)^2$ (fr. 6b) or 2187:2048 ÷ 256:243 (testimonium A26). Ptolemy (*Harm.* I.11, 25.18–26.2) offers the epimoric approximation 65:64 for the interval by which six tones exceed an octave; this is another way to define the *komma*, but Ptolemy does not use the term. The same interval is implicit in *Sect. can.* prop. 9 (whose point Ptolemy is arguing in *Harm.* I.11), but it is neither named nor fully quantified there.

are of a different order from the *diesis*, *apotomē* and *komma*: unlike the latter, they cannot be found 'in numbers', as Greek mathematicians put it. In other words, they are irrational,[116] and this is one of the reasons fr. 6b has been judged spurious.[117] But the fact that this group of intervals will yield a combination which divides the tone exactly in half (*diesis* + *schisma*) has also been taken as significant, especially since the impossibility of dividing an epimoric ratio equally may not have been proven by Philolaus' time.[118]

All this can be done without a *kanōn*. It is true that two of the intervals can be constructed relatively easily on a monochord (the *diesis* and *apotomē*; the *komma* less easily and less exactly because of its size), but the question is rather whether the monochord is presupposed at any point in the scheme. As we shall see, the numbers in testimonium A26 suggest that Philolaus may have known the ratio of the *diesis*, but this would have been the result of calculation, not of experimentation on a monochord; ratios such as 256:243 (unlike 2:1 or 3:2) cannot be extracted empirically from the instrument. Nor is it reasonable to suppose that the smaller intervals were constructed on the monochord as a test. We saw in chapter 1 that it is possible to determine string lengths for irrational intervals using steps available to fourth-century geometers, but these methods are not attested in the fifth century. Furthermore, while it is possible to bisect the *komma* geometrically, it is difficult to believe that the minuscule interval which results from this procedure (the *schisma*) could have been sounded on a monochord with sufficient precision. So far from being presupposed, the monochord would be an albatross in such a method.

Testimonium A26 presents one further, and very different, way of thinking about musical intervals. Here the epogdoic tone is expressed both as the relationship between 243 and 216

[116] That is, there exist no ratios of *arithmoi* (positive integers greater than the unit) in which the *schisma* ($\sqrt{531{,}441:524{,}288}$) and *diaschisma* ($\sqrt{256:243}$) can be expressed.

[117] Huffman (1993: 364–7), who considers the fragment seriously, ultimately prefers to see it as spurious.

[118] Boethius attributes such a proof to Archytas (*Mus.* III.11, 285.7–8). Barker (2007: 275–8) argues that the halving of the epogdoic tone between *messa* and *trita* may have been an important aspect of Philolaus' cosmological application of the intervals of *harmonia*: an attempt, as it were, to pinpoint the exact centre of the symmetrical structure of the universe.

(243:216 = 9:8) and also as the number 27, which is the arithmetical difference between the two terms of the ratio (243 − 216 = 27); the tone is also the relationship between 27 and 24 (27:24 = 9:8) and the number three (27 − 24 = 3). Boethius says that the significance of 27 for Philolaus was that it is the cube of the first odd number (3). The *diesis* is expressed in similar fashion both as 256:243 and as 13 (256 − 243 = 13). Hence the *apotomē* is 14 (because 13 + 14 = 27), and the *komma* is the unit (14 − 13 = 1). Intervals are thus treated both as ratios and as numbers, as though they could be added and subtracted. This presentation stems not from an inability to think in a mathematically consistent way about musical intervals, but, as Burkert says, from a tradition in which numbers themselves are meaningful, and in which the numerical terms of a ratio are considered as significant as the ratio itself.[119] We find another manifestation of this tradition much later in Aristides Quintilianus, who lists the properties of the first twelve numbers in his *De musica* (III.6).

The numbers of testimonium A26 can all be calculated from the ratios of fr. 6a. Moreover, presenting these numbers on a monochord (see fig. 2.1) would do nothing to emphasise the significance Philolaus is reported to have given them, since it would only be plainer in this context that if a tone somehow 'is' 27 because

[119] Burkert 1972: 400; see also Huffman 1993: 369. Burkert was inclined to see this aspect of Philolaus' programme as evidence for a picture of fifth-century Pythagoreanism in which harmonic analysis was less about mathematics or physics than it was about number-mysticism. Huffman (e.g. 1999: 81), who preferred to see more science in Philolaus than Burkert did, used the numbers of testimonium A26 to argue against its authenticity (1993: 374). Mueller, noting that the approach to number found here is hardly unique to Philolaus among those in what he called 'the Pythagorean tradition of Greek music', for whom 'the sense of the cosmic power of pure numbers and the willingness to indulge in meaningless numerical manipulation is always present', still singled out Philolaus for the 'pure nonsense' of his discussion and his 'apparent confusion between numerical relations or ratios and absolute numbers' (1997: 293). Modern readers' approaches to the testimonium depend very much on what they think constitutes the 'scientific' or even the 'mathematical'. Archytas, a generation after Philolaus, recognised a branch of learning (*mathēma*) 'concerned with numbers' (περὶ . . . ἀριθμῶν, fr. 1 ap. Porph. *In Harm.* 56.9), but there is no reason to assume either that such a science in Philolaus' day was limited to the sort of arithmetic we find in Euc. *El.* V and VII–IX, or that the two approaches to number in Philolaus testimonium A26 would necessarily have been seen in the fifth century as nonsensical, or meaningless, or even inconsistent. Our definition of 'harmonic science' may have to be sufficiently broad and sufficiently flexible to include both analysis of ratios and manipulations of their terms.

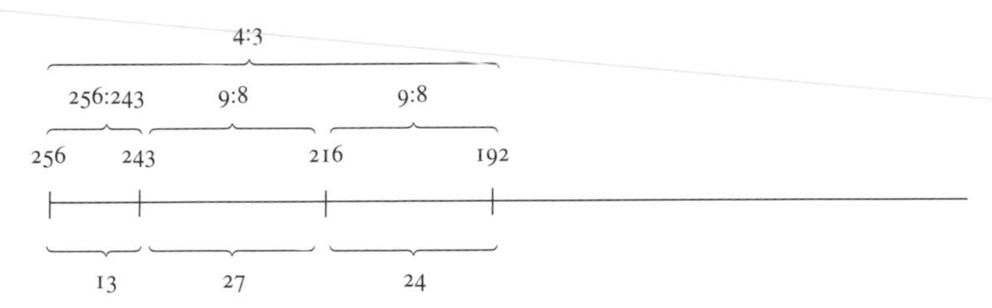

Fig. 2.1 The intervals of Philolaus testimonium A26 set out on a *kanōn*: ratios are above the string, distances below.

243 – 216 = 27, the tone of identical size which lies above 216 'is' only 24, and the one below 256 'is' 32, and so on.[120] The *kanōn*, with its linear, rational way of representing intervals, works against the grain of testimonium A26.

We can find, then, no reason to imagine that the monochord had any part in Philolaus' work, and several reasons why the most inclusive reading of the surviving fragments suggests a way of thinking about the intervals of *harmonia* which are either not easily compatible with the generation of intervals on the monochord, or completely alien to it. With the concord ratios already demonstrated by Hippasus, the instrumental example of the lyre ready to hand, and an approach to number which may have gone well beyond the mathematics of ratio, Philolaus was capable of constructing a detailed account of cosmic *harmonia* without hinting that the ratios of musical intervals can be found in lengths of the same string. If some of the items I have discussed here were not components of that account, the case for the monochord in the fifth century only weakens.

4 Archytas

If one is seeking a probable context for the monochord before Duris and the *Sectio canonis*, the extant fragments of Archytas of Tarentum seem at first glance more promising than those of Philolaus. Here, at last, is an author who investigated musical phaenomena from several scientific approaches, apparently for their own sake rather than as part of a cosmological discourse as

[120] See Burkert 1972: 396 and Barker 2007: 272 n. 19.

Philolaus had done. Ptolemy says that among the Pythagoreans, Archytas was 'the most dedicated to the study of music',[121] and this assessment is borne out by what survives of his work. We have a fragment which attempts to explain the causes of pitched sound, replete with instrumental examples (fr. 1), a fragment on the three mathematical means 'in music' (fr. 2), and a testimonium attributing three tetrachordal divisions to him, complete with ratios (A16). More encouraging still is the fact that the authenticity of these three important passages is no longer a matter of serious doubt.[122] If it were possible to show that these texts tell us things about Archytas' work in acoustics and harmonics which are most reasonably explained by assuming the presence of the monochord, then we would be in a position to speculate that the instrument had come into use by at least the middle of the fourth century, if not earlier.[123] But here, too, we shall find that there is nothing whatever to suggest that the monochord had any part in early fourth-century harmonics, and in fact some reasons to suspect that it had not.

Acoustics and the kalamos

Of the three passages singled out above, the first two may be discounted in fairly short order. The Archytan fragment on acoustics (fr. 1) comes from the opening lines of a book entitled either *On Mathematics* or *Harmonics* quoted at some length by Porphyry and more briefly by Nicomachus.[124] Its starting premise is that sound occurs when things strike each other; it goes on to argue that there is a causal relationship between the force and speed of this striking and the pitch of the sound produced. Archytas uses

[121] μάλιστα τῶν Πυθαγορείων ἐπιμεληθεὶς μουσικῆς, *Harm.* 30.9–10 (trans. Barker 1989: 304).

[122] Authenticity questions are treated comprehensively by Huffman (2005: 91–100, a general overview; 112–14, fr. 1; 166, fr. 2; 406–7, testimonium A16).

[123] On Archytas' dates, see Huffman 2005: 5; his 'best estimate' places Archytas' birth between 435 and 410, and his death between 360 and 350. Huffman's first chapter gives a full account of what is known about Archytas' life, writings and reception.

[124] Περὶ μαθηματικῆς: Porph. *In Harm.* 56.4; ἁρμονικὸς ⟨λόγος⟩: Nicom. *Ar.* I.3.4 (6.17). The fullest version of the fragment is at Porph. *In Harm.* 56.4–57.27, but the two sources are in serious conflict. I follow Bowen (1982), Huffman (1985, 2005) and Barker (1989: 39 n. 42) in preferring Porphyry's version to that transmitted by Nicomachus, and in treating the fragment as genuine. For text, translation and commentary, see Huffman 2005: 103–61.

several examples to illustrate this, including the whirling of a stick through the air, our voices when speaking or singing, missiles when thrown (Porph. *In Harm.* 57.3–7), and finally a series of musical and quasi-musical instruments:

ἀλλα μὰν καὶ ἔν γα τοῖς αὐλοῖς τὸ ἐκ τῶ στόματος φερόμενον πνεῦμα ἐς μὲν τὰ ἐγγὺς τῶ στόματος τρυπήματα ἐμπῖπτον διὰ τὰν ἰσχὺν τὰν σφοδρὰν ὀξύτερον ἆχον ἀφίησιν, ἐς δὲ τὰ πόρσω βαρύτερον, ὥστε δῆλον ὅτι ἁ ταχεῖα κίνασις ὀξὺν ποιεῖ, ἁ δὲ βραδεῖα βαρὺν τὸν ἆχον. ἀλλὰ μὰν καὶ τοῖς ῥόμβοις τοῖς ἐν ταῖς τελεταῖς κινουμένοις τὸ αὐτὸ συμβαίνει· ἡσυχᾷ μὲν κινούμενοι βαρὺν ἀφίεντι ἆχον, ἰσχυρῶς δ' ὀξύν. ἀλλὰ μὰν καὶ ὅ γα κάλαμος αἴ κά τις αὐτῷ τὸ κάτω μέρος ἀποφράξας ἐμφυσῇ, ἀφήσει ⟨βαρέαν⟩ τινα ἁμῖν φωνάν· αἰ δέ κα ἐς τὸ ἥμισυ ἢ ὁπόστον μέρος αὐτῷ, ὀξὺ φθεγξεῖται. τὸ γὰρ αὐτὸ πνεῦμα διὰ μὲν τῶ μακρῶ τόπω ἀσθενὲς ἐκφέρεται, διὰ δὲ τῶ μείονος σφοδρόν.

In auloi too, if the breath travelling from the mouth goes into the finger-holes near to the mouth it emits a higher-pitched sound, because of its vigorous force, but a lower one if it goes into the holes that are further away. Hence it is clear that a swift movement makes a sound high-pitched, while a slow one makes it low-pitched. The same thing happens, once again, with the *rhomboi* that are whirled about in the mystery cults: moved gently they give out a low-pitched sound, moved powerfully a high-pitched one. Similarly, again, with the *kalamos*: if one blocks up its lower end and blows into it, it will give out a low-pitched sound, but if one blows into its half-length or any other part of it, it will utter a high pitch. For if the breath travels through a long passage it comes out weakly, but if the same breath travels through a shorter passage it comes out more vigorously.[125]

Of prime significance to the present discussion is the fact that Archytas employs simple instrumental demonstrations in an attempt to explain the causes of pitched sound. This would seem to be the place to include the monochord, but Archytas does not oblige; more glaring still is his failure to mention stringed instruments of any kind.[126] Now it is possible that Archytas did mention stringed instruments in a part of the discussion which no longer survives. But Porphyry says that he continued, at least in the lines which originally followed the end of the quotation above, with 'other facts about how the motion of the voice is

[125] Porph. *In Harm.* 57.14–23, trans. Barker 1989: 41–2. The text printed here is that of Düring, whose emendations include ἐκφέρεται in the last sentence; Huffman restores the MSS reading (φέρεται), and translates accordingly: 'For the same breath is carried weakly through a long distance and strongly through a shorter distance' (2005: 106).

[126] Huffman also notes this absence, and specifically the absence of the *kanōn* (2005: 142).

intervallic' (ἄλλα περὶ τοῦ διαστηματικὴν εἶναι τὴν τῆς φωνῆς κίνησιν, 57.24–5) before making the summary statement that high notes move more quickly, and low notes more slowly (57.26–7). Furthermore, the acoustic point he is making is that speed and force of impact determine pitch – a point which would be difficult, if not impossible, to illustrate (at least in the manner Archytas does here) with chordophones of any sort.[127] It is a difficulty which will turn out to have implications extending well beyond the scope of Archytas' project in fr. 1. A summary of the argument will help to make the point clearer.

(1) Sound is caused by impact (*plaga*, 56.12);[128]
(2) at the point of perception, faster and more vigorous impacts are perceived as high-pitched, and slower and less vigorous impacts are perceived as low-pitched (56.21–57.2); and this connection between force/speed and high pitch can be seen in sticks (57.3) and *rhomboi* (57.18) when these are whirled through the air, and in the human voice (57.5–7, 9–14);[129]
(3) there is also a connection between force/speed and distance, which is most obvious in the case of missiles (57.7–9); the implication that the force of an impact decreases according to the distance an object travels (though left unstated in as many words) is most demonstrable in the case of auloi (57.14–17), where that object is conceived to be the breath itself (τὸ ἐκ τῶ στόματος . . . πνεῦμα, 57.14–15; cf. 22–3);
(4) if force/speed diminishes as distance increases (3), and if there is a causal connection between force/speed and pitch (2), then there will also be a similar relationship between pitch and distance, which is most evident in the example of the *kalamos* (57.20–2), in which distance now becomes a measurable attribute of pitch.

What Archytas does not do is make use of the measurability of distance to demonstrate a connection between intervals of pitch and ratios of numbers. The *kalamos* example certainly furnishes such an opportunity, and because (as we know from Ptolemy's testimony) Archytas produced tetrachordal divisions in ratio form (A16), it is possible that he went on to make the

127 For a parallel difficulty in Ptol. *Harm.* I.3, see Barker 2000: 48–50.
128 This assumption becomes common after Archytas: cf. πληγή in Pl. *Ti.* 67b3; Arist. *De an.* 419b10; [Arist.] *Pr.* XIX.39 (921a); *Sect. can.* 148.5; πλῆξις at Nicom. *Harm.* 242.20 and Adrastus ap. Theo. Sm. 50.6; see also [Arist.] *De audib.* 800a.
129 Barker (2002: 31) suggests a philological reason for the association between force, speed, volume and high pitch in Archytas fr. 1.

connection explicit; this conclusion may even have been the purpose which his acoustical argument was ultimately meant to serve.[130] If so, its importance can hardly be overemphasised: in the absence of a quantitative theory of pitch as rate of vibration, Archytas' line of thinking, flawed as it is, would have represented the only developed link between acoustics and harmonics in his generation, and thus the only way of attaching numbers to notes in a way which accounted both for the propagation of sounds and for the relationships between their pitches.

The example of the *kalamos* is obviously crucial to such an argument: it is the only instrument in Archytas' list in which the distance–pitch correlation is unambiguously quantifiable. The word is a generic term for reed, of which the many species were used in antiquity for everything from thatching and bedding to fishing and surgery; as a naturally hollow tube, it lent itself to many uses, and the word came to mean, among other things, a reed pen (Latin *calamus*). But here Archytas is thinking of its use as a component of the *syrinx* or panpipes, in which all the *kalamoi*, roughly identical in diameter, were bound together and cut to the same length, and each blocked with wax at the point necessary to produce the desired attunement.[131] What the *kalamos* can do which the monochord cannot is connect steps (2) and (3) of Archytas' acoustic argument: it takes a faster and more forceful breath to sound a reed blocked halfway than one blocked at its end. In other words, *nētē* requires more force and speed (2) and a shorter distance (3) than *hypatē*. The *kalamos* is what permits the inference of step (4), something which no stringed instrument can do – nor, interestingly, can Hippasus' discs. From fragment 1, it appears that the *kalamos* may well have been Archytas' *kanōn*.

[130] See Bowen 1982: 96; Barker 2006: 306 n. 13.

[131] On the *syrinx* in the context of the concord ratios, see part 1 of this chapter. The method of tuning the instrument is described in a discussion of the concord ratios at [Arist.] *Pr*. XIX.23 (919b), where the *kalamoi* are referred to simply by the names of the notes they sound: 'Again, those who tune panpipes cram the wax into the end of *hypatē*, but they fill *nētē* up halfway. And they obtain the fifth by a hemiolic distance [or 'interval': *diastēma*] and the fourth by an epitritic distance in a similar fashion' (ἔτι οἱ τὰς σύριγγας ἁρμοττόμενοι εἰς μὲν τὴν ὑπάτην ἄκραν τὸν κηρὸν ἐμπλάττουσι, τὴν δὲ νήτην μέχρι τοῦ ἡμίσεος ἀναπληροῦσιν. ὁμοίως δὲ καὶ τὴν διὰ πέντε τῷ ἡμιολίῳ καὶ τὴν διὰ τεττάρων τῷ ἐπιτρίτῳ διαστήματι λαμβάνουσιν).

It is worth pointing out that Archytas does not present his instrumental examples in the language of experiment.[132] His way of introducing the *kalamos* provides the clearest example of this. The sentence is phrased as a condition: 'if one blocks up its lower end and blows into it, it will give out a low-pitched sound' (αἴ κά τις αὐτῷ τὸ κάτω μέρος ἀποφράξας ἐμφυσῇ, ἀφήσει ⟨βαρέαν⟩ τινα ἁμῖν φωνάν, 57.20–1). The same type of condition (future more vivid) later became one of the prevailing syntactical forms for the *protasis* (enunciation) of mathematical proofs; four propositions of the *Sectio canonis* begin with a *protasis* of this form (1, 2, 4, 5): e.g. 'if a multiple interval put together twice makes some interval, this interval too will be multiple' (ἐὰν διάστημα πολλαπλάσιον δὶς συντεθὲν ποιῇ τι διάστημα, καὶ αὐτὸ πολλαπλάσιον ἔσται).[133] An important difference between Archytas' usage and that of the *Sectio* is that whereas the *protasis* of a proposition enunciates a general theorem which has then to be proven in a specific case, Archytas here states that the general argument already enunciated will be exhibited in the specific instrumental example introduced. But one can no more reasonably suggest from this that Archytas' method is empirical than that the early propositions of the *Sectio* are empirical. The future more vivid condition is the language of demonstrative challenge.

Three means in music

Fragment 2, which is also preserved by Porphyry later in the same work,[134] is a discussion of three means (arithmetic, geometric and

[132] Although Archytas' examples conform in some respects to some elements of Gera's definition of a 'thought experiment' (2000: 21–2), I do not think that they (quite) satisfy her definition, but the comparisons are interesting and instructive. Of particular relevance is the form of the condition used in one of Gera's examples of a thought experiment presented in a text approximately contemporary with Archytas, the anonymous early fourth-century *Dissoi Logoi*: 'if one were to send a newborn Greek infant off to Persia and raise him there, without his hearing Greek, he would speak Persian; and if one were to bring a Persian baby to Greece, he would speak Greek (αἴ τις εὐθὺς γενόμενον παιδίον ἐς Πέρσας ἀποπέμψαι καὶ τηνεῖ τράφοι, κωφὸν Ἑλλάδος φωνᾶς, περσίζοι κα· αἴ τις τηνόθεν τῇδε κομίξαι, ἑλλενίζοι κα, 6.12)' (Gera 2000: 23). The future less vivid condition here provides a useful grammatical point of contrast to Archytas' future more vivid.

[133] *Sect. can.* prop. 1, trans. Barker 1989: 194.

[134] *In Harm.* 93.6–17; text, translation and commentary in Huffman 2005: 162–81; translation and commentary in Barker 1989: 42; discussion in Michel 1950: 387–91, Barker 2007: 302–3.

subcontrary or harmonic) which can be found 'in music'. Just how they can be found there is not spelled out in the passage Porphyry quotes, but Porphyry is clearly not interested in that issue; he quotes the passage not for its definition of the three means, but for Archytas' use of the word *diastēma* (interval), confusions over the usage of which term proliferated from at least the time of Eratosthenes.[135]

The fragment offers a definition of each mean, and specifies the consequent differences in the sizes of the intervals between the greater and lesser terms.[136] The application to harmonics is not hard to see: if we imagine a set of interval-ratios such as 24:12:9:8:6, the geometric mean of the extremes (12) will lie an octave below the highest note and an octave above the lowest note (24:12:6); the arithmetic mean of the lower octave (9) will lie a fourth from the bottom and a fifth from the top if higher numbers go with lower pitches (12:9:6); the subcontrary or 'harmonic' mean (8) will lie a fifth from the bottom and a fourth from the top (12:8:6). The subcontrary mean is significant to music, then, because it gives a mathematical basis for the division of an octave into two tetrachords separated by a tone; this justifies the new name 'harmonic'.[137] But once the ratios of the fundamental concords are known, as they had been at least since Hippasus, no observations of musical phaenomena are necessary: this is a numbers game. If Archytas included an instrumental demonstration of the application of these means to music (and this is a generous speculation), his *kalamos* would have sufficed.

[135] On this see ch. 4. [136] The means are defined in ch. 1 n. 37.

[137] On the name, and the merits of the claim that it was Archytas himself who coined it (suggested by the way he introduces it: ἁ δ' ὑπεναντία, ἃν καλοῦμεν ἁρμονικάν, 'the subcontrary ⟨mean⟩, which we call 'harmonic', 93.13–14), see Huffman 2005: 170–7. Note that by contrast to the scales of Greek music, our modern equal-tempered scale is built entirely on the geometric proportion, and that in it no note is the arithmetic or harmonic mean of any other. All octaves are 2:1, and therefore each note is the geometric mean of the notes that form an octave above and below it; similarly, since the octave is divided into twelve equal semitones (each the twelfth root of 2:1), every note is the geometric mean of the two that lie an equal number of semitones on either side of it.

Tetrachordal divisions

We come finally to Archytas' divisions themselves (testimonium A16).[138] This collection of ratios is what has suggested to many scholars that the monochord was available to Archytas. Ptolemy, our source for the divisions, does not appear to make such an assumption, but he does set out Archytas' tetrachords in tables (*kanonia*) of string lengths both in *Harmonics* I.13 and in a later chapter (II.14). Obviously this presentation facilitates their construction on the monochord, and that is part of Ptolemy's purpose. But he also includes string lengths for Aristoxenus' tetrachords in the tables of II.14. Presenting tetrachords as string lengths means thinking of them as ratios, and we know that Aristoxenus rejected this approach to harmonic analysis. We cannot then infer from Ptolemy's tabular presentation of Archytas' ratios that they were originally conceived or demonstrated on a monochord.

We can, however, examine the ratios themselves and ask whether their formulation requires the use of such an instrument. It is now generally agreed that the divisions are informed both by observation of the tuning procedures of early fourth-century musicians (the tuning of instruments of the lyre family specifically) and by predetermined mathematical priorities, although the precise nature of this combination is still a matter of disagreement.[139] The balance between the concerns of theory and practice will be easier to see with the passage before us:

τρία μὲν τοίνυν οὗτος ὑφίσταται γένη, τό τε ἐναρμόνιον καὶ τὸ χρωματικὸν καὶ τὸ διατονικόν· ἑκάστου δὲ αὐτῶν ποιεῖται τὴν διαίρεσιν οὕτως. τὸν μὲν γὰρ ἑπόμενον λόγον ἐπὶ τῶν τριῶν γενῶν τὸν αὐτὸν ὑφίσταται καὶ ἐπὶ κζʹ, τὸν δὲ μέσον ἐπὶ μὲν τοῦ ἐναρμονίου ἐπὶ λεʹ, ἐπὶ δὲ τοῦ διατονικοῦ ἐπὶ ζʹ, ὥστε καὶ τὸν ἡγούμενον τοῦ μὲν ἐναρμονίου γένους συνάγεσθαι ἐπὶ δʹ, τοῦ δὲ διατονικοῦ ἐπὶ ηʹ. τὸν δὲ ἐν τῷ χρωματικῷ γένει δεύτερον ἀπὸ τοῦ ὀξυτάτου φθόγγου λαμβάνει διὰ τοῦ τὴν αὐτὴν θέσιν ἔχοντος ἐν τῷ διατονικῷ. φησὶ γὰρ λόγον ἔχειν τὸν ἐν τῷ χρωματικῷ δεύτερον ἀπὸ τοῦ ὀξυτάτου πρὸς τὸν ὅμοιον τὸν ἐν τῷ διατονικῷ τὸν τῶν σνςʹ πρὸς τὰ σμγʹ. συνίσταται δὴ τὰ τοιαῦτα

[138] Ptol. *Harm.* I.13; text, translation and commentary in Huffman 2005: 402–28 (on which see also Barker 2006: 307–9); translation and commentary in Barker 1989: 43–52; further discussion in Barker 2007: 292–302.

[139] See especially Winnington-Ingram 1932. Barker's position (1989: 46–52; 2000: 11–13, 120–8) is followed in general and challenged in particulars by Huffman (2005: 410–25); Barker's interpretation is adjusted in 2006: 307 and 2007: 294–302.

τετράχορδα κατὰ τοὺς ἐκκειμένους λόγους ἐν πρώτοις ἀριθμοῖς τούτοις. ἐὰν γὰρ τοὺς μὲν ὀξυτάτους τῶν τετραχόρδων ὑποστησώμεθα ,αφιβʹ, τοὺς δὲ βαρυτάτους κατὰ τὸν ἐπίτριτον λόγον τῶν αὐτῶν ,βιςʹ, ταῦτα μὲν ποιήσει τὸν ἐπὶ κζʹ πρὸς τὰ ,αϡμδʹ καὶ τοσούτων ἔσονται πάλιν ἐν τοῖς τρισὶ γένεσιν οἱ δεύτεροι ἀπὸ τῶν βαρυτάτων. τῶν δ' ἀπὸ τοῦ ὀξυτάτου δευτέρων ὁ μὲν τοῦ ἐναρμονίου γένους ἔσται ,αωϙʹ. ταῦτα γὰρ πρὸς μὲν τὰ ,αϡμδʹ ποιεῖ τὸν ἐπὶ λεʹ λόγον, πρὸς δὲ τὰ ,αφιβʹ τὸν ἐπὶ δʹ· ὁ δὲ τοῦ διατονικοῦ γένους τῶν αὐτῶν ἔσται ,αψαʹ. καὶ ταῦτα γὰρ πρὸς μὲν τὰ ,αϡμδʹ τὸν ἐπὶ ζʹ ποιεῖ λόγον, πρὸς δὲ τὰ ,αφιβʹ τὸν ἐπὶ ηʹ· ὁ δὲ τοῦ χρωματικοῦ καὶ αὐτὸς ἔσται τῶν αὐτῶν ,αψϙβʹ· ταῦτα γὰρ λόγον ἔχει πρὸς τὰ ,αψαʹ ὃν τὰ σνςʹ πρὸς τὰ σμγʹ. ὑπογέγραπται δὲ καὶ ἡ τούτων τῶν ἀριθμῶν ἔκθεσις ἔχουσα οὕτως.

He posits three genera, the enharmonic, the chromatic and the diatonic, and he makes his division of each of them in the following way. He makes the 'following' ratio the same in all three genera, 28:27; the middle one in the enharmonic 36:35 and in the diatonic 8:7, so that the 'leading' interval in the enharmonic turns out to be 5:4, in the diatonic 9:8. In the chromatic genus he locates the note second from the highest by reference to that which has the same position in the diatonic. For he says that the second note from the highest in the chromatic stands to the equivalent note in diatonic in the ratio of 256 to 243. Such tetrachords, on the basis of the ratios set out, are constituted in their lowest terms by the following numbers. If we postulate that the highest note of each tetrachord is 1,512, and the lowest, in epitritic ratio with this, is 2,016, this latter term will make the ratio 28:27 with 1,944, and that will be the quantity of the second note from the lowest in all three genera. As to the second note from the highest, that in the enharmonic genus will be 1,890, since that makes with 1,944 the ratio 36:35, and with 1,512 the ratio 5:4. The equivalent note in the diatonic genus will be 1,701, since that makes with 1,944 the ratio 8:7, and with 1,512 the ratio 9:8. The equivalent note in the chromatic genus will be 1,792, since that has a ratio to 1,701 as is 256 to 243. The table of these numbers is set out below.[140]

enharmonic ἐναρμόνιον		chromatic χρωματικόν		diatonic διατονικόν	
,αφιβʹ	1512	,αφιβʹ	1512	,αφιβʹ	1512
ἐπὶ δʹ	5:4		(32:27)	ἐπὶ ηʹ	9:8
,αωϙʹ	1890	,αψϙβʹ	1792	,αψαʹ	1701
ἐπὶ λεʹ	36:35		(243:224)	ἐπὶ ζʹ	8:7
,αϡμδʹ	1944	,αϡμδʹ	1944	,αϡμδʹ	1944
ἐπὶ κζʹ	28:27	ἐπὶ κζʹ	28:27	ἐπὶ κζʹ	28:27
,βιςʹ	2016	,βιςʹ	2016	,βιςʹ	2016

[140] Ptol. *Harm.* 30.17–31.18, trans. Barker 1989: 44, 304.

These tetrachords are simple enough to construct mathematically. Practically, however, the matter is more complex. Andrew Barker has raised the difficulty of how Archytas could have known in the first place that the enharmonic *lichanos* he heard musicians commonly play was not a ditone below *mesē*, but the marginally sharper and 'sweeter' 5:4 major third, about which Aristoxenus later complained.[141] In answering this question Barker notes that if Archytas had used a monochord to measure this upper enharmonic interval, he would have required an instrument capable of distinguishing the 5:4 major third from the 81:64 (= $(9{:}8)^2$) ditone with enough precision to allow him to identify the major third as the interval familiar from musical practice. This difference is minuscule (81:80), smaller even than the difference between six epogdoic tones and a true octave (the *komma*, which Ptolemy approximated to the epimoric ratio 65:64).[142] Barker accordingly concludes that the monochord, were it in use by this time ('which is disputable', he adds) is not likely to have been 'a sufficiently accurate device' for the task, and this, among other concerns, leads him to seek another explanation (1989: 50).

If the monochord – which was (later, in any case) a means of demonstrating predetermined interval-ratios rather than determining them by measurement – could not have assisted Archytas in assigning ratios to the intervals of his tetrachords, it remains to be shown how he was able to content himself that his ratios offered, at least in some ideal sense, a true representation of the scales familiar to him. Barker's answer is that observation of musicians provided the key. A lyre can be tuned by the method of concordance (up a fifth, down a fourth, up a fifth, and so on, or the inverse) to a version of the diatonic genus in which the two upper intervals of each tetrachord are epogdoic tones, and the lowest is a *leimma* (*diesis* in Philolaus' terminology).[143] This attunement is often called the 'Pythagorean' diatonic; Ptolemy calls it the

[141] *El. harm.* 23.11–22; Barker 1989: 50–1. On the 'sweetness' of this interval, see Barker 2007: 297–8. For a synopsis of the structure of the scale assumed in the following discussion, see the Introduction.

[142] See n. 115 above.

[143] The method of concordance is discussed in ch. 1 part 1.

'ditonic' (διτονιαῖον) diatonic because of its two epogdoic tones ($9:8 \times 9:8 \times 256:243$).[144] It is often assumed that Philolaus fr. 6a presupposes a tetrachord of this form, and that may be true;[145] but it is worth noting that Philolaus only says that the magnitude of the fourth (*syllaba*) is two epogdoics and a *diesis*, not that the fourth was actually subdivided in this way in real music, nor indeed in what order these three intervals were placed in such an attunement. But even if Philolaus was not thinking of tetrachordal divisions in his discussion of 'the magnitude of *harmonia*' in fr. 6a, it is a reasonable assumption that Archytas was familiar with the ditonic diatonic. The method of concordance made it a starting point in the tuning process,[146] and Plato employed it in the harmonisation of the world soul (*Timaeus* 36a–b).

Once a lyre was tuned in this way, Barker suggests, the *tritē* string, which now lay at $81:64$ ($= (9:8)^2$) below *nētē*, was flattened slightly until the musician had satisfied himself that it lay a minor third above *mesē*.[147] Since *mesē* is a fifth below *nētē*, and since *paranētē* has been tuned by concordance to a tone below *nētē*, *mesē* must be a perfect fourth below *paranētē* ($3:2 \div 9:8 = 4:3$). If *tritē*, which had been $9:8$ below *paranētē* and $32:27$ ($= 9:8 \times 256:243$) above *mesē*, has now been flattened to a pitch still lying a large tone below *paranētē* and a minor third above *mesē*, then Archytas' reasonable guess (and it may have been a consciously approximate one) was that the large tone above the new *tritē* was $8:7$, and the minor third separating it from *mesē* was $7:6$; these together make up the fourth that separates *mesē* from *paranētē* ($8:7 \times 7:6 = 4:3$).[148]

144 'Pythagorean': Barker 1989: 48; διτονιαῖον: Ptol. *Harm.* 40.19.

145 See for instance Huffman 2005: 415.

146 It is possible that this was its only practical role, and that it was not in fact used as a performance attunement on any instrument in the fourth century: Barker 1989: 51, Huffman 2005: 418.

147 Barker 1989: 51.

148 On the importance of this 7:6 minor third see Winnington-Ingram 1932: 206–7. I omit here any discussion of the important question of what mathematical priorities were at play in Archytas' selection of ratios to approximate the adjusted intervals he observed (specifically, the question of how much weight Archytas gave to epimoric ratios in his scheme). The issue is fully discussed by Huffman and Barker (see n. 139 above), and does not bear directly on the present argument; my aim here is simply to show that Archytas did not require a monochord to posit the ratios recorded by Ptolemy.

A further calculation will determine that *paramesē* stands at 28:27 below the new *tritē* (7:6 ÷ 9:8 = 28:27), and once *parhypatē* has been flattened to a perfect fifth below *tritē* the attunement is complete (9:8 × 8:7 × 28:27). Thus by positing the adjustment of only one note in each tetrachord, Barker is able to suggest a way in which to read the ratios of Archytas' diatonic as a deliberate mathematical guess informed by observation of real musicians tuning their lyres. A similar set of tuning procedures can be posited to suggest a basis for the ratios of his chromatic and enharmonic tetrachords as well.[149]

All this relies on a number of assumptions about how lyres were actually tuned in Archytas' lifetime, and there are great uncertainties on this point. But as a conjecture, what it offers is a way of accounting for the ratios of Archytas' tetrachords from the resources we can be sure he had at his disposal (certain mathematical operations, and certain types of instrument). We could certainly speculate that Archytas went on to set out his divisions on a monochord in order to ensure that they sounded like the attunements which he was seeking to quantify. Equally, however, we could speculate that the monochord had not yet been invented.

[149] Ptolemy tells us that Archytas located his chromatic *lichanos* and *paranētē* at 256:243 below their diatonic equivalents (31.2–6). This could be achieved by first tuning the instrument to Archytas' diatonic by the procedure above; next, by tuning the *parhypatē* string to a fourth below *paramesē*, and then flattening *paranētē* to a fifth above this *parhypatē*, the new chromatic *paranētē* will have been made to lie a tone above *paramesē*, and consequently 32:27 below *nētē* and 256:243 below the diatonic *paranētē* (9:8 × 256:243 = 32:27). Since *tritē* has already been located (by the diatonic attunement procedure above) 28:27 above *paramesē*, it can be found by calculation to be 243:224 below *paranētē* (9:8 ÷ 28:27 = 243:224), and when *parhypatē* and *lichanos* have been retuned at a fifth below *tritē* and *paranētē* respectively, the instrument will match Archytas' chromatic.

The enharmonic could be produced by starting with the upper tetrachord tuned to Archytas' diatonic and the lower tetrachord tuned to the 'ditonic' diatonic. *Paranētē* would then be retuned a fifth above *parhypatē*. This new *paranētē*, standing for the moment two whole tones below *nētē*, must then be sharpened slightly to produce a 'sweetened' major third; Archytas will have assigned this interval the ratio 5:4 (the nearest epimoric smaller than the ditone: $(9:8)^2$ = 81:64 and 5:4 = 80:64). By calculation the new enharmonic *paranētē* will be found to lie 36:35 above *tritē*, which has already been located 28:27 above *paramesē*. Again, retuning *lichanos* and *parhypatē* a fifth below *paranētē* and *tritē* respectively will complete the procedure and the instrument will sound an enharmonic from which Archytas could conceivably have derived his ratios.

Given Archytas' reputed talent for gadgetry,[150] it seems likely that a device like the monochord, plainly superior to the *kalamos* in a demonstrative role, would have appealed to him sufficiently to have presented a disconcerting challenge to his theory of acoustics. Had he encountered the *kanōn*, Archytas might well have revised the work from which our fr. 1 survives.

Conclusion

The argument of this chapter has so far been entirely negative: that none of the evidence for mathematical harmonics in the work of the two earliest Pythagoreans to write about the subject can be used to argue for the presence of the monochord in the fifth or fourth centuries. From this negative conclusion, however, we can make some positive suggestions (and they can be no more than that). Let us first take it as an assumption *e silentio* that the monochord was invented after 360 BC.[151] On this basis we might tentatively construct a picture of what could be called 'pre-canonic' mathematical harmonics. Its achievements are in many ways quite respectable: the relationship between concordant intervals and their ratios were demonstrable (Hippasus); this set of ratios could be invested with cosmic significance in order to explain the structure of the universe (Philolaus); the behaviour of various instruments could be used to propose an explanation of how pitched sound arises, one which may even have gone on to derive ratios from distances (Archytas fr. 1); the mathematical properties of the intervals in musical scales could be investigated, both abstractly (fr. 2) and as a means of quantifying the intervals in real music (testimonium A16). If all this could be done without a monochord, then we ought to consider what the monochord brought to its science, and to what extent its presentation of musical intervals determined both the questions and the answers pursued by later harmonicists. This will be the

[150] Aristotle mentions a children's noise-maker named after him (*Pol.* 1340b25–8); Aulus Gellius even reports (quoting Favorinus) that Archytas constructed a wooden dove that flew (*NA* X.12.8–10).

[151] Archytas probably died between 360 and 350 (see n. 123 above).

subject of chapter 5. But first we must turn our attention to the *Sectio canonis*, where the instrument first appears in a demonstrative role, and ask the opposite question: how did the advances in harmonics, acoustics and scientific method in the fourth century prepare the way for the apodeictic use of the *kanōn* in the final propositions of the *Sectio*?

3

THE MONOCHORD IN CONTEXT

'Scientific progress goes "boink"?'.[1]

Scarcely more than half a century separates the end of Archytas' career from the earliest surviving written evidence for the monochord.[2] The conclusions of the previous chapter suggest that the instrument first came into use during this interval, but no fourth-century authors whose work included musicological discussions – Plato, Aristotle, Aristoxenus and Theophrastus are the most important – give any more indication of its existence than Archytas himself. Aristoxenus, for example, whose treatise on harmonics was written within a few decades of Duris' local history of Samos, does not offer so much as a clue; since the instrument facilitated an approach to harmonics whose fundamental tenets Aristoxenus considered extraneous to the science, and given his adversarial temperament, one might have expected a snide remark at the least. This makes its sudden emergence in Duris fr. 23 and the *Sectio canonis* all the more surprising: the monochord is catapulted into a starring role before anything else is heard of it. Perhaps this is no more than should be expected; why, after all, should a gadget deserve mention until there is something to say about it?

The aim of this chapter is to show how the monochord's dramatic entry on the literary stage around 300 BC was prepared by a number of important advances in harmonics, acoustics and mathematical argumentation in the fourth century. From our texts, the *kanōn* appears simultaneously to spring fully formed from the head

[1] Hobbes, in Bill Watterson, *Scientific Progress Goes 'Boink': A Calvin and Hobbes Collection* (Kansas City, Missouri, 1991), 55.

[2] On Archytas' dates, see ch. 2 n. 123; Duris fr. 23 and the *Sectio canonis* provide a *terminus ante quem* for the monochord of c. 300 (see ch. 2 part 2).

of Euclid and from the thigh of Simos, but there is evidence to be gathered about its obscure gestation. Our earliest account of its use (as opposed to its mere existence) is in *Sectio canonis* propositions 19–20, where it is integrated into a rigorously demonstrated argument designed to derive the primary theorems of mathematical harmonics from first principles. Our first task, then, will be to examine the argumentative and demonstrative background to the *Sectio*; here we shall find an inheritance which includes both demonstrative uses of instruments in harmonics and the logical framework of deductive proof in mathematics (this is the subject of part 1 of this chapter). The *Sectio* also has debts to the harmonic developments of the fourth century; these are most evident in the selection of the 'ditonic' division (9:8 × 9:8 × 256:243) of Plato's *Timaeus* (instead of any of those offered by Archytas, for example) and in the rigidly arithmetical treatment of musical intervals (part 2). Thirdly, the *Sectio* is indebted to fourth-century acoustics; here we shall see the influence not only of Archytas (most prominently in the starting premise that sound is caused by impact), but also of accounts of quantitative acoustics which allowed for demonstrations on stringed instruments (part 3). Finally, we shall consider the ways in which Euclid (or his imitator) makes use of this complex inheritance in deploying the *kanōn* at the conclusion of the little treatise that inaugurated 'canonical' harmonics (part 4).

Before we begin, a few words need to be said about the treatise itself. Its transmission is complicated by its survival in three traditions: directly, where it is ascribed to Euclid in the majority of manuscripts, to Cleonides in a minority; indirectly in partial quotation by Porphyry in his commentary on Ptolemy's *Harmonics*; and the first half of it via a Latin version in Boethius' *De institutione musica*. The text has suffered much at the hands of modern scholarship, not only on account of its transmission, but also on the basis of its content. Cleonides' authorship has (rightly) been excluded altogether, Euclid's authorship has been doubted, and the treatise itself has been topped and tailed to such an extent that, on the most critical reading, only the first sixteen propositions are left as possible, but not certain, third-century material, the introduction and last four propositions having been excised as later accretions.[3] The work was known to several authors in antiquity besides Porphyry, who cites its title twice (*Kanonos katatomē*, 'Division of the monochord') with attribution to Euclid both times, and quotes the first sixteen propositions and the introduction separately without attribution.[4] Boethius does not attribute it to anyone, and his Latin version (*Mus*. IV.1–2) includes only the introduction and first nine propositions.[5]

The very fact that Porphyry quotes its title, in Andrew Barker's view, goes some way to making up for his omission of the actual division of the monochord (contained in the disputed propositions 19–20), and based on this and other factors, he has recently advanced a defence of the traditional dating of the treatise (c. 300 BC) and of its integrity as a single work from a single hand, whether that be Euclid's or not.[6]

Firstly, he argues, the fact that Porphyry quotes only the first sixteen propositions with the attribution to Euclid cannot be used to prove that the introduction was not part of the same document; the lack of attribution counts no more against Euclidean authorship than it does for it (2007: 369). Secondly, Porphyry's quotation of the title *Kanonos katatomē* would be difficult to account for if the final propositions containing the division were not part of the treatise as he knew it, despite the fact that he does not quote them himself (370). Furthermore, the canonic division of propositions 19–20 is extraneous to Porphyry's discussion in this part of his *Commentary*, and so the omission, on its own, can hardly be used to excise the last two propositions. Thirdly, the claim that the

[3] The relevant discussions are in Jan 1895 (including an edition of the text); Tannery 1904a; Barbera 1977, 1984b and 1991 (with new editions and translations of the text in its three traditions); Mueller 1980; Barker 1981, 1989 (with translation), 1991 and 2007; Levin 1990; Bowen 1991b. See also Winnington-Ingram 1932: 198; Düring 1934: 177; Burkert 1972: 375 n. 22; Barker 1978b: 338–40; Fowler 1999: 138–46.

[4] Title and authorship: Porph. *In Harm*. 92.29–30 (followed by brief quotations from props. 6 and 4 introduced by the phrase 'for Euclid says') and 98.19; introduction: 90.7–22; props. 1–16: 99.1–103.25. Part of Ptolemy's discussion of the Pythagoreans' postulates on concords (*Harm*. I.5) relies on the arguments of *Sect. can.* props. 3, 10–12 and 16 (see Barker 1989: 286 and 2007: 367). Proclus gives Euclid an *Elements of Music* (Στοιχειώσεις κατὰ μουσικήν, *In Euc.* 69.3) which may refer to what we know as the *Sectio canonis* (see Barbera 1991: 7).

[5] For a full account of the textual transmission of the *Sectio*, see Barbera 1991. On Boethius' version, see also Bower 1989.

[6] Barker 2007, ch. 14.

introduction is irrelevant to the content of the first sixteen propositions is challenged by the argument of the introduction itself: beginning from the connection between motion and sound, it progresses by careful and methodical steps to the conclusion that concords are either multiple or epimoric in ratio, a statement on which much of the rest of the document is based, in one form or another (370–8). As for the alleged problems with propositions 17–20, Barker argues that anomalies in the note-names given there indicate an early date, despite some differences in discursive form between these propositions and the first sixteen (391–406). His arguments cannot be conclusive,[7] but they do suggest strongly that attempts to dismember the treatise and assign certain portions of it a date later than Euclid's are unwarranted. He proposes that the *Sectio* (more or less as we have it) was composed not long after 300 BC as a single document by a single author, perhaps Euclid.

If Barker is right (and I find his arguments very persuasive), then we may be justified in viewing the entire document, from the opening sentences of the introduction to the end of the division itself in proposition 20, as a unified project whose objective, indicated by its title, is finally introduced at the end of the work: the division of the monochord is, as Barker puts it, 'what it is "all about"' (370).

Like the genuine works of Euclid, the *Sectio* is innovative not so much in its content as in its systematic organisation of material already current – that is, in its presentation of this material within an axiomatised deductive system where each proposition relies only on postulates already laid down or on theorems already demonstrated. This, at any rate, is the expectation raised by the author's use of the Euclidean mathematical form and style. But a logical error in proposition 11 creates a rupture in the deductive framework of the treatise, and this has often been cited as grounds for denying its Euclidean authorship.[8] Proposition 11 attempts to prove that the intervals of the fourth and fifth are epimoric. The method is (a) to construct two fourths between named notes, then

[7] As he himself admits (Barker 2007: 365, 367).

[8] The flaw was first pointed out by Tannery (1904a); cf. Fowler 1999: 138, 146 (who does not connect prop. 11 with the question of the treatise's authorship).

(b) to state as a consequence that the combined interval, a double fourth, is discordant, and (c) to conclude from this that it is not multiple. The author then relies on proposition 5 and conclusion (c) to state (d) that if the combined interval is not multiple, the component intervals are not multiple either. But (e) since the component intervals, fourths, are concordant, they must then be epimoric; this relies on premises stated in the introduction.

The error is conclusion (c), on which relies not only step (d), but also proposition 12 (the proof that the octave interval is duple).[9] Conclusion (c) itself is in fact correct: the interval of the double fourth (16:9) is not multiple. But it is not logically valid, since the introduction established only that all concords must be either multiple or epimoric, not that all discords must *not* be multiple or epimoric. (Examples of discordant intervals which *are* multiple are easy to find: e.g. the double octave plus major third – a discord no matter which ancient definition of concord one adopts – whose ratio is 5:1.) In fact, the association of concords with multiple and epimoric ratios is not even given in the introduction as a proof, but only as a 'reasonable' conclusion from mathematical and perceived evidence.[10]

These difficulties notwithstanding, the treatise is otherwise largely careful, methodical and successful in its argumentation, and is clearly indebted, as is immediately apparent from the style of its prose and the formulaic composition of its proofs, to an intellectual tradition that includes Euclid himself.[11] It is also clearly polemical in its aims: in the course of its argument, several shots are fired over the bows of the Aristoxenian school (e.g. prop. 18). The purpose of the text, in certain passages at least, appears to be to

[9] This then creates a logical domino effect throughout the treatise, since prop. 14 (the proof that the octave is less than six tones) relies on prop. 12; prop. 15 (the proof that the fourth is less than two and a half tones, and the fifth less than three and a half tones) relies on prop. 14. Prop. 18 (the proof that the *parhypatai* and *tritai* do not divide the *pyknon* equally) relies directly on prop. 11. The construction of the scale-system in props. 19–20 then relies on many of the mathematical relationships established in these proofs (e.g. by constructing octaves by means of duple ratios of string length). The error of prop. 11 thus compromises the logical validity of the canonic division which is the final goal of the treatise.

[10] Fowler 1999: 139; Barker 2007: 375–8. The 'reasonable' (*eikos*) assumption is made in the concluding sentence of the introduction (149.20).

[11] Netz (1999a) offers a careful analysis of the structure of Euclidean proofs and, more generally, of the formulaic language of Greek mathematics.

use formal argument to counter recent charges against mathematical harmonics.[12] Thus by its integration into such a project, the monochord becomes part of the apparatus of mathematical demonstration, and also a weapon with which to defend the applicability of this method to harmonic science.

The role of the monochord in the polemic of mathematical harmonics against the non-rational approach (by which I mean merely 'not thinking about musical intervals as ratios of numbers') of Aristoxenus was essential in drawing the lines of argument between the two traditions and crystallising the debate for future generations. But both the instrument itself and its use in the *Sectio* owe much to advances in demonstrative argument and the employment of instruments by harmonic theorists of various persuasions in the fourth century. The first task of this chapter is to consider these antecedents in order to see more clearly how this watershed treatise employs them in a careful and methodical argument in which the *kanōn* proves to be indispensable.

1 Demonstration, harmonics and the *Sectio canonis*

It may be useful to begin by distinguishing between harmonic demonstrations which could be classified as *epideictic*, and those which were more strictly *apodeictic*. By the former I mean declamatory or rhetorical displays of musicological learning, in which an instrument could be brought out rather like a stage prop at the appropriate moment, to provide a demonstration of the thesis under discussion.[13] By the latter, on the other hand, I refer to formal scientific demonstrations of a kind methodologically indebted to Aristotle's *Analytics*; these are deductive proofs derived from first

[12] Proposition 18 demonstrates that 'the *parhypatai* and the *tritai* do not divide the *pyknon* into equal intervals'. The *pyknon* ('compressed ⟨interval⟩') was the small composite interval at the bottom of an enharmonic or chromatic tetrachord (*hypatē* to *lichanos* or *paramesē* to *paranētē*); see Aristox. *El. harm.* 24.11–14, 50.15–19. Aristoxenus had described the two small intervals which comprised the *pyknon* as being of equal size (*El. harm.* 50.28–51.11). 'The task of proposition 18 . . . like that of proposition 16, is mainly polemical rather than constructive; it shows that the work of a rival school had been based on a mistake' (Barker 1989: 204 n. 57). On prop. 16, see ch. 1 part 1.

[13] On epideictic displays see especially Lloyd 1979, ch. 2: 'Dialectic and demonstration'. On music as a subject of *epideixis* in the fourth century see Barker 2007: 68–78.

principles (the 'apodeictic proofs' described in chapter 1 part 1). The former may often involve uncontroversial theses; the latter, on the other hand, may advance new positions antithetical to those held by scientific opponents, or offer proof for positions commonly (or even unanimously) taken to be true but not previously proven. We shall find that this distinction can be applied to harmonic demonstrations as early as the fourth century, and in a moment I shall outline some of the reasons why this approach is helpful in tracing the history of the monochord. It should be pointed out, however, that the categories themselves were not rigidly defined in antiquity; both the nouns (*epideixis* and *apodeixis*) and the verbs (*epideiknumi* and *apodeiknumi*) were sometimes used interchangeably, and an argument that began life as an apodeictic demonstration could reappear in the content of an *epideixis*.[14]

The distinction is useful nonetheless because it helps to place the appearance of the monochord in *Sectio canonis* proposition 19 within the broader context of harmonic demonstration, in both the epideictic and apodeictic senses, in the fourth century. To get a better sense of the relevance of this distinction for monochord use specifically, let us first look ahead for a moment. An ideal (if late) example of what I have termed epideictic demonstration is the type of musical activity implied by Ptolemy's comments on the insufficiencies of the *kanōn*, and some improvements to it introduced by the first-century AD musicologist Didymus,[15] at *Harmonics* II.13–14. Ptolemy's discussion presupposes the playing of melodies on the monochord (a task for which it is not as well suited as 'performance' instruments, he notes), and he records certain changes, proposed by Didymus, which facilitated this practice. We know from portions of Didymus' musicological writings quoted by Porphyry that he was a serious musical scientist, much concerned with problems of method (Porphyry quotes him for his discussion of the differences between harmonicists who placed varying degrees of emphasis on either perception or reason), though unarguably more interested in the musical systems preserved in the writings of his

[14] It should also be pointed out that some demonstrations could proceed without argument at all (anatomical demonstrations, for example).

[15] Discussed at greater length in chs. 5 and 6.

predecessors than in those of the music of his own time.[16] Andrew Barker has argued that Didymus' monochord-informed researches were designed to facilitate a reconstruction of the music of Aristoxenus' day, which he could then present to a philhellenic audience eager to hear with their own ears the music of the by-gone age which their aristocratic *paideia* had taught them to revere.[17]

In this context the monochord fulfils an epideictic function; this of course does not preclude the possibility of Didymus' musicological pursuits having included a more apodeictic role for the *kanōn*, nor did it prevent him from criticising 'all those who even now are commonly said to make use of a non-rational procedure (*alogos tribē*)' because 'as far as was at all possible they offered no demonstration (*apodeixis*) and referred nothing to reason'.[18]

Social contexts for musical epideixis*: late examples*

Plutarch gives us an idea of the social context for the type of musicological *epideixis* in which Didymus made melody with his monochord. Like Didymus' audience, Plutarch's was made up of educated Greeks or Hellenising Romans, the focus of whose scholarly discussions lay some four or five centuries distant in the classical age of the Athenian polis. In an anti-Epicurean essay Plutarch berates his long-dead opponents for their attention to the pleasures of musical entertainment rather than to the learned dinner-discussions of serious musical scholars:

τί λέγεις, ὦ Ἐπίκουρε; κιθαρῳδῶν καὶ αὐλητῶν ἕωθεν ἀκροασόμενος εἰς τὸ θέατρον βαδίζεις, ἐν δὲ συμποσίῳ Θεοφράστου περὶ συμφωνιῶν διαλεγομένου καὶ Ἀριστοξένου περὶ μεταβολῶν καὶ Ἀριστοτέλους περὶ Ὁμήρου τὰ ὦτα καταλήψῃ ταῖς χερσὶ δυσχεραίνων καὶ βδελυττόμενος;

What's this, Epicurus? To hear singers to the cithara and performers on the aulos you go to the theatre at an early hour, but when at a banquet Theophrastus holds forth on concords, Aristoxenus on modulations, and Aristotle on Homer, you will clap your hands over your ears in annoyance and disgust?[19]

[16] The fragments are preserved at Porph. *In Harm.* 26.6–29, 27.17–28.26.
[17] Barker 1994a: 71–2.
[18] Didymus ap. Porph. *In Harm.* 26.14–15, 26.10–11 (trans. Barker 1989: 242).
[19] Plut. *Non posse* 1095e, trans. Einarson and De Lacy 1967: 77. Both here and in the next quotation I print 'aulos' for the translators' 'flute': the aulos, a double-reeded

Later in the same discussion Plutarch elaborates further the range of topics on which learned guests could be expected to 'hold forth' at such *symposia*:

ποῖος γὰρ ἂν αὐλὸς ἢ κιθάρα διηρμοσμένη πρὸς ᾠδὴν ἢ τίς χορὸς 'εὐρύοπα κέλαδον ἀκροσόφων ἀγνύμενον διὰ στομάτων' φθεγγόμενος οὕτως εὔφρανεν Ἐπίκουρον καὶ Μητρόδωρον ὡς Ἀριστοτέλη καὶ Θεόφραστον καὶ Δικαίαρχον καὶ Ἱερώνυμον οἱ περὶ χορῶν λόγοι καὶ διδασκαλιῶν καὶ τὰ διαύλων προβλήματα καὶ ῥυθμῶν καὶ ἁρμονιῶν; οἷον διὰ τί τῶν ἴσων αὐλῶν ὁ στενότερος ὀξύτερον ὁ δὲ εὐρύτερος βαρύτερον φθέγγεται· καὶ διὰ τί τῆς σύριγγος ἀνασπωμένης πᾶσιν ὀξύνεται τοῖς φθόγγοις, κλινομένης δὲ πάλιν βαρύνεται, καὶ συναχθεὶς πρὸς τὸν ἕτερον βαρύτερον, διαχθεὶς δὲ ὀξύτερον ἠχεῖ· καὶ τί δήποτε τῶν θεάτρων ἂν ἄχυρα τῆς ὀρχήστρας κατασκεδάσῃς ὁ ἦχος τυφλοῦται, καὶ χαλκοῦν Ἀλέξανδρον ἐν Πέλλῃ βουλόμενον ποιῆσαι τὸ προσκήνιον οὐκ εἴασεν ὁ τεχνίτης ὡς διαφθεροῦντα τῶν ὑποκριτῶν τὴν φωνήν· καὶ τί δήποτε τῶν γενῶν διαχεῖ τὸ χρωματικόν, ἡ δὲ ἁρμονία συνίστησιν;

For what aulos or cithara attuned to vocal music or what chorus sending forth 'a rolling thunder from melodious throats' could so have enthralled the mind of Epicurus and Metrodorus as the minds of Aristotle and Theophrastus and Dicaearchus and Hieronymus were enthralled by discussion of choruses and the production of plays and by questions about double-auloi and rhythms and scale-systems? For example: why of auloi of equal length does the narrower have the higher pitch, the wider the lower? And why, when the *syrinx* is drawn back, are the notes all raised in pitch, but when it is released again, they are lowered? And why, when one pipe is brought close to the other, does it have a lower tone, but a higher when the pipes are drawn apart? And why, when chaff is spread over the orchestra of a theatre, is the resonance muffled, and when Alexander wanted to make the proscenium at Pella of bronze, did the architect demur, as he would thus have spoiled the effect of the actors' voices? And why of the genera does the chromatic relax the hearer, the enharmonic make him tense?[20]

Such questions, very much in the style of the pseudo-Aristotelian *Problems*,[21] will no doubt have been answered – at

pipe, should be distinguished from ancient reedless flute-type instruments such as the *plagiaulos* or *phōtinx*; see e.g. West 1992, ch. 4. LSJ's 'pipe, flute, clarionet' is corrected in the 1996 supplement to 'oboe-type musical instrument, reed-pipe, aulos'.

[20] Plut. *Non posse* 1096a–b, trans. Einarson and De Lacy 1967: 77–81. I have written 'scale-systems' for the translators' 'harmonies'. Plutarch probably means to employ the term *harmoniai* here as Plato does (e.g. at *Phlb.* 17d, where they are also called *systēmata*); *harmonia* in this sense is an arrangement of intervals, often more specifically an 'arrangement of intervals within the octave' (see e.g. Aristox. *El. harm.* 36.31).

[21] See especially books XI and XIX, acoustical portions of which are discussed in part 3 of this chapter.

least at dinner-parties, real or imagined – with a view to entertaining the other guests, rather than presenting rigorously demonstrated arguments. (Indeed, it is difficult to imagine how rigorously demonstrated arguments could be made about most of Plutarch's sample questions.) This is not to say that the answers were not persuasive; an argument can be persuasive without being apodeictic. Sympotic *epideixis* is sometimes persuasive (when it is persuasive) primarily by being a sort of potted *apodeixis*, in which rigorous argument is digested and regurgitated as table-fare.

The musical discussions in Athenaeus' *Deipnosophistae* are of a piece with this picture: in book XIV Masurius, 'an excellent and clever man on every subject' who 'devotes himself unceasingly to music, and even plays instruments', asserts, citing Heraclides Ponticus, 'that Phrygian should not even be called a *harmonia*, and no more should Lydian'.[22] Later he cites Polybius' contention against Ephorus' claim 'that music . . . was brought into use among men as a device for deception and cheating'.[23] Both of these statements are elaborated and defended, but not with any degree of rigour; the point of the exercise is the convincing display of learning that each guest can bring to the dinner-table.[24]

Fourth-century harmonic epideixis

I do not mean to suggest that Plutarch and his contemporaries should be taken as reliable witnesses to the finer details of musical *epideixis* among the educated aristocracy of their fourth-century models. What they do offer, however, is an indication of the extent to which the harmonic *apodeixeis* of serious scholars like Claudius Ptolemy can be seen alongside the more informal *epideixeis* presented by the polymathic drinkers and diners of Second Sophistic dialogues. This dichotomy parallels an earlier one, which is most visible in the contrast between the formal demonstrations of the *Sectio* and the displays of contemporary harmonic experts. One of Plutarch's fourth-century symposiast-scholars, Theophrastus, in a

[22] Ath. 624c–e, trans. Barker 1984: 280–1.
[23] Ath. 626a, trans. Barker 1984: 284. [24] See Lukinovich 1990: 268.

passage of his *Characters* (5.6–10), presents a sketch of a distastefully ostentatious person, whose excessive displays of wealth include, among many other things,

παλαιστρίδιον κονίστραν ἔχον καὶ σφαιριστήριον. καὶ τοῦτο περιὼν χρηννύναι τοῖς σοφισταῖς, τοῖς ὁπλομάχοις, τοῖς ἁρμονικοῖς ἐνεπιδείκνυσθαι· καὶ αὐτὸς ἐν ταῖς ἐπιδείξεσιν ὕστερον ἐπεισιέναι ἐπειδὰν ἤδη συγκαθῶνται ἵν' ὁ ἕτερος εἴπῃ τῶν θεωμένων πρὸς τὸν ἕτερον ὅτι 'τούτου ἐστὶν ἡ παλαίστρα.'

a little palaestra with a sanded area for wrestling and a room for boxing practice. He goes around offering this arena to sophists, drill-sergeants and music lecturers (*harmonikoi*) for them to perform in (*enepideiknusthai*). And he arrives at these performances (*epideixeis*) after the spectators are already seated, so that they will say to each other 'This is the owner of the palaestra.'[25]

Jeffrey Rusten translated *harmonikoi* 'musicians',[26] but Diggle's 'music lecturers' is preferable on several fronts. Each of the three groups (*sophistai*, *hoplomachoi*, *harmonikoi*) is apparently welcome to draw an audience to the man's performance space because this will show off his wealth to the greatest possible effect. It is in keeping with his varied tastes in other things (5.6–9) that those to whom he lends his little palaestra form something of a spectrum, even though they are not listed in that order: the *hoplomachoi* present the most physical sort of display, and the *sophistai* the least; the *harmonikoi* presumably fall somewhere in between the two. It is a programme designed to attract the widest range of spectators, in order to illustrate that their host is a patron of the widest array of arts, from the military to the musical to the rhetorical.

Furthermore, one noun describes the activities of all three types of performer: they give *epideixeis* for their audiences. An *epideixis* is a performance or a 'show' (as Rusten translates it), but it is not the same sort of show as the *thea* mentioned at 5.7. The latter is a spectacle or a theatrical performance; the former is a declamation, demonstration or display.[27] The use of *epideixis* and *epideiknumi* to describe military displays is common,[28] and

[25] *Char.* 5.9–10, trans. Diggle 2004: 81. *Char.* 5.1–5 is about 'the obsequious man' (ὁ ἄρεσκος), but 5.6–10 belongs to a different character.
[26] Rusten, Cunningham and Knox 1993: 73. [27] LSJ s.vv. θέα, ἐπίδειξις.
[28] Thuc. III.16, Pl. *Lach.* 179e.

the name *hoplomachos* ('instructor in fighting with weapons'[29]) makes clear that this was not a blood-sport spectacle of fighters, but a demonstration of the art of fighting in heavy armour.[30] Similarly, the *epideixeis* of sophists are well documented.[31] Thus the two ends of the spectrum suggest that the 'shows' all contained some pedagogic elements: the purpose was presumably not merely to entertain, but also to display knowledge or skill so as to attract potential students. In this context, then, we might even render *harmonikoi* 'musical sophists'. It is true that a musical performance can be referred to in the language of *epideixis*,[32] but it would be surprising if performing musicians were mentioned here by a term used elsewhere to describe specialists in musical theory. The word *harmonikoi* had already come into use to mean 'experts in *harmonia*' by Theophrastus' time. We noted in chapter 2 that it is this word which permits an unequivocal reading of *kanōn* as 'monochord' in Duris fr. 23, where *harmonikos* is the epithet of the *kanōn*-thieving Simos. Its use here thus makes the most sense if understood in the same way. The wealthy man caricatured in 5.6–10 is one whose ostentation is of a distinctly cultured bent, and what we are being led to imagine taking place in his private venue is more of an illustrated public lecture series than a popular entertainment programme.[33]

Harmonikoi *on stage*

Theophrastus' sketch, while it helps to clarify one possible context for demonstrations by *harmonikoi* in the late fourth century, tells us nothing about what such *epideixeis* involved. What happens when these people get on stage? What do they do, and what are

29 LSJ rev. suppl. (1996) s.v.

30 On the *hoplomachoi* and the sophistic character of their displays, see especially Wheeler 1983.

31 Pl. *Phdr.* 235a, *Grg.* 447c; Isoc. 4.17, 5.17. Who counted as a sophist was, of course, problematic.

32 Plato uses the verb *epideiknumi* not only of poetic recitation (*rhapsōidia*) but also of accompanied song (*kitharōidia*), tragedy, comedy and even puppet shows at *Leg.* 658b. For a later example of the verb used in a musical sense, cf. Ael. *VH* IX.36; for an ambiguous case, see n. 37 below.

33 Compare Lloyd's discussion of the *epideixeis* of medical experts given before lay audiences in the fifth and fourth centuries BC (1979: 86–98).

their demonstrations like? Most importantly, for present purposes, do their demonstrations involve instruments?

A few preliminary clues to these questions can be gleaned from the opening lines of the pseudo-Platonic dialogue *Sisyphus*. Socrates speaks first, chiding Sisyphus for arriving late ('we waited for you a long time yesterday, too, Sisyphus') and telling him that he missed an excellent *epideixis* put on the previous day by a certain Stratonicus. This Stratonicus is a 'brilliant fellow' (*anēr sophos*), who demonstrated (the verb is *epideiknumi*) many excellent things 'in both word and action' (387b).

At first glance it might seem an unlikely source; the dialogue is almost certainly not by Plato, although it may have been composed by one of his students near the end of his life.[34] Secondly, there is no reference to music or harmonic theory here. But it just so happens that this Stratonicus was a well-known musician and teacher of music and musical theory. He was a contemporary of Plato's, and died around 350 BC.[35] His claims to fame (many of them surely inflated, particularly those which make him a *prōtos heuretēs*) show just how well he would fit in as one of Theophrastus' musical sophists: he reportedly introduced 'many-notedness' (*polychordia*) into kitharistic music; he was famous for his sharp wit, which was not infrequently aimed at rival musicians; he was the first musical theorist to make a *diagramma*, probably meaning a diagram on which various scales could be mapped out in overlapping fashion.[36] From the *Sisyphus* passage it seems he also gave *epideixeis*, to which he was evidently able to draw audiences, who may even have paid to hear him;[37] and no matter how seriously Socrates means the compliment about Stratonicus'

[34] The *Sisyphus* has long been regarded as spurious. Thesleff (1989: 9) preferred to classify it as 'semi-authentic', to allow for the possibility of Plato's influence even if the composition itself was the work of a student. D. S. Hutchinson gives some reasons for maintaining a mid-fourth-century date and assuming the author to be a 'follower of Plato' (in Cooper 1997: 1706–7).

[35] Burkert 1972: 203; West 1992: 217, 367–8; Barker 2007: 75–7.

[36] Phaenias fr. 32, Ath. 347f–352d; Burkert 1972: 372 n. 12, West 1992: 367–8.

[37] 'When he was giving an *epideixis* in Rhodes and no one applauded, he walked out of the theatre with the words "Since you do not give what costs nothing, how can I expect to receive a fee from you?"' (ἐν Ῥόδῳ δ' ἐπίδειξιν ποιούμενος, ὡς οὐδεὶς ἐπεσημήνατο, καταλιπὼν τὸ θέατρον ἐξῆλθεν εἰπὼν 'ὅπου τὸ ἀδάπανον οὐ ποιεῖτε, πῶς ἐγὼ ἐλπίζω παρ' ὑμῶν ἔρανον λήψεσθαι;', Ath. 350b). But Gulick may be justified in translating *epideixis* as 'recital' here (1930: 85).

brilliance, *sophos* is precisely what a sophist aims to be. There are no details about his presentation, but from the small hints here it appears to have been a lecture on music interspersed with musical examples: he demonstrated 'many splendid things in both speech and action (καὶ λόγῳ καὶ ἔργῳ)'. Since both of these terms are applied to a single *epideixis*, the previous evening's activities seem not to have constituted (just) a kitharistic recital.

The specific *epideixis* in question is certainly fictional: Stratonicus' performing career may not even have begun before Socrates' death in 399. Its narrative function is merely to provide Sisyphus with a reason to say what important civic business detained him the previous day, and it is the theme of political deliberation, not of musical *epideixis*, that takes centre stage for the rest of the short dialogue. But as a dramatic hook to open a dialogue, it can only have served its purpose if what it described was accurate in type and did not distract from the focus of the work at its very outset.

The fact that many of our stories about Stratonicus mention the kithara (either his own performances on it, or his critiques of those by others or their students) suggests that when Socrates says 'action', he means a musical illustration or example performed on the instrument.[38] If we could be sure of this, it would tell us that some sort of public musicological demonstrations with stringed instruments were being carried out within a decade or so of Archytas' death. But the *Sisyphus* gives only a likelihood, not by any means a certainty, and we can do better. More useful clues can be found in one of the Hibeh papyri (I.13), in a fragment of what appears to be a sophistic diatribe (itself an *epideixis*, of sorts) against performing harmonicists, copied in the mid-third century BC and recovered from Graeco-Egyptian mummy-cartonnage in 1902. Grenfell and Hunt, who excavated and published the papyri, suggested an early date for the content of *PHib*. I.13, even going so far as to support Blass's fanciful conjecture that the speech was by Hippias, attacking Damon. More recently it has been argued that the fragment is a product of the early fourth century, to be dated

[38] The stories are preserved by Athenaeus (see n. 36 above).

possibly around 380 BC.[39] The papyrus is (like most) mangled in places, but what we have is largely legible, and presents precisely the sort of details we are after:

[πολλα]κις επηλθε μοι θ̣αυμασαι ω ανδρες [Ελληνες
[ει α]λλοτριας τιν[ες] τας επιδειξεις των ο̣[ικειων τε
[χν]ων ποιουμεν[οι] λανθανουσιν υμας λ̣[εγοντες γαρ
[ο]τ̣ι̣ αρμονικοι εισι και προχειρισαμενοι ω[ιδας τινας
ταυτας συγκρινουσιν τωμ μεν ως ετυχεν
κατηγορουντες τας δε εικηι εγκω[μιαζ]οντες
και λεγουσι μεν ως ου δει αυτους ου[τε ψ]α̣λ̣τας
ουτε ωιδους θεωρειν περι μεγ γαρ τ̣[αυτ]α ετεροις
φασιν παραχωρ[ε]ιν : αυτων δ̣ε̣ ιδιον [ει]ναι το θε
ωρητικον μερος φαινονται δε περι μεν ταυτα
ων ετεροις παραχωρουσιν ου μετριως εσπουδακο
τες εν οις δε φασιν ισχυειν εν τουτοις σχ[εδια]
ζ̣οντες.

It has often occurred to me to be surprised, men [of Greece],[40] at the way certain people construct demonstrations (*epideixeis*) not belonging to [their own areas of expertise], without your noticing. [For they say that] they are *harmonikoi*, and they pick out [various songs] and judge them against one another, condemning some, quite at random, and unsystematically extolling others. Again, they say that it is not their business to think about instrumentalists and singers: these matters, they say, they leave to others, while their own special province is the theoretical branch. Yet they actually display an immoderate enthusiasm for the things they leave to others, while improvising haphazardly in the areas where they say their own strength lies.

[εις τ]ουτο δε ερχονται τολμης ωστε [ολον τον βιο]ν̣ κ̣α̣[τα
[τριβ]ε̣ι̣ν̣ εν ταις χορδαις : ψαλλοντες μ̣ε̣ν̣ [πολυ χ]ε̣ι̣[ρον τω]ν
[ψαλ]τ̣ω̣ν̣ : αιδοντες δε των ωιδων : συνκρινοντες δε
[του τ]υ̣χ̣οντος ρητορο̣ς̣ παντα παντω[ν χει]ρ̣ον ποιουντες
[και π]ερι μεν των αρμ[ο]νικων καλουμ̣[ενω]ν εν οις δη
φ[ασι]ν διακεισθαι πως : ουθ ηντινα φων̣[ην] ε̣χοντες λεγε̣ι̣ν :
εν[θο]υσιωντες δε : και παρα τον ρυθμ[ον δε] π̣α̣ιοντες
το υποκειμενον σανιδιον αυτοι̣ς̣[αμα τοις] α̣π̣[ο] του
ψ[αλ]τηριου ψοφοις.

These people have the effrontery to waste [their entire life][41] on strings. They play on strings [much worse than real instrumentalists], they sing much worse

[39] See particularly Avezzù 1994, Barker 2007: 69 n. 1.
[40] Square brackets indicate editorial conjectures to fill gaps in the papyrus. Avezzù suggests 'of Athens' here (1994: 113).
[41] [a lot of time] Avezzù.

than real singers, and in their critical comparisons they do everything worse than any orator (*rhētōr*) one might come across. As to what is called 'harmonics' (*ta harmonika*), with which they say they have a special familiarity, they have nothing articulate to say, but are carried away with enthusiasm: and they beat the rhythm all wrong, on the wooden bench where they sit, [simultaneously] with the sounds of the *psaltērion*.[42]

This is a caricature, of course, and cannot be taken as a faithful portrait; the author is presumably a sophist himself, and his polemic is intended to distort and discredit the practice of his rivals. But his damnatory assessments of the *epideixeis* given by these self-styled *harmonikoi* can only hit home if the criteria for judgement are taken at face value: they cannot be called incompetent instrumentalists if they never played for their audiences, and they cannot be criticised as bad orators if their presentations included nothing like a speech. Their speeches must have included some theoretical discussions about music (they concentrated on the 'theoretical branch' of music, and claimed familiarity with 'harmonics'), but they must also have played stringed instruments: the vocabulary is clear, varied and repeated (*chordai*, 'strings', the verb *psallō*, 'play ⟨a stringed instrument⟩', and its cognate noun *psaltērion*, 'stringed instrument').

Empirical procedures with stringed instruments in the fourth century

What these passages indicate, then, is an identifiable epideictic context for harmonic demonstrations with strings by the mid-fourth century. Nothing in the *Sisyphus* or the Hibeh papyrus suggests either that these demonstrations included ratios, or that the *kanōn* was being used. A *psaltērion* is literally a 'plucked instrument', but the word is often used to mean 'harp',[43] and an approximately contemporaneous description of an empirical procedure involving a stringed instrument in book VII of the *Republic* (531a–c) is detailed enough to rule out ratio-based harmonics in that instance.

[42] *PHib.* I.13.1–13, 23–31, trans. Barker 1984: 184–5.
[43] West 1992: 74; see also the 1996 supplement to LSJ, s.v.

Two types of harmonic theorists are mentioned in the passage, and in making Glaucon initially misunderstand Socrates, Plato is able to distinguish one from the other with a certain degree of care. (The word *harmonikos*, incidentally, is not used to describe either group; Plato does use the term elsewhere.[44]) Socrates criticises some of those who work on harmonics for their devotion to the sort of pursuit he labels 'pointless' (*ateles*, 530e), claiming that 'they measure heard concords and notes against one another, and so labour to no purpose' (τὰς γὰρ ἀκουομένας αὖ συμφωνίας καὶ φθόγγους ἀλλήλοις ἀναμετροῦντες ἀνήνυτα . . . πονοῦσιν).[45] Glaucon assumes that Socrates is referring to theorists of a decidedly empirical sort:

νὴ τοὺς θεούς, ἔφη, καὶ γελοίως γε, πυκνώματ' ἄττα ὀνομάζοντες καὶ παραβάλλοντες τὰ ὦτα, οἷον ἐκ γειτόνων φωνὴν θηρευόμενοι, οἱ μέν φασιν ἔτι κατακούειν ἐν μέσῳ τινὰ ἠχὴν καὶ σμικρότατον εἶναι τοῦτο διάστημα, ᾧ μετρητέον, οἱ δὲ ἀμφισβητοῦντες ὡς ὅμοιον ἤδη φθεγγομένων, ἀμφότεροι ὦτα τοῦ νοῦ προστησάμενοι.

'Yes, by the gods,' he said. 'Their behaviour is quite ridiculous, when they name some things "*pyknōmata*" and incline their ears as if hunting out a sound from next door, some of them asserting that they can still just hear a sound in between, and that that is the smallest interval, by which measurement is to be made, while others take issue with them, saying that the notes sounded are already the same, each group putting their ears ahead of their mind.'[46]

Socrates then sets Glaucon straight, and in the process we hear more about each of the two types of harmonicist Plato is at pains to distinguish:

σὺ μέν, ἦν δ' ἐγώ, τοὺς χρηστοὺς λέγεις τοὺς ταῖς χορδαῖς πράγματα παρέχοντας καὶ βασανίζοντας, ἐπὶ τῶν κολλόπων στρεβλοῦντας· ἵνα δὲ μὴ μακροτέρα ἡ εἰκὼν γίγνηται, πλήκτρῳ τε πληγῶν γιγνομένων καὶ κατηγορίας πέρι καὶ ἐξαρνήσεως καὶ ἀλαζονείας χορδῶν, παύομαι τῆς εἰκόνος καὶ οὔ φημι τούτους λέγειν, ἀλλ' ἐκείνους οὓς ἔφαμεν νυνδὴ περὶ ἁρμονίας ἐρήσεσθαι. ταὐτὸν γὰρ ποιοῦσι τοῖς ἐν τῇ ἀστρονομίᾳ· τοὺς γὰρ ἐν ταύταις ταῖς συμφωνίαις ταῖς ἀκουομέναις ἀριθμοὺς ζητοῦσιν, ἀλλ' οὐκ εἰς

[44] *Phdr.* 268d–e. [45] *Resp.* 531a, trans. Barker 1989: 55.
[46] *Resp.* 531a–b, trans. Barker 1989: 55.

προβλήματα ἀνίασιν, ἐπισκοπεῖν τίνες σύμφωνοι ἀριθμοὶ καὶ τίνες οὔ, καὶ διὰ τί ἑκάτεροι.

'You are talking,' I said, 'about those worthy persons who bully the strings and interrogate them with torture, racking them on the *kollopes*. But I must not spin out the image too long by talking about the blows inflicted by the plectrum, about accusation and denial and the strings' false pretences, and so I shall abandon the image and say that I do not mean these people, but those whom we said just now we would question about *harmonia*. They do the same as those concerned with astronomy: they seek the numbers in these heard concords, but do not rise to problems, to investigate which numbers are concordant and which are not, and why each are so.'[47]

Glaucon's theorists are certainly no Pythagoreans.[48] Their concern to find the smallest perceptible interval empirically and use it as a type of theoretical 'harmonic unit' by which to measure larger intervals is reminiscent of the activities of those whom Aristoxenus labelled *harmonikoi*.[49] They may even be the sort of theorists who roused Aristoxenus' indignation by their use of what he called 'compression diagrams' (*El. harm.* 28.1), which apparently made use of a minimal unit (the enharmonic *diesis* or quarter-tone) in order to map out several overlapping musical systems on one diagram: Glaucon's use of the word *pyknōmata* is a plausible hint that the procedure to which he refers is somehow related to what Aristoxenus calls *katapyknōsis* ('compression').[50]

Like some of Aristoxenus' *harmonikoi*, Glaucon's theorists made use of instruments in their empirical researches; for the latter it is stringed instruments, while in the case of the former it was auloi.[51] But there is no chance of the strings in question at *Republic* 531b belonging to the *kanōn*.[52] Socrates' image of slaves being tortured for their testimony on the rack only works as a metaphor

47 *Resp.* 531b–c, trans. Barker 1989: 56.

48 Levin (1994: 157) assumes that they are Pythagoreans, but does not argue the point. She is right that they seek 'a minimum unit measure with which they could fix the size of all the intervals in their melodic vocabularies'; but a minimum intervallic unit expressed as a ratio (the customary analytical language of Pythagorean harmonics) is impossible to extract from strings by twisting the *kollopes*. See West 1992: 240.

49 *El. harm.* 2.30; see Barker 1989: 124–5.

50 *El. harm.* 38.3, 53.4–5; Arist. *Metaph.* 1016b18–24. See also Barker 1978a: 8–16 and 1989: 72 n. 16, 132 n. 34.

51 Aristox. *El. harm.* 39.4–43.9.

52 Barker 1989: 56 n. 4.

because Glaucon's theorists are trying to obtain their minimal intervallic unit by changing the tension, not the length, of the strings. Instead of the movable bridges of monochords, the tuning devices here are the *kollopes* of lyre-type instruments, thick leather 'collars' on the crossbar, round which the strings were wound. They could be twisted one way or the other to tighten or slacken the strings, tuning them up or down.[53] In most of the lyres depicted on classical period vase-paintings the *kollopes* appear as lumps or bulges along the crossbar.[54] In Socrates' vivid analogy they stand for the rack, on which slaves were tortured to extract forensic information. Glaucon's harmonicists, he says, 'bully' the strings, 'interrogate them with torture', 'racking [or twisting: στρεβλοῦν-τας] them on the *kollopes*'.[55] What we are being led to imagine, on the musical side of the metaphor, is a group of enthusiastic persons gathered in heated debate around a lyre or kithara on which two strings, perhaps, have been tuned nearly to the same pitch; the flatter of them is being gradually sharpened by the twisting of its *kollops*, until the interval between it and its neighbour is as small as possible without becoming imperceptible.

It will be clear, then, that Socrates' play on the twisting of the *kollopes* makes little sense in the context of the monochord. Since the function of the *kollops* was to allow a string to be tuned and retuned by altering its tension, any minimal interval obtained in this way will not be precisely quantifiable by any method available to Plato's contemporaries: it will be impossible to say exactly how many such intervals measure the tone, for example. This is only one of the reasons why the exercise is 'pointless' (*ateles*).[56] A *kanōn* might have a *kollops* (some of Ptolemy's did), but the construction of intervals was done with the movable bridge, since it was only

[53] Bélis 1985; Roberts 1981; Anderson 1994: 174–5. A similar tuning mechanism is found on many modern African lyres: see Lawergren 1984: 163 fig. 20; Wegner 1984: 104; Plumley 1966: 18–19; Omondi 1984: 265 fig. 5.

[54] Many examples are visible in vase-paintings reproduced in Paquette 1984 and Maas and Snyder 1989.

[55] *Resp.* 531b, trans. Barker 1989: 56; my addition.

[56] For Socrates the most important reason is that any pursuit of harmonics, or any of the other *mathēmata*, which does not take account of the most fundamental questions – which things fall into which classes and why – is pointless labour: it has no real purpose (*telos*); it does not aim the mind toward knowledge of the good.

by comparing lengths of string that intervals could be rationally quantified in antiquity.[57]

Plato's metaphorical vocabulary for the activities of Glaucon's theorists suggests a harmonics more akin to that of Aristoxenus, where pitch is conceived as tension (*tasis*).[58] Socrates' theorists, by contrast, are clearly Pythagoreans: he has named them earlier in the passage (530d) in quoting Archytas' tag about astronomy and harmonics being 'sister sciences',[59] a notion with which he is in agreement. It is noteworthy that when he quotes Archytas he credits 'the Pythagoreans' rather nebulously, rather than Archytas himself specifically, with the idea he praises. This, combined with the fact that one of the two criticisms levelled at this group – that of seeking the numbers in perceived concords – hits home on Archytan territory, suggests that Archytas is in some sense representative of what Plato is here lumping together as Pythagorean harmonics. (Aristoxenus, writing several decades later, reports Pythagorean harmonicists as 'saying that it is certain ratios of numbers and speeds relative to one another in which high and low pitch arise'.[60] This summary fits what we know of Archytas' approach to the subject.[61])

What we learn from this passage in the *Republic*, then, is that stringed instruments were being used not only in harmonic *epideixeis* by the mid-fourth century, but also in empirical procedures of a sort methodologically incompatible with both ratio-based harmonics and the monochord. Furthermore, Plato mentions no instrumental procedures of any sort, empirical or demonstrative, in connection with Socrates' Pythagorean harmonic theorists. They incur Socratic criticism not for 'putting their ears ahead of their

57 Some of the more advanced demonstrations described in Ptolemy's *Harmonics* (II.16 and III.1) require the construction of many-stringed *kanones* with *kollaboi* (= *kollopes*) at one or both ends of each string. The function of the *kollaboi* is to make it easier to maintain the strings' pitches when a single bridge is being moved under them; intervals are still constructed and conceived as ratios of length, not as differences of tension.

58 Aristox. *El. harm*. 3.30, 8.29, 8.31, 9.6, and in many other places; *tasis* is Aristoxenus' usual word for 'pitch'.

59 Cf. Archyt. fr. 1 ap. Porph. *In Harm*. 56.10 with Pl. *Resp*. 530d.

60 φάσκοντες λόγους τέ τινας ἀριθμῶν εἶναι καὶ τάχη πρὸς ἄλληλα ἐν οἷς τό τε ὀξὺ καὶ τὸ βαρὺ γίγνεται, Aristox. *El. harm*. 32.24–6.

61 See ch. 2 part 4.

mind' like their empiricist counterparts, but for failing to 'investigate which numbers are concordant and which are not, and why each are so'.

Mathematical apodeixis *and harmonics*

This background is important to the *Sectio canonis*, with its ratio-based theory and its demonstrative use of a stringed instrument. But the *Sectio* is not a speech, and the author's notion of proof is more rigorous than that suggested by the epideictic displays of fourth-century *harmonikoi*. It is in this document that the vehicle of argumentation provided by Aristotelian *apodeixis*, already adopted in the conventions of Euclidean mathematical discourse,[62] is used, for the first time as far as we can tell, to provide a tightly concatenated series of arguments for the basic doctrines of mathematical harmonics. Whereas the epideictic rhetoric of earlier musical (perhaps even earlier monochord-based) demonstrations may have been persuasive, the apodeictic method inherited from Aristotle provided more than mere persuasion; it was designed to lead the student irreversibly toward one set of interlocking conclusions, and show that none of them could exist in any other way. It was a method by which, in mathematics at least, one was to be led to the recognition of absolute truths.[63] The connections of this apodeictic method to rhetorical *epideixis* on the one hand, and to mathematics – and especially to the form of the Euclidean mathematical treatise – on the other, hinge on the logical work of Aristotle.[64] G. E. R. Lloyd was the main champion of the view, now generally accepted as correct, that the formal axiomatised deductive system of Aristotle's *Analytics* had its origins in the political and legal necessities of convincing and proving, and of 'criticising and undermining rivals'.[65]

[62] See Mueller 1969. [63] Michel 1950: 67.
[64] Michel 1950: 50–1; Lloyd 1973: 36–9, Netz 1999a: 293.
[65] Lloyd 1990b: 97. The main arguments in Lloyd are 1979, ch. 2, and 1990b, ch. 3; for general acceptance of his thesis see Netz 1999a: 292–3; Rihll 1999: 9, 12–13. Netz, summarising Lloyd 1990b, ch. 3, writes that 'the development of rigorous arguments in both philosophy and mathematics must be seen against the background of rhetoric, with its own notion of proof. It was the obvious shortcomings of rhetoric which led to the bid for incontrovertibility, for a proof which goes beyond mere persuasion . . . So there is,

Jonathan Barnes noted another aspect of Aristotelian *apodeixis* in his attempt to account for the lack of any satisfactory examples of the method in Aristotle's own treatises. Demonstration, he argued, was not intended as a 'research' method, but as a pedagogical one. It answers, in Barnes' view, the question not of how scientific knowledge is to be gained in the first place, but of how it is to be passed on from teacher to student.[66] This squares well with later conceptions of demonstration: Archimedes, for example, distinguishes clearly between heuristic and demonstrative methods.[67] It also fits neatly with Lloyd's general conclusions about the origins of Aristotelian deductive logic, since teachers of various sorts – rhetorical, medical and others – used their expertise in epideictic oratory as an essential tool in the securing of students, who would be more likely to attach themselves to a teacher's circle if his gifts of persuasion were compelling enough to win their respect.[68]

There is no doubt about the place of the *Sectio* in this tradition: the treatise is clearly based on the Euclidean *Elements*, both in the form of its composition and in its reliance on theorems proved there,[69] and the *Elements*, which Netz calls 'the most explicitly pedagogic Greek mathematical text',[70] was influenced by the logical works of Aristotle. It is not surprising, then, to see both adversarial and pedagogic concerns in the *Sectio*. We may even go so far as to suppose that the *Sectio*, too, was intended as a pedagogic

first, an activity of great prestige for the Greeks: that of making compelling arguments. And there is one type of argument which is more compelling than others, which leaves less room for controversy than others. This is mathematics' (1999a: 309–10). Michel would have agreed: 'La grande affaire est de "prouver", d'acculer l'adversaire à la contradiction, de l'emprisonner dans l'absurde' (1950: 50).

66 'Teaching is the proper function of demonstrative argument' (Barnes 1975: 80). The origins of Aristotle's theory of demonstration are therefore to be located in pedagogic conversation (81).

67 See Lloyd's example of Archimedes' distinction between a mechanical method (for discovery) and a proof at *Method* praef. (426–30 Heiberg): Lloyd 1973: 45–6; 1979: 122 with n. 330; 1990b: 89–91.

68 See Lloyd's remarks on the uses of demonstration by physicians in securing a clientele for themselves (1979: 89–96).

69 Prop. 2 relies on a proof given in *Elements* VIII.7; prop. 3 is dependent on *Elements* VII.33 (see Barbera 1991: 127) and VIII.8; and prop. 9 on *Elements* VII.2. It may be noted in addition that the *Sectio* makes use of *reductio* arguments (props. 3–4), a type of indirect proof much used in the *Elements*. See also McKirahan 1978: 207.

70 Netz 1999a: 262; see also Rihll 1999: 54.

document – a sort of *Elements* for the science of harmonics, though much shorter than its mathematical model.[71]

The 'same domain' rule and the subordination of harmonics

It is a curious feature of this inheritance that the polemical arguments of the treatise are dependent on Aristotle in two ways: there is, on the one hand, the direct route (the Peripatetic tradition of apodeictic proof); on the other, there is Aristotle's own enunciation of what is often called the 'same domain' rule. Though broken by Aristotle himself, this rule gave rise to the non-mathematical approaches to harmonics championed by his students, Aristoxenus and Theophrastus.

Aristotle lays out the 'same domain' rule in *Posterior Analytics* I.28: 'A science is one if it is about one domain (*genos*)' (μία δ' ἐπιστήμη ἐστὶν ἡ ἑνὸς γένους, 87a38); 'there is evidence for this when you come to the undemonstrables – for they must be in the same domain as what is demonstrated' (τούτου δὲ σημεῖον, ὅταν εἰς τὰ ἀναπόδεικτα ἔλθῃ· δεῖ γὰρ αὐτὰ ἐν τῷ αὐτῷ γένει εἶναι τοῖς ἀποδεδειγμένοις, 87b1–3).[72] On this view of what constitutes a science, any attempt to link (undemonstrable) common notions about the properties and groupings of ratios with those of musical intervals, however much the two may resemble each other, would not be a scientifically unified pursuit. But this is not the whole story. Aristotle provided an exception which allowed him to treat certain sciences, such as harmonics and optics, as 'subordinate' to branches of mathematics, rather than as entirely separate domains (*An. post.* I.7):

οὐκ ἄρα ἔστιν ἐξ ἄλλου γένους μεταβάντα δεῖξαι, οἷον τὸ γεωμετρικὸν ἀριθμητικῇ . . . διὰ τοῦτο τῇ γεωμετρίᾳ οὐκ ἔστι δεῖξαι ὅτι τῶν ἐναντίων μία ἐπιστήμη, ἀλλ' οὐδ' ὅτι οἱ δύο κύβοι κύβος· οὐδ' ἄλλῃ ἐπιστήμῃ τὸ ἑτέρας, ἀλλ'

[71] And hardly as exhaustive in its scope: see McKirahan 1978: 208. If Proclus was thinking of the *Sectio* when he mentioned a Euclidean *Elements of Music* (see n. 4 above), then ancient readers may have read the treatise similarly. But we cannot be sure.

[72] The translations are those of Scholz (1975: 53), who summarises the passage in this way: 'a science in Aristotle's sense is thus a sequence of undemonstrable and demonstrable sentences referring to the same domain of things'.

ἢ ὅσα οὕτως ἔχει πρὸς ἄλληλα ὥστ' εἶναι θάτερον ὑπὸ θάτερον, οἷον τὰ ὀπτικὰ πρὸς γεωμετρίαν καὶ τὰ ἁρμονικὰ πρὸς ἀριθμητικήν.

It is not possible to give a demonstration by shifting from one class (*genos*) of things to another: for instance, one cannot demonstrate something in geometry by means of arithmetic . . . Hence it is impossible to demonstrate by means of geometry that opposites are studied by a single science, or even that the product of two cubes is a cube: nor can any other science demonstrate something belonging to a different one, except where the subjects are so related that one is subordinated to the other, as optics is subordinated to geometry and harmonics to arithmetic.[73]

Harmonics thus subordinated is the sort Aristotle calls 'mathematical harmonics', as opposed to 'hearing-based harmonics'; those who practise it know 'the reason why' as opposed merely to 'the fact that', because they 'are in possession of the demonstrations (*apodeixeis*) of the causes' (*An. post.* 79a1–4). Aristotle calls them 'mathematical harmonicists' (*Top.* 107a15–16).[74] The project of the *Sectio* is the product of this view of harmonics: as Richard McKirahan notes, it is a science which 'uses the principles and conclusions of arithmetic to give a mathematical account of certain observed musical facts'.[75] But Aristoxenus and Theophrastus, both staunch Aristotelians in their own ways, disallowed the exception of *Posterior Analytics* I.7. Aristoxenus even shaped his harmonic science with explicit reference to Aristotelian *apodeixis*, and maintained the 'same domain' rule to exclude any kind of mathematical harmonics.[76]

Some of his harshest criticism is reserved for those who, he says, 'delivered oracular utterances on individual topics, without explanation (*aitia*) or demonstration (*apodeixis*), and without

73 75a38–9, b12–17, trans. Barker 1989: 71. Scholz also points out that the restrictions of *An. post.* I.28 are not universal: 'it should not be inferred from this either that Aristotle did not know about, or that he did not appreciate, the possibility of an isomorphism between two materially different domains' (1975: 59 n. 26). On Aristotle's conception of the 'subordinate sciences', see especially McKirahan 1978.

74 On these passages, see the Introduction.

75 McKirahan 1978: 210. See his comments on the *Sect. can.* (208) and his careful definition of Aristotle's conception of 'mathematical harmonics' (210–11).

76 See Aristox. *El. harm.* 32.19–30, 44.15–20. Theophrastus is nowhere so explicit on this point, but in general his argument separates harmonic theory from mathematics in a similar fashion (the relevant fragment (716) is quoted by Porph. *In Harm.* 61.16–65.15; translation and commentary: Barker 1989, ch. 6; revised text and translation by Barker in Fortenbaugh *et al.* 1992: II.562–73; see also Lippman 1964: 157–61; Sicking 1998; Barker 2007, ch. 15).

even properly enumerating the perceptual data'.[77] When Aristoxenus uses the term *apodeixis* he does so in full knowledge of its Aristotelian overtones, as is clear from the next sentence:[78]

ἡμεῖς δ' ἀρχάς τε πειρώμεθα λαβεῖν φαινομένας ἁπάσας τοῖς ἐμπείροις μουσικῆς καὶ τὰ ἐκ τούτων συμβαίνοντα ἀποδεικνύναι.

We, on the other hand, try to adopt initial principles (*archai*) which are all evident to anyone experienced in music, and to demonstrate (*apodeiknunai*) what follows from them.[79]

Unlike his opponents, Aristoxenus intends to present his arguments as properly demonstrated proofs, derived from *archai* by *apodeixis*.[80] The *harmonikoi* he attacks have made assertions which they have not bothered to back up by logical argument. Aristoxenus has already dealt with Pythagorean music theorists earlier in the same paragraph,[81] and his remarks here are directed at a group of musical thinkers whom he regarded as his predecessors: these are the bumbling incompetents he designates with the term οἱ καλούμενοι ἁρμονικοί, 'so-called harmonicists' (40.26), which Barker paraphrases 'these people who dignify themselves with the honourable title of students of ἁρμονική, but in fact do not deserve it'.[82] Aristoxenus appears to have identified their musicological activities, in broad outline at least, with his own, however unsatisfactorily they carried them out; the Pythagoreans, on the other hand, pursue a project whose accounts are 'altogether extraneous, and totally in conflict with the appearances' (ἀλλοτριωτάτους . . . καὶ ἐναντιωτάτους τοῖς φαινομένοις).[83]

77 *El. harm.* 32.29–30, translation mostly Barker's (1989: 149). See also 5.22, 6.10, 6.14, 6.24, 6.26.

78 Aristoxenus mentions Aristotle by name twice in as many pages immediately before this passage (30.16, 31.11); both times it smacks of name-dropping. The strong influences of Aristotle's philosophy on Aristoxenus' thinking are evident throughout the *El. harm.*: see e.g. 44.3–15; Bélis 1986.

79 *El. harm.* 32.31–33.1, trans. Barker 1989: 149–50. See also 32.19–20.

80 The promise is fulfilled in book III, on which see Gibson 2005: 68–72, Barker 2007: 197–208.

81 *El. harm.* 32.18–29. For a similar dismissal of Pythagorean researches – this time in acoustics – as irrelevant, cf. *El. harm.* 12.4–13.

82 Barker 1978a: 7. See also Barker 1989: 127 n. 8 for a list of Aristoxenus' references to his predecessors.

83 *El. harm.* 32.27–8, trans. Barker 1989: 149.

In directing the polemical propositions of the *Sectio* (14, 15, 16 and 18) squarely against the ramifications of this apodeictic but non-mathematical approach to the science, its author is presenting a rival document to Aristoxenus' *Elementa harmonica*: it, too, is a sort of harmonic *Elements*, but one which vindicates Aristotle's own views of harmonics, using his own apodeictic method (the very argumentative authority invoked by Aristoxenus), against the errors of two of his less mathematical students.

2 Finding a harmonic syntax: the 'ditonic' division

The construction of the division in *Sectio canonis* propositions 19–20, as I have noted already, employs a series of intervals whose pedigree was already well established by the end of the fourth century. The intervals themselves appear in Philolaus fr. 6a,[84] though not in the form of a tetrachordal division, and Plato uses them in his harmonisation of the world soul in the *Timaeus*. Arranged as a tetrachord, the intervals constitute the 'ditonic' or 'Pythagorean' diatonic (9:8 × 9:8 × 256:243), in which the two tones (9:8) are the upper intervals, and the *leimma* is the lowest. (I shall refer to this division simply as 'ditonic'.)

It is curious that the author of the *Sectio* selects this scale in particular, since there were evidently alternatives; Archytas' are the only extant examples in ratios, but there may perhaps have been others.[85] Because the opening sentences of the introduction, which state a connection between impact (*plēgē*) and pitched sound (*phthongos*), seem to acknowledge a dependence on the acoustics of Archytas, it is possible that the author knew Archytas' work directly.[86] Archytas' influence on the *Sectio* is indisputable,[87] but

[84] Discussed in ch. 2 part 3.

[85] Ptolemy (*Harm.* II.13), while preparing the ground for his tabulated summary of both his own attunements and those of his predecessors, qualifies his reference to 'those that have previously been handed down' (τῶν ἄνωθεν παραδοθεισῶν) with 'those, at any rate, that we have come across' (ὅσαις γοῦν ἐνετύχομεν, 69.10, trans. Barker 1989: 344). Whether any alternative tetrachords were proposed in ratio form by any pre-Euclidean author other than Archytas is impossible to say.

[86] Cf. *Sect. can.* 148.5–6 with Archyt. fr. 1 ap. Porph. *In Harm.* 56.11–12. *Sect. can.* prop. 3 may rely on a proof by Archytas (testimonium A19 = Boeth. *Mus.* III.11); see Huffman 2005: 451–70.

[87] Barker 1989: 40 n. 45, 191 n. 1, 195 n. 12.

it may have been transmitted through (and moderated by) that of Plato; R. P. Winnington-Ingram even suspected the *Sectio* of being the product of an Academic hand (1932: 198).

The selection of the ditonic scale for the operations of propositions 17–20 goes some way to confirming the strength of the influence of Platonic ideas about harmonic theory. On the other hand, the author also relies on the 'method of concordance' (*lēpsis dia symphōnias*) to generate epogdoic tones and *leimmata* by constructing fourths upward and fifths downward in alternation.[88] This was a practical tuning technique, used by Greek lyre and kithara players. But the simplicity and economy of this method of generating the intervals of the ditonic division apparently did not recommend it to Plato, since the harmonisation of the world soul in the *Timaeus* is achieved without it. Likewise, the procedures of *Sectio canonis* proposition 20, in which it would have been useful, avoid it.[89]

The value of the ditonic division, from a Platonic point of view, is constituted by more than mere ease of manipulation. At *Republic* 530–1 Plato argues that any pursuit of harmonics which involves no analysis of the causes of concord and discord is purposeless, and the point (as we have seen already) is driven home by the extra attention he devotes to the passage by allowing Glaucon to be confused about the precise target of Socrates' attack. But purposelessness may be only half of Plato's objection to the type of harmonics whose practitioners are preoccupied with the measurement of notes and intervals: he may also mean that practising *mathēmata* in this way is an interminable pursuit. 'They measure heard concords against one another, and so labour *endlessly* (*anēnuta*), just like the astronomers' (531a); *anēnuta*, I suggest, has a double sense here, meaning both 'ineffectually' and

[88] *Sect. can.* prop. 17; Winnington-Ingram 1932: 198; Barker 1989: 203 n. 54. On the method in Aristoxenus and the *Sectio*, see ch. 1 part 1.

[89] *Ti.* 36b; *Sect. can.* prop. 20 (165.5–8); Barker 1989: 207 n. 73. Given Plato's concern that harmonics be sufficiently abstracted from musical practice, it is hardly surprising that he avoids using the method of concordance (i.e. its mathematical equivalent, whereby one would multiply or divide the terms by 3:2 and 4:3 in alternation, instead of deploying 9:8 independently) in the *Timaeus*.

'never-endingly'.[90] Plato's point is that if the task of harmonics (even mathematical harmonics) is one of accurately attaching numbers to the perceived intervals of musical practice, it not only misses its purpose (its *telos*), but also has no natural point of completion (*telos*): since there can exist an indefinite plurality of attunements of the octave,[91] defining the task of harmonics as the accurate quantification of musical scales would condemn the harmonicist to a labour as never-ending (*ateles*) as the water-carrying of the Danaïds in Hades.[92]

Numbers to tune the world by

If the quantification of the perceived intervals of practical music was ruled out by Platonic conceptions of harmonics, it remained to find a mathematical platform on which to base harmonic investigations – a division, that is, whose benefits to the theorist were unrelated to whether or not it was currently in vogue among musicians. Plato's selection of the ditonic division for the harmonisation of the world soul both satisfied the demands of *Republic* VII and fixed the scale as the 'default' division of much of later harmonic theory.[93] It satisfied the conditions set out in the *Republic* on two counts. Firstly, it was not (simply) a practical attunement, and secondly, its ratios can be generated by a very small number of very simple mathematical operations. Though at one time it may have been current in musical practice, Archytas' implicit rejection of the ditonic division in the construction of his own tetrachords suggests that it was not necessarily so at any time in the fourth century.[94] That is not to say that it was not used by musicians; the

90 I follow the reading of Burnet and Slings (i.e. the better attested ἀνήνυτα rather than the less well-attested ἀνόνητα). See LSJ s.v. ἀνήνυτος ('never-ending'), which Plato sometimes uses to mean ἀνήνυστος ('ineffectual'). Similarly, the ἀνήνυτον ἔργον of Penelope's weaving and unweaving at *Phaedo* 84a is both 'endless' and 'futile' (compare e.g. the translations of Hugh Tredennick (1954, in Hamilton and Cairns 1961) and Harold North Fowler (1914)).

91 Aristoxenus, for one, allowed that there was an unlimited number of possible attunements (*El. harm.* 26.24–7). In a similar way, the cataloguing of visible stars is an astronomical task whose finitude is determined only by that of human faculties of perception.

92 Δαναΐδων ὑδρεῖαι ἀτελεῖς, [Pl.] *Axiochus* 371e.

93 The importance of the *Timaeus* commentary tradition as a context for subsequent discussions of harmonic subjects cannot be overemphasised.

94 It may not have been a practical attunement at all: see ch. 2 n. 146.

very fact that it is generated by the method of concordance meant that it would have been one of the first things a young musician was taught, and consequently it held a place which no other attunement could rival.[95] Other *accordature* may have been based on it (e.g. those of Archytas), but it remained the foundation, and its practical primacy gave it universal recognition and a central place in Greek musical theory.

Yet it is not the practicality of the division which Plato approved; his emphasis on abstraction is far too strong for that.[96] A further hint of his reasoning can be gleaned from *Timaeus* 35–6, in which the divine craftsman (*dēmiourgos*) constructs the world soul out of a long strip of soul-stuff, which he divides according to the ratios of the ditonic scale:

ἤρχετο δὲ διαιρεῖν ὧδε. μίαν ἀφεῖλεν τὸ πρῶτον ἀπὸ παντὸς μοῖραν, μετὰ δὲ ταύτην ἀφῄρει διπλασίαν ταύτης, τὴν δ' αὖ τρίτην ἡμιολίαν μὲν τῆς δευτέρας, τριπλασίαν δὲ τῆς πρώτης, τετάρτην δὲ τῆς δευτέρας διπλῆν, πέμπτην δὲ τριπλῆν τῆς τρίτης, τὴν δ' ἕκτην τῆς πρώτης ὀκταπλασίαν, ἑβδόμην δ' ἑπτακαιεικοσιπλασίαν τῆς πρώτης· μετὰ δὲ ταῦτα συνεπληροῦτο τά τε διπλάσια καὶ τριπλάσια διαστήματα, μοίρας ἔτι ἐκεῖθεν ἀποτέμνων καὶ τιθεὶς εἰς τὸ μεταξὺ τούτων, ὥστε ἐν ἑκάστῳ διαστήματι δύο εἶναι μεσότητας, τὴν μὲν ταὐτῷ μέρει τῶν ἄκρων αὐτῶν ὑπερέχουσαν καὶ ὑπερεχομένην, τὴν δὲ ἴσῳ μὲν κατ' ἀριθμὸν ὑπερέχουσαν, ἴσῳ δὲ ὑπερεχομένην. ἡμιολίων δὲ διαστάσεων καὶ ἐπιτρίτων καὶ ἐπογδόων γενομένων ἐκ τούτων τῶν δεσμῶν ἐν ταῖς πρόσθεν διαστάσεσιν, τῷ τοῦ ἐπογδόου διαστήματι τὰ ἐπίτριτα πάντα συνεπληροῦτο, λείπων αὐτῶν ἑκάστου μόριον, τῆς τοῦ μορίου ταύτης διαστάσεως λειφθείσης ἀριθμοῦ πρὸς ἀριθμὸν ἐχούσης τοὺς ὅρους ἓξ καὶ πεντήκοντα καὶ διακοσίων πρὸς τρία καὶ τετταράκοντα καὶ διακόσια. καὶ δὴ καὶ το μειχθέν, ἐξ οὗ ταῦτα κατέτεμνεν, οὕτως ἤδη πᾶν κατανηλώκει.

This is how he began to divide. First he took away one part from the whole, then another, double the size of the first, then a third, hemiolic with respect to the second and triple the first, then a fourth, double the second, then a fifth, three times the third, then a sixth, eight times the first, then a seventh, twenty-seven times the first. Next he filled out the double and triple intervals, once again cutting off parts from the material and placing them in the intervening gaps, so that in each interval there were two means, the one exceeding [one extreme] and

[95] Ptolemy also vouches for its centrality in the musical context of his own century (*Harm.* I.16).

[96] This is not to say that Plato's view was unaffected by the obvious practical associations of the ditonic division. As Barker notes, this advantage may have offered him 'an additional encouragement' in his exclusive selection of this division as the intervallic context for the harmonic constructions of his own work (1989: 51).

exceeded [by the other extreme] by the same part of the extremes themselves, the other exceeding [one extreme] and exceeded [by the other] by an equal number. From these links within the previous intervals there arose hemiolic, epitritic and epogdoic intervals; and he filled up all the epitritics with the epogdoic kind of interval, leaving a part of each of them, where the interval of the remaining part had as its boundaries, number to number, 256 to 243. And in this way he had now used up all the mixture from which he cut these portions.[97]

The procedure described here, which makes no use of the method of concordance, generates the series of numbers 1, 2, 3, 4, 9, 8, 27, which results from the conflation of two geometric series: that of doubles (1, 2, 4, 8) and that of triples (1, 3, 9, 27). When combined, they present a straightforward set of 'harmonic' ratios, whose corresponding musical intervals include all those which Greek harmonic theory unanimously considered concordant: the octave, octave plus fifth, double and triple octaves, fifth and fourth; besides these the tone was located between 9 and 8. The entire series, conceived musically, encompasses four octaves and a major sixth.

Thus while the harmonic construction of the world soul has no recourse to the practical methods of attunement current among musicians, it takes full advantage of the fact that its numbers can be generated by mathematically simple operations: one needs only two geometric series, combined and divided again by the insertion of two means and an interval already generated by the combination of the two series (the epogdoic tone); the *leimmata* (256:243) are simply the by-products after two epogdoic tones have been taken from a perfect fourth. The ditonic division created by this process has the advantages of simplicity and clarity, which were prominent concerns of Greek mathematics from Plato onward.[98]

Plato's use of the ditonic division in the *Timaeus* therefore conforms to, and is supported by, the concerns expressed in *Republic* VII. It is not a division which is tied to any particular music; it is not the notes which are conceived as concordant or discordant, but the numbers themselves.[99] It is worthwhile recalling

[97] Pl. *Ti.* 35b–36b, trans. Barker 1989. [98] Michel 1950: 64.

[99] The task of the harmonicist is 'to consider which numbers are concordant, and which are not' (ἐπισκοπεῖν τίνες σύμφωνοι ἀριθμοὶ καὶ τίνες οὔ, *Resp.* 531c).

here that in *Republic* VII harmonics comes, after astronomy, at the end of Plato's list of subjects to be included in the mathematical curriculum of the guardians of his ideal state, following arithmetic, plane geometry and stereometry. Astronomy and harmonics are said to be 'sister sciences' because they are both sciences of motion – one visible, the other audible (530c–e). But they are not to be distinguished from the first three *mathēmata* as sciences with sensible (as opposed to non-sensible) objects of study, for Plato's criticisms make clear in each case that the way they are practised in his time does not allow either to be a 'study that would draw the soul away from the world of becoming to the world of being',[100] which is the goal of all five *mathēmata* in Plato's curriculum. As stereometry deals with solids, astronomy deals with solids in revolution (528a–b), and as arithmetic treats numbers and their fundamental properties, harmonics treats numbers in relation to one another, as either concordant or discordant.[101] Thus for Plato harmonics is a study of numbers, not of notes, since the qualities of number 'appear to lead to the apprehension of truth'.[102] Eric Havelock's comments on Plato's view of astronomy in *Republic* VII are equally appropriate to harmonics:

> The visible heavens are to function only as a paradigm from which to elucidate the universal behaviour of bodies, expressed in equations which 'are' and do not 'become' or change . . . Invisible astronomy [or, we might say, 'inaudible harmony'] becomes a device for thinking in terms of what (a) is purely abstract and (b) can be stated in a timeless syntax as that which always 'is' and never 'is not'.[103]

The ditonic has to be modified to suit current practice, but current practice will change. The ditonic division is therefore beyond the realm of attunements which 'become' or 'change'. Whereas the business of quantifying the intervals of time-bound music will ensnare the theorist in a task as endless and futile (*anēnuton*) as Penelope's weaving and unweaving,[104] the tasks of the Platonic

[100] μάθημα ψυχῆς ὁλκὸν ἀπὸ τοῦ γιγνομένου ἐπὶ τὸ ὄν, *Resp.* 521d, trans. Shorey (1930) in Hamilton and Cairns 1961.
[101] See Mueller 1980: 112.
[102] ταῦτα δέ γε φαίνεται ἀγωγὰ πρὸς ἀλήθειαν, *Resp.* 525b, trans. Shorey (see n. 100).
[103] Havelock 1963: 260. Cf. Pl. *Ti.* 27d–28a.
[104] See n. 90 above.

harmonicist are meaningful and finite at least partly because they employ the ditonic division as a 'timeless syntax', whose numerical structures always 'are' and never 'are not'. It is a division which, since it is 'based on pure concordances and the most economical application of proportion-theory', as Barker observes, Plato was able to regard as 'truly "harmonious" at a metaphysical level' (1989: 51).

That the author of the *Sectio canonis* is heir to this tradition is evident. We need not, however, go so far as Winnington-Ingram's assumption that the treatise was 'an Academic document' (1932: 198). For while the author of the *Sectio* obviously accepted the ratios (and, in part, the procedures[105]) of Plato's harmonics, he did not adopt a kind of Platonic 'inaudible harmony', nor could the harmonics of the *Sectio* be called 'a device for thinking in terms of what . . . is purely abstract'. Yet if the project was to provide harmonics (illustrated by canonic division) with a scientific foundation so that it could be treated, like astronomy, within the scope of mathematics,[106] then the expression of its features in a 'timeless syntax' is a great advantage. A scale whose properties are determined by the unchanging behaviour of the simplest numbers is the ideal language in which to couch an attack on the conclusions of rivals to the harmonics of the mathematical tradition. What it allows for, most crucially, is the demonstration of the *necessity* of the conclusions reached in the propositions about such contentious issues as the number of tones in the perfect fourth (prop. 15) or the octave (prop. 14, leaning on prop. 9), and the indivisibility of the *pyknon* into equal intervals (prop. 18). The ditonic division, because of its clear and direct link to the properties of the simplest number-ratios, and the ease of demonstrating this link with the monochord, is the perfect 'syntax' for arguments which aim at necessary and universal conclusions.

[105] Such as the construction of a tone downwards (from *nētē hyperbolaiōn* to *paranētē hyperbolaiōn*) in prop. 20 by dividing the length of string needed to sound *nētē hyperbolaiōn* into eight parts and then extending its length by an amount equal to one of these eighths. As Barker observes (1989: 207 n. 73), this procedure is the same as that employed at *Ti*. 36b.

[106] See Barker 1981: 15.

The Euclidean package and its contents

The selective inheritance from Archytas, Plato, Aristotle and Euclid evident in the *Sectio* helps to suggest some answers to a question that arises from the material considered in chapter 1: since by the end of the fourth century Aristoxenus had already elaborated an entire system of harmonic analysis founded on the assumption that irrational intervals (the semitone, for example) exist not only theoretically but in musical practice, and given that it was possible for a theorist familiar with Euclidean geometry to construct irrational intervals on the monochord, why did no one (apparently) attempt it? Why did those who adopted the mathematical approach to harmonics, like the author of the *Sectio*, insist on working within an uncompromisingly rational framework?

One answer seems to be that adopting ratios as an intervallic language in the first place meant adopting an intellectual package, whose contents also included handling instructions. One of the most important of these was evidently the proof attributed to Archytas that no ratio of numbers will divide an epimoric equally (the argument of *Sect. can.* prop. 3). Another was Aristotle's 'same domain' rule, the exception to which allowed for arithmetical demonstrations of propositions in its subordinate science of harmonics, but did not allow for harmonic demonstrations based on the separate domain of geometry. The second of these then closed off the only avenue open to the harmonicist once the first had been proven: if rational bisection of epimoric intervals is impossible, and if irrational bisection is ruled out on methodological grounds, one is left with rational approximations. The ditonic division contains the first and most basic of these, the 256:243 *leimma*, whose very name signals that it is a 'remnant', not in the first-round draft pick of intervals. The only alternative to this expedient is to ditch the rational approach to harmonic analysis altogether, deny that the fourth 'is' epimoric in any meaningful sense, and posit its division into two tones and a semitone instead. The positions adopted by Aristoxenus and the author of the *Sectio* are thus polarised as much by which package they appropriate as by the instruments they use, and the use of the monochord in the *Sectio* is constrained by the rules on the packing slip, all of which had been written well before the end of the fourth century.

3 Acoustics and the *Sectio canonis*

If, as I have suggested, the *Sectio* owes its harmonic debts to Plato, and its methodological and argumentative debts to Aristotle, then a break from this inheritance was required in the field of acoustics. For the *kanōn* to be used as the *Sectio* uses it – and I shall argue that the instrument is crucial to the writing of the treatise: it is, in effect, its *sine qua non* – requires a solution to the dilemma which kept Archytas to his auloi and *kalamos*: if pitched sound has its origins in impact (*plēgē*), and if faster and more vigorous impacts produce higher-pitched sounds, then connections between numerical ratio and musical interval will be framed in terms of either speed or distance of travel,[107] and the monochord cannot be used to provide quantitative demonstrations of either of these.

The acoustic assumptions of Plato and Aristotle offered no workable solutions to this dilemma. Aristotle was aware of the problems inherent in the Archytan acoustical model but could not resolve them;[108] for Theophrastus, they were sufficient reason to reject the quantitative approach to harmonics altogether.[109] Thus the essential steps of preparing the way for the monochord, contained in the *Sectio*'s introduction, had to be supplied by other conceptions of how sound is produced. The necessary change to the Archytan model, while starting from the same premise (that pitched sound originates in impacts), was to think of pitches not in terms of variable speeds of travel, but in terms of variable rates of vibration. This is not quite what modern physics calls 'frequencies'; rather, it means thinking of the relative quantities in musical intervals as arising from the relative rates of lateral oscillation in

[107] These are the assumptions of Archyt. fr. 1.

[108] See, for example, Pl. *Ti.* 67a–c, 80a–b; Arist. *Gen. an.* 786b, *De an.* 420a–b. Aristotle was disturbed by the obvious implication of the pitch-velocity theory that when two sounds of different pitch are emitted simultaneously, the higher-pitched ought to reach the hearer sooner than the lower-pitched, something which can be disproved by simple observation (*Sens.* 447a12–b21; 448a9–13, a20–b2).

[109] Theophr. ap. Porph. *In Harm.* 63.20–64.7.

a string, a phenomenon which, while not directly measurable in antiquity, could at least be crudely observed. Such a shift in acoustical thinking appears not to have taken place until after Aristotle, but his scientific legacy may have prepared the way for it, since the only possibly pre-Euclidean expressions of the theory occur in texts which were the product of his school.[110]

In his commentary on Ptolemy's *Harmonics*, Porphyry quotes part of a book on acoustics, the *De audibilibus* (*On Things Heard*), which he attributes to Aristotle (67.24–77.18). It is unlikely that the treatise was in fact written by Aristotle, and a number of other possible authors have been suggested, most notably Strato of Lampsacus, who succeeded Theophrastus as head of the Lyceum around 287 BC.[111] In this case the *De audibilibus* would have been written too late to have influenced the composition of the *Sectio* (assuming, of course, that the latter dates from around 300 BC), but although this attribution is plausible, it is far from conclusive, and an earlier date may well be possible.

What the treatise (such as we have it) offers is a collection of assertions and observations about the production of musical and vocal sounds. Its author is more at pains to distinguish the causes of difference in timbre than of pitch, but from the few and occasionally ambiguous remarks on the latter it is clear that the notions of pitch-causation in the *De audibilibus* can be located – ideologically, if not chronologically – somewhere between Aristotle and the *Sectio*. Sounds arise 'as the result of impacts made by the breath or by strings';[112] 'the impacts made on the air by strings are many and separate, but because of the smallness of the time between them the ear is unable to detect the gaps, and hence the

110 The rate of vibration theory expressed by Heraclides (ap. Porph. *In Harm.* 30.1–31.21) could only be considered an influence on the *Sect. can.* if this Heraclides were assumed to be the elder Heraclides Ponticus (fl. c. 361–322 BC), as both Düring and Guthrie assume (see the introduction to ch. 2). If, on the other hand, this Heraclides belongs to the first century AD, then we should see him as (possibly) dependent on the *Sectio*, rather than the other way around. The fragment is translated by Barker (1989: 235–6); for a summary of its content see Creese 2008b.

111 Gottschalk 1968; on this and other proposals see Barker 1989: 98. Solomon (2000: 12 n. 58), following Düring, accepts the treatise as the work of Aristotle.

112 διὰ τὰς τοῦ πνεύματος καὶ τῶν χορδῶν γινομένας πληγάς, *De audib.* 800a, trans. Barker 1989: 99.

sound seems to us single and continuous';[113] 'for in all concords the impacts of the air belonging to the higher notes occur more frequently, because of the swiftness of the movement'.[114]

H. B. Gottschalk thought that this constituted a theory of pitch-determination in which more frequent impacts are the cause of higher pitch, but Barker disagreed, arguing instead that the acoustics in the *De audibilibus* operates on the now familiar hypothesis that pitch is constituted by the speed of a sound's transmission.[115] Where it does differ from earlier accounts is that whereas for Archytas the variable of force/speed was undifferentiated, the author of the *De audibilibus* makes some effort to distinguish the separate functions of its two constituent elements: 'swiftness of the breath makes the voice high-pitched, and force makes it hard' (ἡ μὲν γὰρ ταχυτὴς τοῦ πνεύματος ποιεῖ τὴν φωνὴν ὀξεῖαν, ἡ δὲ βία σκληράν, 803a).[116] And this distinction carries consequences, for as Archytas made the connection between force/speed and distance by analogy with missiles (which travel more swiftly to a greater distance the more forcefully they are thrown), claiming, in the case of auloi and the *kalamos*, that the thing which travelled with greater or lesser force/speed was the breath itself,[117] the present author, by separating the two, deals only with speed as a causal factor of pitch. For him it is not the breath travelling at variable speeds which accounts for differences of pitch, but the transmission of a series of discrete impacts through the stationary, but flexible, medium of the surrounding air. Pitch variation is still a consequence of the speed at which these impacts are transmitted, and hence, rather indirectly, a higher-pitched note will also happen to be constituted from more frequent impacts. The number of references to strings in the *De audibilibus* suggests that problems of how to account for their behaviour within a quantitative acoustic theory were being puzzled out at the time of its composition, and we may even suspect that the monochord was behind

[113] αἱ δὲ πληγαὶ γίνονται μὲν τοῦ ἀέρος ὑπὸ τῶν χορδῶν πολλαὶ καὶ κεχωρισμέναι, διὰ δὲ σμικρότητα τοῦ μεταξὺ χρόνου τῆς ἀκοῆς οὐ δυναμένης συναισθάνεσθαι τὰς διαλείψεις, μία καὶ συνεχὴς ἡμῖν ἡ φωνὴ φαίνεται, 803b, trans. Barker 1989: 107.

[114] πλεονάκις μὲν γὰρ ἐν πάσαις ταῖς συμφωνίαις ὑπὸ τῶν ὀξυτέρων φθόγγων αἱ τοῦ ἀέρος γίνονται πληγαὶ διὰ τὸ τάχος τῆς κινήσεως, 804a, trans. Barker 1989: 107.

[115] Gottschalk 1968; Barker 1989: 107 n. 40.

[116] Trans. Barker 1989: 105. [117] Archyt. fr. 1 ap. Porph. *In Harm.* 57.14–15.

these efforts. But the treatise as it stands, and with the focus of its concerns directed elsewhere, falls short of providing an acoustical foundation for the *Sectio canonis*.

Another text in which attempts are made to explain the acoustics of stringed instruments comes, like the *De audibilibus*, from a Peripatetic source. The pseudo-Aristotelian *Problems* also suffers from uncertain authorship and vexed chronology. It is certainly Aristotelian in inspiration, but more probably a compilation of work by various Aristotelian scholars in the late fourth century than a work of the philosopher himself.[118] Thus it is possible, but hardly certain, that the relevant acoustical portions of the text predate the *Sectio*, and we should again be justified only in treating it as an indication of the type of acoustical speculations which were being undertaken around the time of the *Sectio*'s composition, rather than as a direct influence. On the other hand, it is not impossible that the *Sectio* was written first, and that its introduction was known to those responsible for the acoustical sections of the *Problems*.

In its approach to acoustics the *Problems* is hardly unified. While the general treatment of problems to do with music keeps more or less consistently to the thesis that higher pitch and smaller size are related (a position Archytas would have supported, and, for that matter, the author of the *De audibilibus* also), there are inconsistencies in the answers to the problems themselves. *Problems* XI.6, for example, operating on the familiar assumption that 'what is swifter is high-pitched' (τὸ γὰρ θᾶττον ὀξύ, 899b),[119] reflects on the Archytan missile-theory and offers an improvement – that it is not the same air which travels from the source of the sound to the point of its perception, but that the air emanating from the source moves the air around it, and that this air is moved in turn, and so on. This 'displacement theory' is not the same as that offered in the *De audibilibus*, nor is it quite as sophisticated, for although it will account for the acoustics of strings, it will not account for the transmission of sounds through solid objects. The transmission theory of the *De audibilibus* satisfies this desideratum: the

[118] On questions of authorship and date, see Hett 1936: vii–x; Barker 1984: 190 and 1989: 85–6.

[119] Trans. Barker 1989: 87.

treatise's many examples include bronze statues resonating under the file, and ships' masts being tapped to detect cracks (802a). But despite this limitation, the displacement theory of *Problems* XI.6 is a step forward from the problematic theory of Archytas, in that it saves the phaenomena of sound transmission from both stringed and wind instruments.

A further step is taken at XIX.39. Here the discussion centres around the special character of the octave ('the pleasantest of the concords'[120]), and what constitutes its uniqueness among musical intervals. Nowhere do speeds of transmission get a mention, but instead we are given the analogy of metrical feet in duple ratio.[121] In the time it takes one to beat twice, the other has beaten only once; likewise with two notes in octave ratio, the higher note strikes the air twice in the time it takes the lower to strike once: 'Now it happens that *hypatē* has the same ending of the periodic movements in its notes: for the second blow on the air made by *nētē* is *hypatē*' (ἔτι δὲ ὑπάτη συμβαίνει τὴν αὐτὴν τελευτὴν τῶν ἐν τοῖς φθόγγοις περιόδων ἔχειν. ἡ γὰρ δευτέρα τῆς νεάτης πληγὴ τοῦ ἀέρος ὑπάτη ἐστίν, 921a).[122]

Although this account does not treat frequency of impact as a direct cause of pitch differentiation, the two are directly linked in a discussion which appears deliberately to avoid reference to theories of speed or force. None of the problems in the treatise is provided with anything like a formulated proof, and the general method employed throughout is one which begins with the statement of the question to be discussed, followed by one or more possible answers (often expressed as questions themselves: hence the common formula διὰ τί . . . ; ἢ ὅτι . . . ; – 'Why . . . ?' 'Is it because . . . ?'), the merits of which are then weighed in light of specific examples. This is hardly the sort of document in which to find a rigorously demonstrated argument that only one acoustical model can explain the transmission of sound from various sources. Apodeictic argument is all about the answers; the *Problems* is all

[120] 921a, trans. Barker 1989: 95.

[121] Cf. Porph. *In Harm.* 37.25–38.2 on the assumptions made by the *kanonikoi* about the similar natures of rhythmic and harmonic ratios.

[122] Trans. Barker 1989: 95. This thesis seems also to underpin one of the proposed answers to problem XIX.42 (921b).

about the questions. Where its value lies for present purposes, however, is in the indication that, once again, strings were an instigating factor behind the reassessment of received acoustic theories.

Other problems in the collection, too, suggest that strings contributed to the developments of XIX.39. At XIX.23 the author notes that 'in triangular *psaltēria* where the tension is equal, the concord of an octave is produced when the length of one string is double, that of the other half'.[123] Earlier in the same problem (on the subject of why *nētē* is double *hypatē*[124]), the author observes that a string will sound an octave interval when its half-length and its whole length are plucked in turn. In XIX.12 there is an even more precise reference to string division: 'in division (*dialēpsis*) there are two *nētai* in *hypatē*' (τῇ διαλήψει δύο νῆται ἐν τῇ ὑπάτῃ γίνονται, 918b).[125]

It is clear, then, that strings were a concern to the author(s) of the *Problems*, and therefore quite possibly to harmonic theorists around the time the *Sectio* was written as well. What instrument these strings were normally part of is far from clear. It is possible, since only octaves are mentioned among the intervals generated by the *dialēpsis* of XIX.12, that the author had an instrument of the lyre family in mind; in this case, the higher note would have been the harmonic which sounds an octave above the fundamental.[126] Or it could have been a lute of some sort, where string stopping would have been the familiar method of generating notes.[127] The monochord may be assumed here, but it need not be, and in any

[123] 919b, trans. Barker 1984: 197.

[124] Double in rate of vibration, that is, not in string length.

[125] Trans. Barker 1984: 191.

[126] There is no other way to generate two notes an octave apart on a single string in an instrument without a finger-board, but no ancient Greek text provides an unequivocal reference to this technique.

[127] Burkert's assumption that this passage provides an indication that 'flageolet tones seem to have been recognised' (1972: 374, following Jan 1895: 84 n.) overlooks the lute as a possible instrumental context. He also cites Theo. Sm. 59.22 (a passage rife with textual problems in the midst of an extensive quotation from Adrastus), but here again it makes more sense to assume a lute than a lyre, since although the first division (*dialēpsis*) generates the octave by pressure on the midpoint of the string (μέσην πιέσας, 59.25), the second generates the fifth from two-thirds of the same string (60.2–3), a result which guarantees that 'pressing' on the string here means stopping it. (Otherwise, two-thirds of the string would have sounded the harmonic an octave and a fifth above the fundamental.) Because the monochord is excluded by name (59.23–4), the only instrument possible here is the lute.

case the more common terms for canonic division are *katatomē* and *diairesis*.[128] It is worth noting that the word *kanōn* occurs in the *Problems*, at XIX.43: the notes of the lyre accompanying the voice are said to act in this context as *kanones*; since they are 'bare' (ψιλοί) in timbre, they 'make a mistake in the song stand out clearly'.[129] The sense of the word seems merely to be 'rulers' – the lyre's notes act, in other words, like an auditory straight-edge for singers. Hence *kanones* has been rendered here as 'yardsticks' or as 'criteria':[130] as a ruler shows the inexactitude in a line drawn freehand, so the notes of the lyre show the inaccuracies in the pitching of notes by the voice.

Thus *Problems* XIX.39, like the *De audibilibus* and the *Sectio* itself, seems to suggest that efforts were being made to include strings in an updated acoustics. This is not surprising: we saw in part 1 of this chapter that stringed instruments were being used by harmonic theorists in epideictic and empirical contexts by the middle of the fourth century. What is left unstated in XIX.39, but evidently implied there, is that a string which vibrates more rapidly strikes the air more frequently, and the opposite. It is precisely this assumption that is adopted in the introduction of the *Sectio*. Brief references to a similar theory occur in Theophrastus' attack on the mathematical harmonicists.[131] But the first attempt at an apodeictic use of the theory appears in the *Sectio*'s introduction, where it is used, rather obliquely it must be admitted, to underpin and justify the use of the monochord in the second half of the treatise. For if the causal determinant of height or depth of pitch in notes is the temporal density of the impacts of a moving body (*Sect. can.* 148.6–149.3), then it follows from this that strings – which produce their various notes by the addition and subtraction of movement

[128] *Katatomē*: *Sect. can.* 148.2, Theo. Sm. 87.4, Nicom. *Harm.* 260.13 and *Ar.* II.27.1 (136.12), Ptol. *Harm.* 32.12 (and in many other instances), Aristid. Quint. 98.21; *diairesis*: used frequently by authors such as Cleonides, Nicomachus, Bacchius, Gaudentius, and in the *Exc. Neap.* (see the index rerum of Jan 1895, s.v. διαίρεσις, διαιρεῖν; cf. also Proclus' use of *diairein* for canonic division, *In Ti.* II.237.18, 27). Philodemus uses the noun *katadiairesis* for canonic division at *De mus.* IV col. 144.14 (= IV.30.14 Kemke). The word *dialēpsis* is used specifically with reference to the *kanōn*, however, by Theon (perhaps quoting Adrastus) at 59.22; see previous note.

[129] 922a, trans. Barker 1984: 202.

[130] 'Yardsticks': Barker 1984: 202; 'criteria': Hett 1936: 411.

[131] Theophr. ap. Porph. *In Harm.* 62.14–16, 63.19–20.

(προσθέσει κινήσεως, 149.5; ἀφαιρέσει κινήσεως, 149.4), either by being tightened (ἐπιτεινομένους, 149.5) and slackened (ἀνιεμένους, 149.3) or by being shortened and lengthened – can be used to demonstrate the ratios of differing densities of impact if the corollary variable of length is used as a measure of the former.

It is important to note that this critical step is not made by the author of the *Sectio*: we are given only the materials, and are required to draw the inferences for ourselves. The foundation which would support the use of the monochord is left formally incomplete, and this has been highlighted as one of the logical weaknesses of the document.[132] But if the thrust of the introduction is to distinguish concordant intervals as corresponding to either multiple or epimoric ratios, and so to establish a basis for the subsequent treatment of audible phaenomena mathematically, then the acoustic justification of the monochord appears as a secondary (though equally necessary) objective. The conclusions are not drawn, but the raw ingredients are set out along the way. The process of argumentation may leave out formally necessary steps (as in the flawed prop. 11), but the author of the treatise does at least state the acoustic theory on which the *Sectio* is based in sufficient detail to justify (at least in his view) his use of the *kanōn*.

4 The *kanōn* and the *Sectio canonis*

Without the *kanōn* itself, the treatise would not have been written: this conclusion is a consequence of regarding the document as a unified project. If, as suggested by the title, the division of the monochord is the goal for which the successive propositions of the *Sectio* prepare the way, then the fact that this division is carried out not abstractly but with the help of the monochord is of great significance. For the division laid out in propositions 19–20 does not consist of an analysis of a scale-structure into its constituent intervallic ratios; it is the opposite, a construction, a method for dividing a single string so as to generate a two-octave scale, a method in which all the intervals can be simultaneously heard and

[132] See especially Barker 1981.

seen, and in which their identity with the ratios already established in the previous propositions will be plainly evident to the mind just as their intervals are plainly audible to the ear.

The instrument is involved in this method from the very first sentence of proposition 19:

ἔστω τοῦ κανόνος μῆκος, ὃ καὶ τῆς χορδῆς τὸ ΑΒ, καὶ διῃρήσθω εἰς τέσσαρα ἴσα κατὰ τὰ ΓΔΕ.

Let there be a length of the *kanōn* which is also the length AB of the string, and let it be divided into four equal parts, at Γ, Δ and E.[133]

The author then goes on to detail the rest of the division of the 'immutable' scale-system (*ametabolon systēma*), identifying each note by letter, by name, and by musical and mathematical relationship to certain other notes in the division. Every step in the construction depends on previous propositions.[134] The procedure may be somewhat haphazard – in several places there are more elegant possibilities (both mathematically and practically) than the ones the author chooses – but what is abundantly clear is that the steps described in propositions 19 and 20 are there to be followed; they are, in effect, an invitation to 'try this at home'.

Furthermore, they are there to be followed specifically on the monochord, and not on some other instrument, where it would be difficult to perform the division in the way it is set out.[135] And once we read the treatise with the monochord in mind – even, perhaps, with a monochord at our elbow – then the propositions which

[133] *Sect. can.* 163.17–18, trans. Barker 1989: 205 (I have retained the Greek letters). Compare the language of representation here with that of Ptolemy (*Harm.* I.8), Diocles (*On Burning Mirrors* prop. 1) and Heron (*Dioptra* 3), discussed in ch. 1 part 2. Here it is the *length* (μῆκος), a geometrical entity, on which the operations of props. 19–20 will be carried out; its application to the string is stated at the outset as a simple equality (ὃ καὶ τῆς χορδῆς τὸ ΑΒ), and it is the neuter 'length', not the masculine '*kanōn*' or the feminine 'string' (χορδή), to which the designation-letters then apply (τὸ ΑΒ).

[134] The intervals of prop. 19 begin with the fourth: it is constructed first as an epitritic ratio between two line segments; the fact that this relationship of string-lengths sounds the fourth is stated as a consequence, relying on prop. 12 (which in turn relies on props. 6 and 11); this having been established, its notes can be named: it is now not simply *a* fourth, but a particular one. The other intervals in props. 19–20 are constructed in similar, logically dependent, fashion.

[135] Barker notes that 'the method is . . . unworkable for anything but the monochord . . . To transfer the method to an eight-stringed lyre, for example, involves complexities in which our author is plainly not interested' (1981: 15).

precede the division itself offer themselves readily to instrumental illustration as well. Propositions 10–16, which attach musical significance to the strictly mathematical conclusions of propositions 1–9, and propositions 17–18, which offer further conclusions based on those of their forerunners, make frequent demands on the musical ear, either for the construction of intervals of specified magnitude, or for the confirmation that a certain interval is concordant or discordant.

Take, for example, proposition 10, the proof that the octave is multiple: the reader is presented with three notes, *nētē hyperbolaiōn*, *mesē* and *proslambanomenos*, which (he is expected to know) are each separated by an octave. These notes are given letters (Α, Β, Γ) by which their correspondence to the lines on an accompanying lettered diagram can be seen – they are thereby visually quantified.[136] The interval ΑΓ (that is, the interval between *nētē hyperbolaiōn* and *proslambanomenos*), a double octave, is then claimed to be concordant. This is true, and the reader could satisfy himself of the fact by humming the notes to himself as he read, just as the claim that two equal intervals must be multiple if when put together they make a whole that is multiple (proven in prop. 2) can be 'checked' by means of mental arithmetic alone. But the author offers additional verifiability by inviting visual and aural comparison – as the diagrams of the first nine propositions offer a visual parallel to the arithmetic they present, so in the subsequent propositions these lines become strings, offering both visual and aural parallels to assist the reader. As we saw in chapter 1, arithmetically limited geometrical entities (line segments representing numerable quantities) become the visual intermediary between numbers and

[136] The presence of diagrams in the earliest copies of the treatise cannot be taken for granted. But I have argued in ch. 1 part 2, partly on the basis of papyrus copies of diagrams from Euclid's *Elements*, that there are some reasons to take the MS diagrams of Ptolemy as a good indication of what readers saw in the second century AD. Since papyri copied within three or four centuries of the composition of the *Elements* show that diagrams were part of its presentation at that time, it seems likely that this was equally true of other works in the Euclidean tradition. The thirty-two figures of Paris papyrus 1 (see ch. 1 n. 98) are concrete evidence for the use of diagrams in the mathematical sciences less than two centuries after Euclid. It is impossible to say whether or not early third-century copies of either the *Elements* or the *Sectio* possessed diagrams, but in the absence of convincing arguments to the contrary, I am inclined to think that both did. (The MS diagrams of the *Sectio* and the diagrammatic language of prop. 3 in particular are discussed in greater detail in ch. 1 part 1.)

notes. They can only do so, however, if length is conceived as a kind of quantity which has acoustical, and not merely mathematical, relevance. It is true that the parallel visual and aural (or mathematical and musical) constructions could be read and understood without an instrument; but the constant appeal to the reader to think in terms of distance-quantities makes the monochord the easiest instrumental accessory to the text itself, and consequently it seems reasonable to suppose that that is precisely what the author had in mind.

Like the diagrams, the monochord is not a hidden research tool – empirical apparatus – whose existence is covered up in the subsequent presentation of the proofs, but part of that presentation itself. The very fact that the *kanōn* is written into propositions 19–20 indicates that the author considered it an important tool for understanding his project. The way it is used there makes it indispensable: without it, the mathematical and musical statements of the construction simply cannot cohere. Reviel Netz has argued that the lettered diagram was a regular feature of the presentation of mathematical proofs (both written and oral) in the time of Euclid.[137] The author of the *Sectio* appears to place the *kanōn* in a similar role, so that it acquires the didactic force of a lettered diagram for the student of mathematical harmonics.

Moreover, the fact that the lines of early mathematical diagrams were often drawn with the aid of ruler and compasses (famously, the only two mathematical tools allowed by Plato) becomes pertinent in light of the relationship between lines and strings in the *Sectio*.[138] Both ruler and monochord – which share the name *kanōn* – perform a similar function: they assist the reader by, as Ptolemy says, 'straightening [*kanonizein*] those things in sense-perception that are inadequate to reveal the truth'.[139] But whereas

[137] His conclusion is that 'Greek mathematical exchanges, as a rule, were accompanied by something like the lettered diagram. Thus an exclusively oral presentation (excluding, that is, even a diagram) is practically ruled out. Two methods of communication must have been used: the fully written form, for addressing mathematicians abroad, and (hypothetically) a semi-oral form, with some diagram, for presentation to a small group of fellow mathematicians in one's own city' (Netz 1999a: 14).

[138] Netz 1999a: 17, Michel 1950: 63–4, Fowler and Taisbak 1999.

[139] *Harm.* 5.12–13, trans. Barker 1989: 278.

the ruler is not strictly necessary once one posits the existence of perfectly straight lines (because lines drawn freehand will still assist the presentation of a proof without introducing logical problems), the monochord is logically necessary even if one reads the *Sectio* without it (because the connection between linear quantities and musical notes cannot be made by humming): harmonic diagrams are only hummable once one has the monochord in mind. The reader who follows the construction of propositions 19–20 on a monochord redraws their diagrams acoustically. The monochord is thus an indispensable diagram-reading instrument in the *Sectio*.

Because the intervals of propositions 19–20 are constructed according to ratios established in propositions 12–13, which in turn rely either directly or indirectly on the introduction and earlier propositions (2, 3, 6, 7, 8, 10, 11), the musical significance of the purely mathematical propositions (1–9) becomes clear by reading backward through the text. Once the monochord has been introduced as a diagram-reading instrument, the lines of the diagrams in propositions 1–9 are made retroactively 'pluckable': that is, they are given a double significance by the addition of harmonic meaning, and hence a kind of 'audibility'. In this respect the monochord reaches beyond its immediate musical context: because it converts ratios to intervals by the interpretation of geometrical diagrams, its argumentative force is not limited to propositions 19–20. On a second reading of the treatise, the monochord can reach back and offer a new layer of interpretation to propositions 1–9. In this way, their relevance to the musical propositions they support (10–20) is prepared in advance, and the monochord's demonstrative function is doubled.

These aspects of the monochord's deployment in the *Sectio* make the text and its instrument a formidable apodeictic ensemble. It is worth recalling here both the distinctly pedagogical nature of some of the *epideixeis* mentioned in fourth-century sources,[140] and the commonly acknowledged pedagogical aims of the Euclidean-style treatise. The apodeictic method is, in large part, an extension

[140] See the discussion of Theophr. *Char*. 5.9–10 in part 1 of this chapter.

of the epideictic, adding rigour to persuasion.[141] But it can also be the basis for pedagogical and epideictic pursuits by its readers. It is in this respect that the appearance of the *kanōn* in the *Sectio* can be said to constitute a watershed moment in the history of mathematical harmonics. It is the first time, as far as we know, that an instrument had been used to complete and complement a chain of argument beginning with silence and ending with all the notes of a two-octave scale-system.[142]

This achievement, imperfect though it is, was made possible by many others: the demonstration of the concord ratios in Hippasus' discs at least a century and a half earlier; the application of these ratios to the immovable notes of the octave by Philolaus; the development of the notion that sounds are caused by impact, and Archytas' attempts to use it to explain difference of pitch; Archytas' proof that an epimoric cannot be divided equally in ratios of numbers; the ditonic division and its role in Plato's *Timaeus*; the 'method of concordance' and its argumentative use by Aristoxenus; Aristotle's formulation of apodeictic demonstration and its application to mathematics in Euclid's *Elements*; the development of the language and practice of the lettered diagram in fourth-century mathematics; the use of stringed instruments for epideictic and empirical purposes by fourth-century harmonicists; the critique of Archytan acoustics by Aristotle and Theophrastus; and, just possibly, some of the solutions offered in the acoustical passages of the pseudo-Aristotelian treatises.

It is because of these achievements that the monochord could now be placed in the pedagogical and polemical role it occupies in the *Sectio*. Here it is instrumental not only in the founding of a new harmonic pedagogy to rival that of Aristoxenus, but in the systematic undermining of the entire approach to harmonic science advocated and exemplified by his work. From this we might expect to find it in prominent use more or less immediately among those interested in the mathematical approach to harmonics. But we do

141 See n. 65 above.

142 The *Sectio* begins with the words: 'if there were stillness and lack of movement, there would be silence' (εἰ ἡσυχία εἴη καὶ ἀκινησία, σιωπὴ ἂν εἴη).

not. Either because too few were interested, or because in polarising harmonics as it did the *Sectio* appeared to close off further avenues of argument, the monochord is not mentioned again in Greek literature for over two centuries. As we shall see in the next chapter, not even being named as the author of 'canonic divisions' necessarily guarantees that one has employed the monochord at all.

4
ERATOSTHENES

With the demonstrative and pedagogic project of the *Sectio canonis* we have arrived at a new phase in the monochord's history. The instrument has been included in a method of argument which aims to instruct and persuade by means of incontrovertible proofs. It has now become part of these proofs, has been written into them as much as the lettered diagrams of Euclidean mathematical propositions. This constitutes a major development not only for the monochord, but for scientific method in Greek harmonics more generally, and consequently the *Sectio* can be seen as a watershed document in both areas. It would be hard to overemphasise the importance of the achievements of the treatise, for, imperfect though it is, it pointed a way forward for later theorists, several of whom evidently knew the work and drew on it in their own projects.

But if we assume from this that the *Sectio canonis* heralds an emerging trend of apodeictic monochord use among Hellenistic harmonic theorists, and hope on this account to find an increasing number of references to the monochord in the literature of the generation that followed its composition, our searches will be in vain. While geometry flourished in the third century BC, harmonics apparently stagnated; unless it is a mere accident of preservation, the subject seems not to have been much written about until the first century AD, in either the Aristoxenian or the mathematical fashion.[1] And even if the *Sectio* was in regular use for the teaching of mathematical harmonics, the subject produced no Archimedes or Apollonius of Perga to build on the achievements of Euclid (or his imitator). The very status of the monochord in the musicology of the period will prove to be ambiguous.

[1] The few authors writing between Eratosthenes and Ptolemy whose surviving work contains evidence about the history of the monochord will be discussed in ch. 5.

In this impenetrable cloud of obscurity we can make out only one author who produced new work in mathematical harmonics which still survives. This is Eratosthenes, the contemporary of Archimedes, who was born in Cyrene around 285 BC,[2] studied in Athens as a young man, and came to Alexandria about 245 to succeed Apollonius of Rhodes as head of the royal Library and to become tutor to the young Ptolemy (IV) Philopator; he remained there until his death in 194. Although his scholarly energies were directed in many widely differing areas (on account of which he acquired the epithet *pentathlos*, 'pentathlete'), his work in chronology, geography and geometry earned him the respect not only of his contemporaries (including Archimedes) but of later authors as well.[3]

Although he appears not to have written a book solely on musical theory (no *Harmonics* or the like is attributed to him), Eratosthenes' interests in the subject are clear from the surviving fragments of his work. His *Platonicus*, which some have identified as a commentary on Plato's *Timaeus*, apparently contained a discussion of musical passages in that dialogue,[4] and he evidently dealt with Platonic cosmology and harmonics in a poem entitled *Hermes*.[5] In addition we have the reports of Nicomachus, who mentions that he produced a canonic division (*kanonos katatomē*),[6] and Ptolemy, who preserves three tetrachordal divisions attributed to him.[7]

It is the last two testimonia which raise expectations that Eratosthenes may be a witness to the use of the monochord in the late third century. Here, if nowhere else in our record for the period, is a credible context for the instrument: a 'division of the monochord' published by an author whose interests in mathematics and

[2] Fraser 1970: 175–6. The biographical tradition of Eratosthenes relies mainly on Strabo and the *Suda*.

[3] Fragments and testimonia preserved by later authors are collected by Hiller (1872) and Powell (1925). 'Pentathlete' was already used to mean 'a person who tries everything' by the fourth century (LSJ s.v.).

[4] For a reconstruction of the *Platonicus*, see Hiller 1870; on whether it constituted a *Timaeus* commentary, or a dialogue in which Plato appeared as a character, see Fraser 1970: 180–1; Heath 1921: II.104; Solmsen 1942; Wolfer 1954.

[5] Fragments of the *Hermes* are published together in Hiller 1872, Powell 1925: 58–63 and *Suppl. Hell.* 397–8. See Solmsen 1942: 199–213 for discussion of the poem and its relation to its most obvious successor, the *Somnium Scipionis* of Cicero. See also Geus 1995 and 2002.

[6] *Harm.* 11 (260.12–17). [7] *Harm.* II.14 (70.5–72.14).

music are well attested. But the question turns out not to be at all straightforward. The first complication is how to treat the testimony of Nicomachus and Ptolemy, a matter much vexed by problems of textual transmission not only between Eratosthenes and our sources, but between Ptolemy in particular and the manuscripts on which modern editions of his work are based. For this reason we must begin with an evaluation of the quality of this evidence, before addressing the question of whether Eratosthenes used the monochord in his work, and if he did, how it was incorporated into his method. His project will turn out to have been innovative, but we shall find no reasons either to take Nicomachus' report of his 'division of the monochord' literally, or to rule out the monochord altogether.

The evidence of Nicomachus

Nicomachus does not preserve the division itself, and his testimony constitutes our only surviving reference to a canonic division by Eratosthenes. (Ptolemy, our source for three sets of Eratosthenean tetrachord-ratios, does not use the words 'canonic' or 'division' to describe them.) Near the end of the penultimate chapter of his short *Handbook* Nicomachus lists a number of projects he plans to postpone and treat more extensively in a later work; one of these, he says, will be his own canonic division:

καὶ προσεκθησόμεθα τὴν[8] τοῦ Πυθαγορικοῦ λεγομένου κανόνος κατατομὴν ἀκριβῶς καὶ κατὰ τὸ βούλημα τοῦδε τοῦ διδασκάλου συντετελεσμένην, οὐχ ὡς Ἐρατοσθένης παρήκουσεν ἢ Θράσυλλος, ἀλλ' ὡς ὁ Λοκρὸς Τίμαιος, ᾧ καὶ Πλάτων παρηκολούθησεν, ἕως τοῦ ἑπτακαιεικοσιπλασίου.

In addition we shall set out the division of the so-called Pythagorean *kanōn*, worked out accurately and completely according to Pythagoras' own intention, not in the manner of Eratosthenes or Thrasyllus, who misunderstood it, but in that of Timaeus of Locri, whom Plato also followed, right up to the twenty-seven-fold ratio.[9]

The sentence raises several disturbing questions about the nature of Nicomachus' sources and the criteria by which he judged canonic

[8] τὴν] τὸν Jan. [9] Nicom. *Harm.* 260.12–17, trans. Barker 1989: 266, slightly modified.

division. These will be postponed until chapter 5, when they can be assessed in the context of the aims and objectives of the science which later came to be called 'canonics'.

The crucial question for the moment, if we are to take from Nicomachus' report that Eratosthenes was one of the more famous (or infamous) of those harmonicists who worked out a *kanonos katatomē*, is whether this *katatomē* actually involved the *kanōn* at any stage, or whether it was purely a construction of abstract theory, produced by mathematical calculations unattended and untested by the ear, meant perhaps to be read in silent contemplation of its mere numbers. That is, was his 'division of the monochord' precisely that, or should we take the phrase less literally, either because it already meant something more abstract by the time Nicomachus applied it to Eratosthenes, or because Nicomachus himself assumed that an author who published tetrachord-ratios must have intended them for demonstration on the *kanōn*? What complicates the matter further is that *kanonos katatomē* is not necessarily Eratosthenes' own label for what he produced, and could have been applied by an intermediary (like Didymus),[10] or perhaps only by Nicomachus himself; we cannot even be sure whose terminology we are evaluating in this instance.

In the case of Thrasyllus, whom Nicomachus names alongside Eratosthenes, the terminology must be taken literally, since Nicomachus' younger contemporary Theon of Smyrna preserves a lengthy and detailed *kanonos katatomē* by him, in which there are explicit references to the instrument's string.[11] But Thrasyllus had been dead less than a century when Nicomachus wrote his *Handbook*; Eratosthenes had been dead nearly three centuries. It is more likely that Nicomachus had access to a copy (as opposed to a summary or a quotation in some intermediate source) of Thrasyllus' book than of Eratosthenes', and since Thrasyllus' division clearly involved the monochord, it would have been natural for Nicomachus to assume that Eratosthenes' did too.

[10] Didymus is a probable source for Ptolemy's knowledge of Eratosthenes (Barker 2000: 129), and may thus have been Nicomachus' immediate source as well.

[11] Theon uses the phrase 'division of the monochord' twice in the context of Thrasyllus, both when introducing (87.4) and when concluding (93.8–9) the quotation. Thrasyllus refers to the monochord's string at 87.13–14; his division will be discussed in ch. 5 part 2.

The evidence of Ptolemy

The only remnant of Eratosthenes' work which bears any resemblance to Thrasyllus' *kanonos katatomē* is the set of tetrachord-ratios recorded in Ptolemy's *Harmonics* (II.14). They are transmitted entirely without context; Ptolemy offered at least some critique of Archytas' tetrachords (*Harm.* I.13–14), but in the case of Eratosthenes' he gives no more than the ratios and an octave's worth of string lengths, tabulated in standardised columns alongside those of other theorists. All we possess, then, is a close-range look at some of the results of Eratosthenes' harmonic activities, without any quotation from the work in which they originally appeared, or any discussion of how they were arrived at. Whatever we might have learned about the overall range and disposition of the numbers in Eratosthenes' original *kanonos katatomē* is lost to us in its presentation in Ptolemy's tables, and the only part of the division that can be ascribed to Eratosthenes with any certainty (and this somewhat qualified in light of the textual problems considered below) are the nine ratios. This hardly seems a promising place to start. But the ratios themselves contain several curiosities, and we shall see that these offer some clues about the aims and the procedures of Eratosthenes' harmonic investigations. Moreover, since Ptolemy includes the divisions of only four other theorists alongside his own – those of Archytas, Aristoxenus, Eratosthenes and Didymus – each division is precious evidence indeed; Eratosthenes is the only representative of the third century on Ptolemy's list.

Textual problems

Before proceeding to the divisions themselves, a word must be said about the text of *Harmonics* II.14, our source. It just so happens that the very chapter where the division-tables are set out contains the first major lacuna in nearly all the extant MSS of Ptolemy's treatise. The lacuna, which begins at 71.7, just after the discussion of the enharmonic divisions and its accompanying table, affects the remaining text of the chapter, in which the chromatic and diatonic tetrachords of the five theorists are set out in prose: thus out of a total of twenty-three interval-sets, eighteen are missing from the text of II.14 as it survives. The enharmonic division of Eratosthenes is reasonably secure, as its numbers are given along with those of the other theorists at the beginning of the chapter (*Harm.* 71.3–4), and can be verified against the numerals in the corresponding table; but whereas all the other missing chromatic and diatonic divisions can be restored from other portions of Ptolemy's text, those of Eratosthenes alone cannot.[12] For these we must rely on the statistically more corrupt, but largely unanimous witness of the tables, which most MSS reproduce without the missing text.[13] James Mountford, whose examination of the textual tradition of this chapter influenced subsequent editorial decisions, concluded that despite some discrepancies among the MSS which provide the missing intervals for Eratosthenes' chromatic and diatonic, there was, in the end, 'no serious doubt about the evaluations of these two scales' (1926: 87).

The division-tables

For reasons which will soon become clear, I have selected the tables presenting Aristoxenus' and Eratosthenes' divisions (see table 4.1). In them, each division is set out (as in the MSS) in the same way to allow easy comparison: an octave (2:1) is constructed on a string 120 units long, giving the outer boundaries of 60 and 120; 60 stands for the highest note (*nētē*), 120 for the lowest (*hypatē*). The octave is subdivided into two tetrachords, each spanning a fourth (4:3); the upper tetrachord lies between 60 (*nētē*) and 80 (*paramesē*), the lower between 90 (*mesē*) and 120 (*hypatē*). The tone that lies between these two tetrachords is epogdoic (90:80 = 9:8). These four 'standing' notes (*hestōtes*) are by definition immovable in Greek musical theory,[14] and hence they remain constant throughout the twenty-three divisions in Ptolemy's tables. What do change are the other two notes in each tetrachord, the

[12] Mountford discusses the problem in detail (1926: 82–7).

[13] Two MSS (one a direct descendant of the other) offer a reconstruction of the chapter's missing text: see Mountford 1926: 87–95. On the greater corruptibility of the Greek numerals in the tables, see Mountford 1926: 85.

[14] Aristox. *El. harm.* 21.31–22.21, 46.20–47.8; Ptol. *Harm.* 28.24.

Table 4.1 *Nine of the tables of Ptol.* Harm. *II.14, showing the divisions of Aristoxenus and Eratosthenes (Düring 1930: 70–3). All fractional quantities are written in sexagesimals, as they are in the Greek;* hence:*

$$\xi\zeta'\ \lambda' = 67\ 30' = 67^{30}/_{60} = 67^{1}/_{2} = 67.5.$$

The enharmonic genera

According to Aristoxenus	*According to Eratosthenes*
60	60
76	76
78	78
80	80
90	90
114	114
117	117
120	120
24 + 3 + 3 = 30	19:15 × 39:38 × 40:39 = 4:3

The chromatic genera

According to Aristoxenus			*According to Eratosthenes*
soft chromatic	hemiolic chromatic	tonic chromatic	
60	60	60	60
74 40′	74	72	72
77 20′	77	76	76
80	80	80	80
90	90	90	90
112	111	108	108
116	115 30′	114	114
120	120	120	120
22 + 4 + 4 = 30	21 + 4$^{1}/2$ + 4$^{1}/2$ = 30	18 + 6 + 6 = 30	6:5 × 19:18 ×20:19 = 4:3

(*cont.*)

Table 4.1 *(cont.)*

The diatonic genera		
According to Aristoxenus		*According to Eratosthenes*
soft diatonic	tense diatonic	
60	60	60
70	68	67 30′
76	76	75 56′
80	80	80
90	90	90
105	102	101 15′
114	114	113 54′
120	120	120
$15 + 9 + 6 = 30$ $7{:}6 \times 38{:}35$ $\times 20{:}19 = 4{:}3$	$12 + 12 + 6 = 30$ $17{:}15 \times 19{:}17$ $\times 20{:}19 = 4{:}3$	$9{:}8 \times 9{:}8$ $\times 256{:}243$ $= 4{:}3$

* Mountford (1926: 86), Düring (1930: 70–3), Barker (1989: 347–50) and Solomon (2000: 99–102) all convert Ptolemy's sexagesimals to common fractions.

middle or 'movable' notes (*kinoumenoi*). Thus each division contains eight numbers: four for the fixed notes (the *hestōtes*) and four for the movable (the *kinoumenoi*). Below some columns are the ratios used to generate the string lengths of that division. So, for example, the second note from the highest in each tetrachord of Eratosthenes' diatonic is sounded by lengthening the amount of string needed to sound the highest note by the ratio 9:8; thus the numbers of the *paranētē* and *lichanos* in this division are greater than those of the *nētē* and *mesē* respectively by nine eighths.

Mathematical absurdities

The obvious anomaly in Ptolemy's collection is the group of divisions attributed to Aristoxenus. Aristoxenus' convention of expressing musical intervals as 'spaces' is here awkwardly maintained alongside the rational framework of the other divisions.[15] Below the columns assigned to Aristoxenus the intervals comprising the fourth are expressed as units of intervallic space, where equal numbers represent intervals of perceptibly equal magnitude. This unit-system is built directly on the work of Aristoxenus, though not entirely on any writing of his which we now possess. He presented the tetrachordal divisions of his *Elementa harmonica* in intervals which could all be defined as parts or multiples of a tone.[16] In this approach, the equal division of the type of intervals which mathematical theorists expressed as epimoric ratios (and consequently regarded as impossible to divide equally by ratios of numbers) poses no theoretical problems. Thus the tense diatonic, for example, is composed of two tones and a semitone, and the enharmonic is made up of a ditone and two quarter-tones (which, according to Aristoxenus, are equal in magnitude).[17] Porphyry says that Aristoxenus also divided the tone, for theoretical purposes, into twelve equal parts, and used this small interval as a unit by which to measure the melodic intervals of his tetrachordal divisions.[18] Thus the fourth, defined in his system as two and a half tones (*El. harm.* 46.2), consists of 30 units,

[15] On intervals as spaces between notes, see Aristox. *El. harm.* 8.13–9.1. The singing voice 'appears to cross a space (*topos*)' between notes; this is what Aristoxenus calls 'intervallic' (*diastēmatikē*), as opposed to continuous, vocal movement (9.14–20). It is this 'intervallic space' (*diastēmatos topos*, 10.16) which he later quantifies in tones and fractions of a tone in his tetrachordal divisions.

[16] *El. harm.* 50.19–52.32. Like the mathematical theorists (cf. *Sect. can.* props. 8, 13), Aristoxenus defined the tone as the interval by which a fifth exceeds a fourth (21.21–31, 45.34–46.8, quoted in ch. 1 part 1); the disagreements were all about how it could be divided, and how it measured larger intervals (cf. *Sect. can.* props. 9, 14, 15 with Aristox. *El. harm.* 56.14–58.6 and Ptol. *Harm.* I.10–11).

[17] Interval series are to be read downward, so that 'ditone and two quarter-tones' means a tetrachord in which the ditone is the highest interval, the quarter-tones the lowest. (Here I follow the convention of Ptolemy rather than that of Aristoxenus, simply for consistency in the present discussion.) For the tense diatonic, see *El. harm.* 51.29–31; for the enharmonic, 50.22–5. According to Aristoxenus, the equal quarter-tones taken together make a semitone, which is precisely half of a tone, which again is the interval by which a fifth exceeds a fourth, and of which there are exactly six in an octave. In the rational conception of intervals espoused by, for example, the author of the *Sect. can.*, these definitions contain contradictions which render them harmonic impossibilities (see ch. 1 part 1).

[18] Porph. *In Harm.* 125.24–126.2 (citing the fourth book of Aristoxenus' lost work Περὶ μελοποιΐας, *On Melodic Composition*), 137.25; cf. Cleonides 192.12–193.2. Adopting the twelfth of a tone as a measuring unit facilitated the division of the tone into three, four or six parts (125.26–126.1). Aristoxenus uses the twelfth of a tone as a measure

the tone 12, the semitone 6 and the quarter-tone 3.[19] This way of expressing the magnitudes of the intervals within a tetrachord is internally consistent, but incompatible with either the rational language of the mathematical theorists or the columns of Ptolemy's tables, where numbers represent lengths of a monochord string.[20]

Because of this, the inclusion of six scales by Aristoxenus in Ptolemy's tables should come as a surprise. Closer inspection of the numbers themselves shows that they have been generated in the simplest possible manner: Aristoxenus' interval-units have been equated with the units of string length in the lower tetrachord of the octave (from *mesē* to *hypatē*, 90 to 120). Aristoxenus' 30-unit tetrachords are then deposited directly into this 30-unit length of string, so that, for example, the enharmonic (ditone + quarter-tone + quarter-tone = 24 + 3 + 3) is sounded by string lengths of 90, 114, 117 and 120 units. Mathematically the procedure is absurd, since the two lowest intervals of Aristoxenus' tetrachord, conceived in his system as equal quarter-tones, are here neither equal nor properly a quarter of a tone ($\sqrt[4]{(9:8)} \neq 120:117 \neq 117:114$). The ratios of the tetrachord are then extracted from these numbers ($114:90 \times 117:114 \times 120:117 = 19:15 \times 39:38 \times 40:39 = 4:3$) and used to divide the upper tetrachord, in which the differences between the string lengths, naturally, no longer correspond to the Aristoxenian 24 + 3 + 3 intervallic units. The same procedure has been employed to convert the rest of Aristoxenus' divisions as well: three chromatics and two diatonics; six divisions in all, and every one of them given string lengths worked out in a way plainly nonsensical from a mathematical standpoint.[21]

in one passage of the *El. harm.* (25.11–26.6), and makes clear that it is a theoretical interval rather than a melodic one.

19 Aristid. Quint. I.9 (18.1–4) doubles the units to 60 per tetrachord and 24 per tone; Ptolemy also does this occasionally in his treatment of Aristoxenus' divisions (e.g. at *Harm.* 29.21–3).

20 It is consistent only within the tetrachord, since the tone defined as $^{12}/_{30}$ of a fourth is smaller than the tone defined as the interval by which a fifth is greater than a fourth ($^{12/30}\sqrt{4:3} < (3:2 \div 4:3)$).

21 Ptolemy appears to criticise a procedure of this very kind in *Harm.* I.9, but then goes on to use this one in II.14 to represent Aristoxenus' systems. A possible explanation for the discrepancy is that Ptolemy was not himself the author of the conversion (see next note).

Eratosthenes the Aristoxenian?

The second surprise in the tables is that the string lengths of the enharmonic and chromatic divisions of Eratosthenes are identical to those assigned to the enharmonic and tonic chromatic of Aristoxenus. For this reason it seems at least possible that it was Eratosthenes himself who carried out the conversion.[22] We know from Porphyry that for all his Platonic interests, Eratosthenes followed Aristoxenus in admitting eight concords, not six as the mathematical theorists did:

Αἰλιανὸς δ' ὁ Πλατωνικὸς Εἰς τὸν Τίμαιον γράφων κατὰ λέξιν λέγει ταῦτα. '... τῶν δὲ συμφωνιῶν ἓξ τὸν ἀριθμὸν οὐσῶν' – ἃς μόνας ὁ Πτολεμαῖος κατηρίθμησε, παρεὶς τὰς λοιπάς· Ἀριστόξενος γὰρ καὶ Διονύσιος καὶ Ἐρατοσθένης καὶ ἄλλοι πολλοὶ ὀκτὼ κατηρίθμησαν – 'ἁπλᾶς μὲν ἐκάλουν οἱ παλαιοὶ τήν τε διὰ τεσσάρων καὶ διὰ πέντε, συνθέτους δὲ τὰς λοιπάς'.

Aelianus the Platonist, writing in his *On the Timaeus*, says the following, in exactly these words. '... Of the concords, which are six in number' – Ptolemy counted only six and omitted the rest; for Aristoxenus, Dionysius, Eratosthenes and many others counted eight – 'the ancients called the fourth and the fifth "simple" and the remainder "composite"'.[23]

What Aelianus and Ptolemy presumably have in mind is a two-octave system, in which the concords are of finite number;[24] for in an abstract theoretical sense, there is an infinite number of concords.[25] In a two-octave system the six concords will be fourth, fifth, octave, octave plus fourth, octave plus fifth and double octave.[26] Yet if Porphyry is right, then Aristoxenus, Dionysius[27] and Eratosthenes must have done their counting within a different range, since under no ancient definitions of *symphōnia*

[22] An interpretation suggested by Barker (1989: 345 n. 112, 346 nn. 116–17; 2000: 252–4); see also West 1992: 239.

[23] Porph. *In Harm.* 96.7–8, 10–14, trans. Barker 1989: 231, with my supplements as necessary (Barker does not include Porphyry's parenthesis).

[24] Barker 1989: 231 n. 94.

[25] Aristox. *El. harm.* 20.12–23.

[26] Bacch. I.11 (293.13–16), Gaud. *Harm.* 9 (338.8–339.2). Gaudentius adds that the two-octave limit is purely for practical reasons (339.2–7).

[27] It is not clear which of the many men of this name Porphyry had in mind. Elsewhere Porphyry quotes from the first book of an *On Likenesses* (Περὶ ὁμοιοτήτων) by 'Dionysius the musician' (*In Harm.* 37.15–20); Düring suggests that the man in question here is Aelius Dionysius, an Atticist lexicographer who lived in the time of Hadrian. See Creese 2008a.

can more than six be found in the space of two octaves. Aristoxenus makes clear statements about the range within which he is willing to do his harmonic thinking: considered in terms of musical practice (rather than 'the nature of melody' (ἡ τοῦ μέλους φύσις), 20.16), he says that the greatest concordant interval is the double octave and a fifth, 'since our compass does not extend as far as three octaves'.[28] This range contains the first six concords and two others, the double octave plus fourth and double octave plus fifth, making up the eight mentioned by Porphyry. The question of how many concords various theorists admitted thus has nothing to do – at least here – with the thorny issue of the octave plus fourth (excluded by some mathematical theorists on account of its ratio);[29] on the contrary, it has to do entirely with range, since all of the theorists mentioned by Porphyry counted the octave plus fourth. Those whose theoretical range was only two octaves admitted six concords; those who extended it a fifth beyond this admitted eight.[30] Porphyry's parenthetical remarks are therefore useful insofar as they place Eratosthenes, by one criterion only, along with Dionysius and 'many others', in an Aristoxenian camp.

If Eratosthenes was, in one sense at least, more Aristoxenian than Ptolemy, then it is perhaps more likely that he is to blame

[28] Aristox. *El. harm.* 20.29–30, trans. Barker 1989: 139. Despite the clarity of this statement, other notions about Aristoxenus were evidently afloat in the murkier waters of later musical theory. Compare, for example, the narrower two-octaves-plus-fourth range which Adrastus attributes to Aristoxenus (containing only seven concords), and the broader range of three octaves and a tone (containing nine), which some of Adrastus' contemporaries apparently used (Adrastus ap. Theo. Sm. 64.2–6). Adrastus hints also that there were many other ranges between the two. The question has to do with various theorists' differing approaches to the topic of *tonoi* ('keys', roughly). Ptolemy's (*Harm.* II.4) requires only two octaves; other theorists set out their *tonoi* as overlapping two-octave scales a semitone apart, and therefore required a greater overall range, containing a greater number of concords (cf. e.g. Aristid. Quint. I.11).

[29] The ratio of the octave plus fourth is 8:3, which is neither multiple nor epimoric (the two mathematical criteria for concordance posited at the end of the introduction to the *Sectio canonis*). According to the most rigidly mathematical harmonicists, it is therefore a discord, even though it may sound like a concord. Ptolemy discusses the matter at *Harm.* I.6.

[30] Any theorists who did not recognise the octave-plus-fourth interval would have omitted the double octave plus fourth as well, since its ratio (16:3) falls into the same (epimeric) category. Thus on this reckoning a two-octaves-plus-fifth range would contain only six concords – but this model cannot be applied to Ptolemy, who is clear about his two-octave working range and his acceptance of the octave plus fourth (*Harm.* I.5–6).

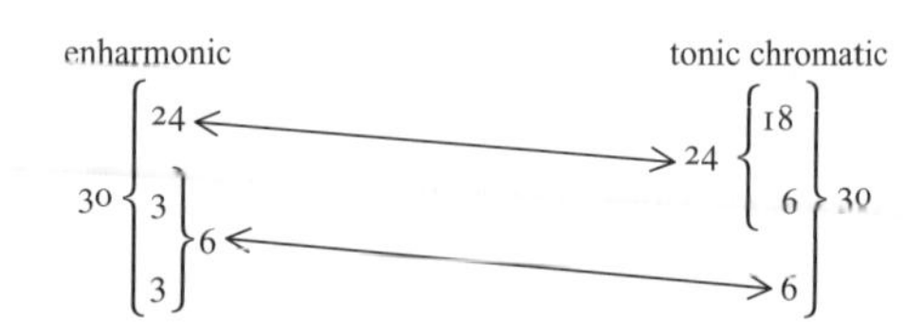

Eratosthenes

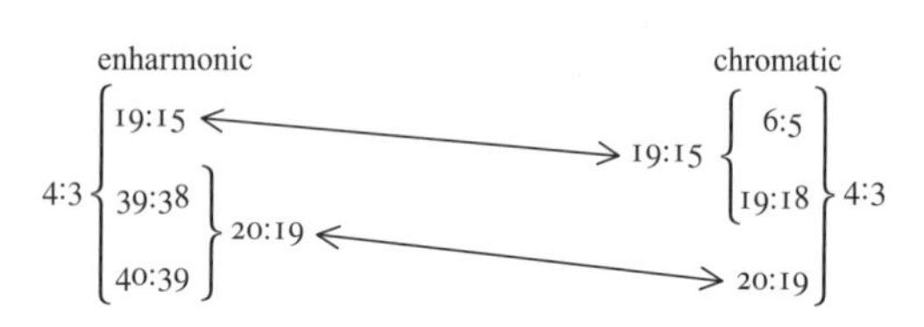

Fig. 4.1 The enharmonic and chromatic tetrachords of Aristoxenus and Eratosthenes. Those of Aristoxenus are expressed in intervallic units, where 12 units = one tone. The fourth (here $2^1/_2$ tones) spans 30 units. Eratosthenes' tetrachords are expressed in ratios, but maintain the same interval-combinations despite inevitable inequalities of magnitude between individual intervals.

than Ptolemy for the muddled conversion attempt of *Harmonics* II.14, not least because Ptolemy's evident disdain for Aristoxenus' way of thinking about musical intervals would have made such a task appear to him unattractive, unnecessary and hardly worth the effort.[31] It is more probable that Ptolemy was guilty of including another author's gravely flawed numbers in his tables because he was not interested in the aims of the project for which they were generated than that he himself worked them out in such a confused manner.[32]

What makes this hypothesis even more likely is the fact that Eratosthenes' enharmonic and chromatic divisions are not only identical with two of those given under Aristoxenus' name in Ptolemy's tables, but are also formulated in such a way as to take account of Aristoxenian theoretical principles. Eratosthenes makes the upper interval of his enharmonic equal to the two highest intervals of his chromatic combined ($24 = 18 + 6$ and $19{:}15 = 6{:}5 \times 19{:}18$), and likewise he makes the two lowest intervals in his enharmonic (referred to together as the *pyknon*) equal to the

[31] See especially Ptol. *Harm.* 30.3–9.

[32] A summary of Barker's position (2000: 254).

lowest interval in his chromatic ($3 + 3 = 6$ and $39{:}38 \times 40{:}39 = 20{:}19$; see fig. 4.1).[33]

The two intervals of the enharmonic *pyknon* are generated according to a procedure for 'halving' ratios attributed to Eratosthenes in the *Excerpta Neapolitana*:

ἔστι δὲ ἡ εὕρεσις τῶν τόνων καὶ τῶν ἡμιτονίων καὶ τῶν διέσεων κατὰ τὸν Ἐρατοσθένην

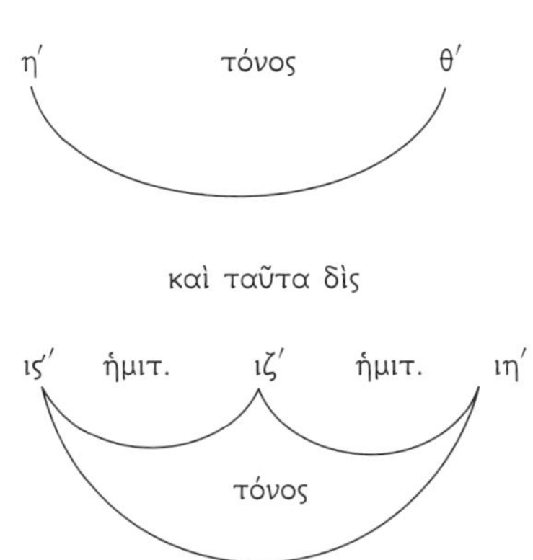

καὶ μεταξὺ εὑρίσκεται ἀριθμὸς ὁ ιζ′.

καὶ ταῦτα πάλιν δίς

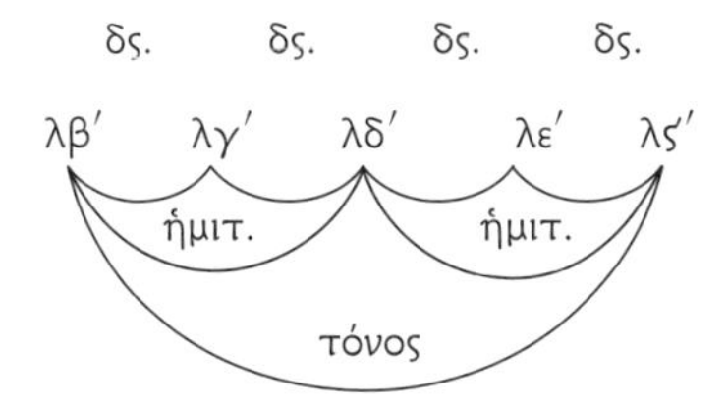

εὑρίσκονται μεταξὺ ὡρισμένα λγ′ λε′.

Then there is the method of finding the tones, semitones and *dieseis*[34] according to Eratosthenes:

[33] Barbera 1977: 302.

[34] The term *diesis* here means the enharmonic *diesis* or quarter-tone, an Aristoxenian usage (see *El. harm.* 25.11–16; it is 'the smallest of all the melodic intervals', 47.1–2).

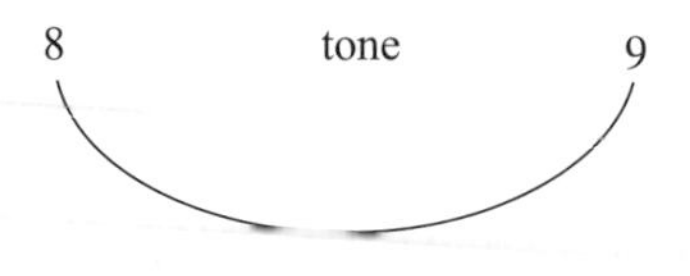

and these multiplied by two =

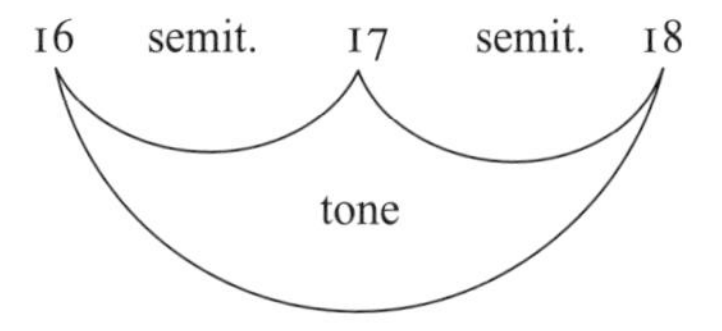

and between them the number 17 is found.

And these multiplied by two again =

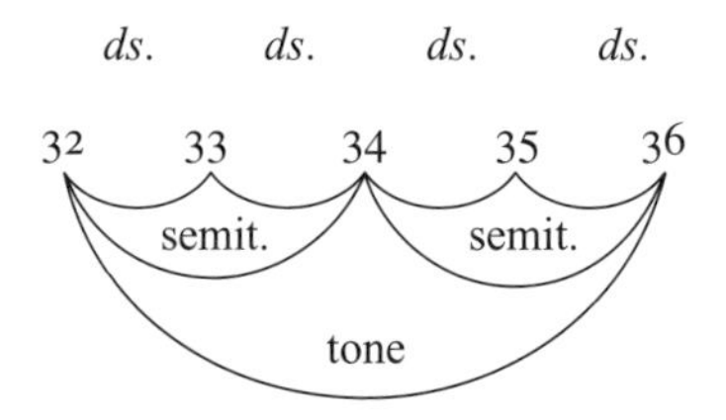

The terms thereby defined, 33 and 35, are found in between.[35]

The difference between what we have here in the *Excerpta Neapolitana* and what we find in the Aristoxenian/Eratosthenean enharmonic of Ptolemy's tables is that in the former the intervals are all generated out of the epogdoic (9:8) tone, whereas in the latter they are derived from the chromatic *pyknon*, a smaller tone (10:9). The procedure, however, is the same in either case (see fig. 4.1): the lowest terms of the ratio to be divided (e.g. the chromatic *pyknon*, 10:9) are first doubled (20:18); the arithmetic mean is then inserted between them (20:19:18), giving a division into two near-equal ratios (20:19 × 19:18 = 10:9). Eratosthenes'

[35] *Exc. Neap.* 19 (416.12–417.6). In the translation the abbreviations semit. = semitone and *ds.* = *diesis* (see previous note).

enharmonic *pyknon* is divided by the same procedure (20:19 = 40:38 = 40:39 × 39:38).

When large intervals are divided in this way, the difference between the component intervals is correspondingly large (the difference between the fifth and fourth, for example, which are produced when this procedure is applied to the octave, is the tone, 9:8). But this difference decreases rapidly along with the magnitude of the interval to be divided, and at some point it becomes musically (though not mathematically) negligible. Tannery believed that Eratosthenes was taking advantage of this in his construction of the chromatic, where, he says, the difference between 20:19 and 19:18 amounts to nothing more than 'une pure fiction mathématique' (1902: 344). It is indeed a minimal difference (361:360, less than a fortieth of a tone); the corresponding difference in Eratosthenes' enharmonic, between 40:39 and 39:38, is positively infinitesimal (1521:1520, about $^1/_{180}$ of a tone).[36] Thus on these criteria Eratosthenes' conversion of Aristoxenus' enharmonic and tonic chromatic is successful, despite its mathematical absurdity: he has managed to preserve the theoretical principles by which the magnitudes of the intervals in the enharmonic and chromatic are related to one another in Aristoxenus' divisions, and his near-equal division of the *pyknon* in these genera is internally inconsistent only in the smallest possible degree.

The problem of the Eratosthenean diatonic

The question which immediately arises from this analysis is why, after proceeding thus far, Eratosthenes apparently rejected the Aristoxenian diatonics in favour of the now standard 'ditonic' diatonic (9:8 × 9:8 × 256:243) of the *Timaeus* and the *Sectio canonis*.[37] The answer to this question has been treated as

[36] In general I avoid the modern logarithmic system of cents (where 1 cent = $^1/_{100}$ of a tempered semitone = $^1/_{1200}$ of an octave = ($\sqrt[1200]{(2:1)}$)) in favour of expressions used by ancient authors (ratios or parts of a tone). This inevitably introduces complications (here I use fractions of an epogdoic tone), and in one instance below where these complications would be significant, I shall break my own rule and use cents for the sake of clarity (as I did in ch. 1 part 1).

[37] See ch. 3 part 2.

straightforward:[38] among other things, Eratosthenes is known to have worked seriously on musical portions of the *Timaeus* and consequently he must have been influenced by Platonic conceptions of harmonic theory, in which the ditonic diatonic was employed as a 'timeless syntax', and the enharmonic and chromatic received less attention (if any).[39] Kathleen Schlesinger was unsatisfied with this picture: why, after all, should Eratosthenes have carried out his project with reasonable success only to abandon it at the final stage in deference to Plato? For this and other reasons she rejected the ratios Ptolemy gives for Eratosthenes' diatonic, claiming that those assigned to Aristoxenus' tense diatonic (17:15 × 19:17 × 20:19 = 4:3) were the ones actually intended by Eratosthenes himself (see fig. 4.2).[40]

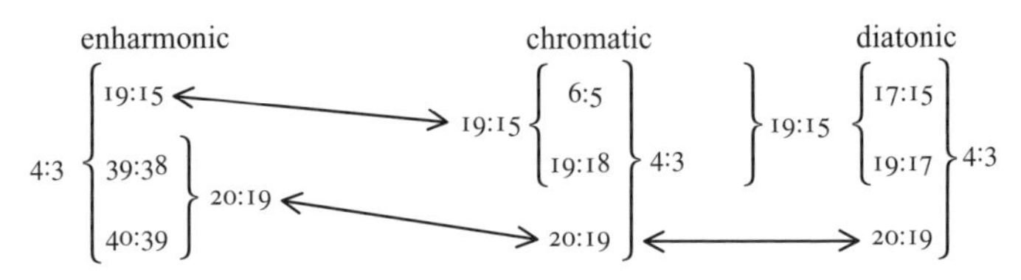

Fig. 4.2 Eratosthenes' tetrachords as proposed by Schlesinger (1933: 94). The annotations showing correspondences between the three genera are my additions.

38 Barker 2000: 254. Barbera (1977: 302) and West (1992: 169–70, 237–9) pass over the discrepancy without comment.

39 In the *Sect. can.* the ditonic division is the basis not only of a diatonic scale-system (props. 19–20), but also of the construction of enharmonic intervals (props. 17–18, where the upper interval is a ditone (81:64 = 9:8 × 9:8) rather than the major third (5:4 = 80:64) of Archytas' enharmonic). Thrasyllus (ap. Theo. Sm. 92.27–93.2), a Platonist who devotes as little effort as possible to the enharmonic, follows the same approach as the *Sectio*.

40 Schlesinger 1933: 94; 'the substitution in Ptolemy's formula must be laid at the door of Pythagorean enthusiasts'. A thorough critique of Schlesinger's article cannot be attempted here. Although she presents much of value, her method is in many points questionable and her conclusions are frequently offered without supporting evidence of any kind (see esp. p. 92 n. 1). Perhaps the most troubling aspect of her method is that she validates her hypothesis about Eratosthenes' diatonic by claiming to have found it in the scales produced by two surviving auloi, without attempting to date the instruments to Eratosthenes' period or suggesting any reason why Eratosthenes should have been intending to record such scales. Her proposed Eratosthenean diatonic is not without merit, however, particularly on the grounds we have considered so far: it follows the pattern set by his enharmonic and chromatic both in its 'halving' procedure and in its preservation of the interval-combinations within the original Aristoxenian tetrachords (see fig. 4.2). For this reason we shall have cause to revisit her hypothesis below.

There is another possibility, however: Eratosthenes may have selected the ditonic diatonic precisely because it *was* the division that best fit his programme of 'translating' Aristoxenus' work. This was Barker's view:

> No doubt the simplicity of its construction recommended it [sc. the ditonic diatonic] to Eratosthenes: the equality of its two upper intervals echoes an obvious feature of Aristoxenus' higher diatonic (and in both systems will be described as 'tones'), and the terms are in fact closer to those assigned to the latter than are those of any other system that Ptolemy lists. Eratosthenes may have construed it as a rationally corrected version of the Aristoxenian division.[41]

On this interpretation, Eratosthenes recognised that what Aristoxenus meant by 'tone + tone + semitone' was what mathematical theorists meant by 'tone + tone + *leimma*'. But since he had already produced a rationally corrected version of the same Aristoxenian tetrachord (17:15 × 19:17 × 20:19), he would also have decided that the ditonic diatonic was a better rational correction than his own. To account for his choice, we would need to answer two questions. First, on what basis did he favour the ditonic diatonic over his own rational conversion of Aristoxenus' tense diatonic? And second, why should he have singled out the diatonic genus as requiring this disruptive shift, and have left the enharmonic and chromatic as they were?

The first question is less straightforward than it might seem. Posing it in the first place involves assuming that if Eratosthenes' priorities were primarily mathematical, then two different mathematical agendas governed his choices; alternatively, mathematical priorities were secondary at this point, in which case his concern would presumably have been to select the set of ratios which most faithfully represented the scale he thought Aristoxenus intended. A simple explanation would be that he set out the rival divisions on a monochord and chose the one that sounded best to him. In this scenario, Eratosthenes' ear trumped his mathematics, and we are immediately faced with the second question: why did he stop at the diatonic? If one set of rational conversions did not measure up under a monochord test, how could he have been confident that the other two were satisfactory?

[41] Barker 1989: 349 n. 125.

The very fact (if it is one) that Eratosthenes produced his own rational conversion of Aristoxenus' tense diatonic in the first place seems to indicate that at least some context existed in which he was not willing to make a simple equation between the latter and the ditonic diatonic. If his choice of the ditonic diatonic in the set of divisions Ptolemy preserves was the result of an attempt to judge objectively between the two rival rational approximations of Aristoxenus' tense diatonic, then it is also an admission of a failure of his own arithmetic. This would be a musical failure, since its success as an arithmetical procedure is unrelated to the impression it made on the ear. If Eratosthenes attempted to answer the question of which set of ratios sounded a scale closest to the one Aristoxenus intended, he would first have needed to determine as exactly as possible the string lengths corresponding to Aristoxenus' intervallic units, and then to compare them with the two sets generated by ratio. On the basis of this comparison he could have accepted the rational division which most nearly approximated the irrational division of Aristoxenus. If he continued this comparison to assess the enharmonic and chromatic as well, then the inconsistent set of tetrachords attributed to him would be the result of a final round of testing; this would answer the second question raised above. All this assumes that the divisions reported by Ptolemy originally formed part of a single, coherent project which included all three genera, and in which the accurate rational 'translation' of Aristoxenus' tetrachords was a primary goal. For reasons I shall elaborate later, I believe that these assumptions may not constitute the best account of the evidence.

What Eratosthenes might have done

The first task, by arithmetic alone, was impossible for Eratosthenes: he could not have calculated exact string lengths for Aristoxenus' intervallic units, since the ratio of the fourth (4:3) cannot receive a geometric mean in numbers. But we can construct a solution for Eratosthenes out of the methods and apparatus available to him. We know from both Eutocius and Pappus that Eratosthenes offered a new mechanical solution to the so-called 'Delian problem', that is, the problem of how to double the cube, which in the Academy had been equated with the problem of how to find two mean proportionals (two geometric means) in continued proportion between two straight lines.[42] This solution involved an instrument called the *mesolabon*, 'mean-finder', which consisted of overlapping rectangular tablets or panels, on which diagonals were inscribed (see fig. 4.3). The tablets were movable: they were set in grooves so as to slide under one another, and by lining them up so that the exposed vertical edge of each overlapped the diagonal of the next in a straight line, one could find the mean proportionals between the lengths of the outside edges of the outermost tablets where these were cut by the straight line:

δύο τῶν δοθεισῶν εὐθειῶν δύο μέσας ἀνάλογον εὑρεῖν ἐν συνεχεῖ ἀναλογίᾳ. δεδόσθωσαν αἱ ΑΕ, ΔΘ. συνάγω δὴ τοὺς ἐν τῷ ὀργάνῳ πίνακας, ἕως ἂν κατ' εὐθεῖαν γένηται τὰ Α, Β, Γ, Δ σημεῖα. νοείσθω δή, ὡς ἔχει ἐπὶ τοῦ δευτέρου σχήματος. ἔστιν ἄρα, ὡς ἡ ΑΚ πρὸς ΚΒ, ἐν μὲν ταῖς ΑΕ, ΒΖ παραλλήλοις ἡ ΕΚ πρὸς ΚΖ, ἐν δὲ ταῖς ΑΖ, ΒΗ ἡ ΖΚ πρὸς ΚΗ· ὡς ἄρα ἡ ΕΚ πρὸς ΚΖ, ἡ ΚΖ πρὸς ΚΗ. ὡς δὲ αὗται πρὸς ἀλλήλας, ἥ τε ΑΕ πρὸς ΒΖ καὶ ἡ ΒΖ πρὸς ΓΗ. ὡσαύτως δὲ δείξομεν, ὅτι καί, ὡς ἡ ΖΒ πρὸς ΓΗ, ἡ ΓΗ πρὸς ΔΘ· ἀνάλογον ἄρα αἱ ΑΕ, ΒΖ, ΓΗ, ΔΘ. ηὕρηνται ἄρα δύο τῶν δοθεισῶν δύο μέσαι.

ἐὰν δὲ αἱ δοθεῖσαι μὴ ἴσαι ὦσιν ταῖς ΑΕ, ΔΘ, ποιήσαντες αὐταῖς ἀνάλογον τὰς ΑΕ, ΔΘ τούτων ληψόμεθα τὰς μέσας καὶ ἐπανοίσομεν ἐπ' ἐκείνας, καὶ ἐσόμεθα πεποιηκότες τὸ ἐπιταχθέν. ἐὰν δὲ πλείους μέσας ἐπιταχθῇ εὑρεῖν, ἀεὶ ἑνὶ πλείους πινακίσκους καταστησόμεθα ἐν τῷ ὀργανίῳ τῶν ληφθησομένων μέσων· ἡ δὲ ἀπόδειξις ἡ αὐτή.

Between two given straight lines to find two means in continuous proportion. Let ΑΕ, ΔΘ be the given straight lines. Then I move the tables in the instrument until the points Α, Β, Γ, Δ are in the same straight line. Let this be pictured as in the second figure [fig. 4.3]. Then ΑΚ:ΚΒ is equal, in the parallels ΑΕ, ΒΖ, to ΕΚ:ΚΖ, and in the parallels ΑΖ, ΒΗ to ΖΚ:ΚΗ; therefore ΕΚ:ΚΖ = ΚΖ:ΚΗ. Now this is also the ratio ΑΕ:ΒΖ and ΒΖ:ΓΗ. Similarly we shall show that ΖΒ:ΓΗ = ΓΗ:ΔΘ; ΑΕ, ΒΖ, ΓΗ, ΔΘ are therefore proportional. Between the two given straight lines two means have therefore been found.

If the given straight lines are not equal to ΑΕ, ΔΘ, by making ΑΕ, ΔΘ proportional to them and taking the means between these and then going back to the original lines, we shall do what was enjoined. If it is required to find more means,

[42] Eutoc. *In Sph. Cyl.* (88.3–96.27); Papp. III.7, 54.22–56.13; Thomas 1939: 256–61, 290–7. See also Cuomo 2000: 137–40; Netz 1999a: 16 with n. 19.

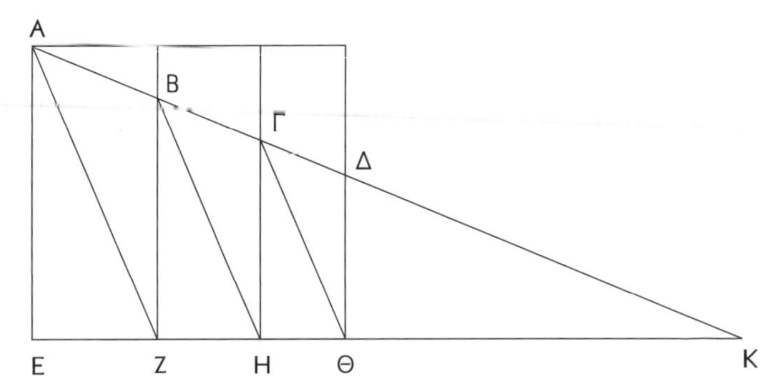

Fig. 4.3 Eratosthenes' *mesolabon* (after Thomas 1939). BZ and ΓH are the mean proportionals of AE and ΔΘ.*

we shall continually insert more tables in the instrument according to the number of means to be taken; and the proof is the same.[43]

Although the instrument was designed as a mechanical (or 'instrumental', *organikē*) solution to the problem of how to double the cube, and as such it required only three panels (to generate two mean proportionals), more could be added if more means were sought. In the epigram to Ptolemy III which accompanied the instrument, apparently inscribed on the column on which it was set up, Eratosthenes even permitted himself the hyperbolical statement that with it one could easily generate an unlimited number of means (μεσόγραφα μυρία).[44] Now since the smallest interval in Aristoxenus' tense diatonic is the semitone (6 units), of which there are five in a tetrachord, a reasonable approximation of the string lengths of his spatial divisions could be reached by finding four mean proportionals between, for example, the 90 and 120 units of string length used in Ptolemy's tables. Eratosthenes could have found the required lengths by setting up his instrument with five tablets and locating the straight line that will pass through all the points of overlap and divide the outer edges of the instrument in 4:3 ratio to one another (see fig. 4.4).

* The similarities between this instrument and Ptolemy's variant on the *helikōn* (*Harm.* II.2) in both form and function, and the possible importance of this in evaluating Ptolemy's handling of the scales he presents in II.14, will be taken up in chapter 6.

43 Eratosth. ap. Eutoc. *In Sph. Cyl.* 94.15–96.9, trans. Thomas 1939: 295.

44 Eratosth. ap. Eutoc. *In Sph. Cyl.* 96.20–1.

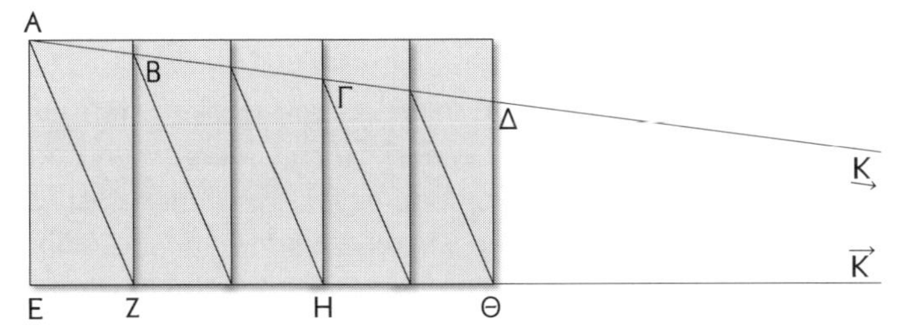

Fig. 4.4 Five-panelled *mesolabon* arranged to divide an epitritic fourth (AE:ΔΘ = 4:3) into five equal semitones. Aristoxenus' tense diatonic (tone + tone + semitone) will be found in the lengths ΔΘ, ΓΗ, BZ, AE.

He could then have continued the process, adding more tablets to find more means as necessary, until he had obtained approximate irrational string lengths for all six of Aristoxenus' tetrachords. The second task, of judging these against the intervals already generated by the rational 'halving' method, would now be a matter of comparing the resulting lengths. The best way for Eratosthenes to have judged whether or not the discrepancies identified in this way were musically (and not just mathematically) significant would have been to compare the notes generated by these lengths – and this can only be done effectively with the help of a monochord.

If Eratosthenes proceeded this far, he will have found, to his satisfaction, that many of his rational conversions of Aristoxenus' tetrachords are very close indeed to the string lengths obtained by the *mesolabon*. In one case the discrepancy is less than 6 cents; in others it is higher, but on the whole the results are reasonably good, particularly in the enharmonic and the three chromatics (see table 4.2).[45]

It will have come as a disappointment to him, however, to find that the inconsistency in his conversions of both diatonics of Aristoxenus is higher than in the other genera: the rational *lichanoi* and *paranētai* are especially flat (over 17 cents) in both cases. Furthermore, the discrepancy between incomposite intervals which

[45] The advantage of using the modern system of cents here is that it provides a standard measure. One disadvantage is that its round numbers are those of our tempered intervals; for the sake of comparison, the epitritic fourth is 498.0 cents, the epogdoic tone 203.9 cents, the *leimma* 90.2 cents, the *komma* 23.4 cents.

Table 4.2 *Comparison of rational and irrational conversions of six Aristoxenian tetrachords into string lengths. In the left-hand column are the string lengths of Ptol.* Harm. *II.14 and the interval magnitudes in both Aristoxenian units and ratios. The middle column gives string lengths for the same scales using an irrational conversion of Aristoxenian units; the* mesolabon *will generate lengths approximating these. Tetrachords are bounded by epitritic (4:3) fourths separated by an epogdoic (9:8) tone. Inner intervals have been calculated so that 1 Aristoxenian unit =* $\sqrt[30]{4:3}$.* *All fractional quantities are expressed, following Ptolemy's usage, in sexagesimals.**1 *The right-hand column gives the interval (in cents) by which the notes in the middle column are sharper (+) or flatter (−) than those in the left-hand column.*

* *There are other ways this could be done, and all (including this one) involve problems. Tempering the entire octave, rather than the fourth, spreads out the error more evenly (a modern piano, if well tuned, illustrates this effectively). But this would require a much larger* mesolabon, *whose results would not be as reliable (an issue which will be taken up below).*

*1 *In Eratosthenes' day, however, fractional quantities were still expressed as sums of parts (*merē*); see e.g.* PHib. *I.27, where the length of the night on a certain date is recorded as* $13 + {}^{1}/_{12} + {}^{1}/_{45}$, *which would be expressed in our (sexagesimal) reckoning as 13 hours 6 minutes and 20 seconds (Fowler 1999: 229–30 with pl. 7).*

enharmonic

Aristoxenus ap. Ptol.	*Aristoxenus by mesolabon-conversion*	*difference in cents*
60	60	0.0
76	75 32′	+10.8
78	77 44′	+6.0
80	80	0.0
90	90	0.0
114	113 17′	+10.8
117	116 36′	+6.0
120	120	0.0
24 + 3 + 3 = 30		
19:15 × 39:38 × 40:39 = 4:3		

(*cont.*)

Table 4.2 *(cont.)*

soft chromatic

Aristoxenus ap. Ptol.	*Aristoxenus by mesolabon-conversion*	*difference in cents*
60	60	0.0
74 40′	74 6′	+13.4
77 20′	76 59′	+7.7
80	80	0.0
90	90	0.0
112	111 8′	+13.4
116	115 29′	+7.7
120	120	0.0
22 + 4 + 4 = 30		
[56:45 × 29:28 × 30:29 = 4:3]		

hemiolic chromatic

Aristoxenus ap. Ptol.	*Aristoxenus by mesolabon-conversion*	*difference in cents*
60	60	0.0
74	73 23′	+14.4
77	76 37′	+8.5
80	80	0.0
90	90	0.0
111	110 5′	+14.4
115 30′	114 56′	+8.5
120	120	0.0
$21 + 4^1/_2 + 4^1/_2 = 30$		
[37:30 × 77:74 × 80:77 = 4:3]		

tonic chromatic

Aristoxenus ap. Ptol	*Aristoxenus by mesolabon-conversion*	*difference in cents*
60	60	0.0
72	71 18′	+16.8
76	75 32′	+10.8
80	80	0.0
90	90	0.0
108	106 57′	+16.8
114	113 17′	+10.8
120	120	0.0
18 + 6 + 6 = 30		
6:5 × 19:18 × 20:19 = 4:3		

(cont.)

Table 4.2 *(cont.)*

soft diatonic

Aristoxenus ap. Ptol.	*Aristoxenus by mesolabon-conversion*	*difference in cents*
60	60	0.0
70	69 17′	+17.8
76	75 32′	+10.8
80	80	0.0
90	90	0.0
105	103 55′	+17.8
114	113 17′	+10.8
120	120	0.0
15 + 9 + 6 = 30		
7:6 × 38:35 × 20:19 = 4:3		

tense diatonic

Aristoxenus ap. Ptol.	*Aristoxenus by mesolabon-conversion*	*difference in cents*
60	60	0.0
68	67 19′	+17.5
76	75 32′	+10.8
80	80	0.0
90	90	0.0
102	100 59′	+17.5
114	113 17′	+10.8
120	120	0.0
12 + 12 + 6 = 30		
17:15× 19:17 × 20:19 = 4:3		

Aristoxenus treats as equal is at its extreme in this genus: the 17:15 'tone' in the highest position of the tense diatonic, for instance, is greater than the 19:17 'tone' immediately below it by 24 cents (i.e. more than a *komma*).[46] The inequality of two notionally equal intervals is less jarring when they are in different locations within a scale (e.g. *mesē* to *lichanos*, *lichanos* to *parhypatē*) than when

[46] A comparison of composite intervals reveals even greater inequalities. The upper interval of Aristoxenus' soft diatonic, for instance, is equal to the two lower ones combined; in Eratosthenes' rational conversion, the upper interval (7:6) is greater than the two lower ones combined (38:35 × 20:19 = 8:7) by 35.7 cents. Eratosthenes' smallest melodic interval (40:39), for comparison, is 43.8 cents.

they are constructed from the same starting pitch.[47] But even given the inaccuracies of determining string lengths by means of the *mesolabon*, and the anomalies of a gut-stringed monochord, differences such as these would probably have been audible. What would certainly have been audible is the remarkable proximity of the ditonic diatonic to the irrational conversion of Aristoxenus' tense diatonic, and its superiority over Eratosthenes' rational conversion as an approximation of the scale Aristoxenus intended (see tables 4.3 and 4.4).

Having offered Eratosthenes such a solution, we find that it appears to answer the two questions raised above: here is a method by which he could have seen and heard that the ditonic diatonic was a more successful rationally corrected version of Aristoxenus' tense diatonic than his own. It would also explain why he thought that only the diatonic required adjustment. The idea that Eratosthenes might have used the monochord in such a project can be entertained partly because we know from Eutocius and Pappus that in at least one other instance he constructed a 'laboratory' instrument in order to solve a specific scientific problem.[48]

If this reconstruction were accurate, it would help to explain an apparent inconsistency in the surviving record of Eratosthenes' tetrachordal divisions. More importantly, however, it would help to build a clearer picture of the way in which harmonic theorists made use of the monochord in the early centuries of its history. If in the Euclidean *Sectio canonis* the instrument had been a tool for demonstrating the primary theorems of mathematical harmonic science, then with Eratosthenes its role would already have been broadened: it would now be employed to facilitate the process of judging between rival sets of intervals, a use to which it was later put by Ptolemy. The value of Eratosthenes' work in harmonics would then be considerable: through his efforts to convert the

[47] Schlesinger gives another example: 'a sensitive ear would experience no difficulty in distinguishing the intervals $\frac{12}{11}$ and $\frac{11}{10}$ [a difference of 14.4 cents] when starting from the same note, but the ear does not readily estimate such differences when the intervals follow one another in melodic succession' (1933: 92).

[48] He was also aware of the instrument's potential as a heuristic tool in very practical contexts: in the epigram he suggests its application to the construction of pens for livestock, pits for grain and cisterns for water (Eutoc. *In Sph. Cyl.* 96.10–15).

Table 4.3 *The ditonic diatonic as a suitable conversion of the Aristoxenian tense diatonic, but not of the soft diatonic. Here the left-hand column gives the diatonic Ptolemy records as Eratosthenes'; the middle column gives the two Aristoxenian diatonics of Ptol.* Harm. *II.14 in irrationally converted string-lengths; the right-hand column gives the difference in cents (middle and right-hand column data calculated as in table 4.2). It will be observed that the difference between the ditonic diatonic and Aristoxenus' tense diatonic as determined by* mesolabon *is considerably less than the error generated by any of Eratosthenes' rational conversions (cf. table 4.2).*

ditonic diatonic v. soft diatonic

Eratosthenes ap. Ptol.	*soft diat. by mesolabon-conversion*	*difference in cents*
60	60	0.0
67 30′	69 17′	−45.1
75 56′	75 32′	+9.1
80	80	0.0
90	90	0.0
101 15′	103 55′	−45.0
113 54′	113 17′	+9.4
120	120	0.0
9:8 × 9:8 × 256:243 = 4:3		

ditonic diatonic v. tense diatonic

Eratosthenes ap. Ptol.	*tense diat. by mesolabon-conversion*	*difference in cents*
60	60	0.0
67 30′	67 19′	+4.7
75 56′	75 32′	+9.1
80	80	0.0
90	90	0.0
101 15′	100 59′	+4.7
113 54′	113 17′	+9.4
120	120	0.0
9:8 × 9:8 × 256:243 = 4:3		

Table 4.4 *The diatonic division-tables of Aristoxenus and Eratosthenes at Ptol.* Harm. *II.14. Numerals in square brackets indicate the string lengths of Aristoxenus' tense diatonic as yielded by the* mesolabon *method. (Eratosthenes' results could not have been so precise, however, since they would have been obtained by measurement rather than by calculation.)*

The diatonic genera			
According to Aristoxenus			*According to Eratosthenes*
soft diatonic	tense diatonic		
60	60		60
70	68	[67 19′]	67 30′
76	76	[75 32′]	75 56′
80	80		80
90	90		90
105	102	[100 59′]	101 15′
114	114	[113 17′]	113 54′
120	120		120
15 + 9 + 6 = 30	12 + 12 + 6 = 30		9:8 × 9:8
7:6 × 38:35	17:15 × 19:17		× 256:243
× 20:19 = 4:3	× 20:19 = 4:3		= 4:3

irrational scales of Aristoxenus into the ratios of the mathematical theorists, his methodological innovations with the monochord would have helped to pave the way for the more sophisticated and rigorously controlled demonstrations of his successors.

Several objections may be raised, however, which in combination make this an unlikely reconstruction of Eratosthenes' procedures. The most important is that although it explains Eratosthenes' selection of the ditonic diatonic, it does not account for his choice of chromatic. For if he had prioritised minimal difference between rational and irrational conversions of the same scales, then we should not expect him to have adopted Aristoxenus' tonic chromatic, since this is the least successfully converted of the three (see table 4.2). The second objection is a practical one: if Eratosthenes had used his *mesolabon*, he would have found it more difficult to test the soft and hemiolic chromatics than the tonic chromatic – harder, in fact, than *any other* Aristoxenian scale. Because of the

Table 4.5 *Number of panels needed to check Eratosthenean conversion of Aristoxenian divisions on the* mesolabon.

Aristoxenian division	number of panels needed
tense diatonic	5
tonic chromatic	5
soft diatonic	10
enharmonic	10
soft chromatic	15
hemiolic chromatic	20

inconvenient magnitude of the lower intervals in Aristoxenus' hemiolic chromatic ($4^1/_2$ units), a total of twenty panels would have been needed in order to complete this comparison (see table 4.5).

If Eratosthenes got as far as setting up a ten-panelled *mesolabon* for testing his enharmonic conversion, he could also have checked the soft diatonic. With only five panels he might have been able to see (though probably not hear) that the tonic chromatic was marginally better than the tense diatonic, but in order to compare it with the other chromatics he would have needed to double the size of his instrument to twenty panels. A second practical difficulty attends the first: the greater the number of panels, the greater the margin of error in the *mesolabon* itself. Thus there comes a point at which the device for checking error becomes more unreliable than it is worth, and for all Eratosthenes' poetic boast of generating an unlimited number of means, it is unlikely that his *mesolabon* was adequate for this task. If his rejection of the tense diatonic was based on string lengths obtained from a five-panelled *mesolabon*, then one would expect his suspicions to have been raised about the tonic chromatic even if he went no further with his tests, and yet he adopted it. Thus not even the simplest application of the *mesolabon* makes consistent sense of Eratosthenes' choices as Ptolemy records them.

Conclusion

I have attempted to provide Eratosthenes with a method that would satisfy the questions raised by an interpretation of his choice of the ditonic diatonic as a rationally corrected version of Aristoxenus' tense diatonic, and I have tried to show why I do not think he employed this method. One of the attractions of this reconstruction was that in rationalising Eratosthenes' selection of an anomalous diatonic, it suggested a role for the monochord in third-century harmonics which goes beyond that of the *Sectio canonis*. It seems, rather, that what I have called the 'Euclidean package' was still intact when Eratosthenes took up harmonics.[49] The inadmissibility of geometrical proofs to the arithmetical domain of harmonics, one of the main components of that package, would then have deterred Eratosthenes from using his own geometrical invention to assist him in a project designed to bring the Aristoxenian and mathematical harmonic traditions closer together.

Rejecting the *mesolabon-kanōn* hypothesis thus leaves two loose ends: first, how do we account for his ditonic diatonic? And second, where does this leave the monochord?

One way of approaching the first loose end is to consider where the tetrachordal divisions might originally have appeared in Eratosthenes' work. The evidence about Eratosthenes gleaned from Ptolemy can be broken down into three items: (1) a complete set of rational conversions of six Aristoxenian tetrachords; (2) a smaller set of three tetrachords attributed directly to Eratosthenes, comprising (a) the Aristoxenian enharmonic and tonic chromatic and (b) the ditonic diatonic. It is possible that Eratosthenes presented all these components in a single discussion. But it is equally possible that they had already been compiled from more than one Eratosthenean text in Ptolemy's immediate source.[50] We might expect that items 1 and 2a originally belonged to the same context, and that 2b was grafted on to 2a at a later stage. A look at the candidates for an original Eratosthenean context tends to support this expectation.

Among the most likely places for a detailed canonic division in Eratosthenes' known works are the *Hermes* and the *Platonicus*. In the *Hermes* it might have been incorporated into the account of the newborn god's ascent, lyre in hand, through the spheres; in the *Platonicus* it might have formed part of a discussion of the scale of

[49] See ch. 3 part 2. [50] See n. 10 above.

Plato's *Timaeus*. In either case, the intervals would almost certainly have been derived from the ditonic division, even if the chromatic and enharmonic were included along with the diatonic. For this reason, it seems improbable that the (non-ditonic) enharmonic and chromatic divisions Ptolemy attributes to Eratosthenes originally formed part of any 'cosmic' scale. And while some authors (such as Nicomachus and Thrasyllus) discussed chromatic and enharmonic divisions in the same context as the *Timaeus* scale, the 'harmony of the spheres' was a consistently diatonic one whenever Greek authors specified its intervals. Since Eratosthenes' diatonic intervals are straightforwardly ditonic (as we receive them, at any rate), they may indeed have been taken from a document such as the *Hermes* or the *Platonicus*, for all we know of these.

Wherever the enharmonic and chromatic intervals came from in Eratosthenes' published output, we may at least be justified in doubting whether they originally appeared in the same document as his diatonic. It is hard to imagine how two such scales, let alone six, could have been accommodated in a poem like the *Hermes*; in the *Platonicus*, if it was a *Timaeus* commentary, it would have required a lengthy digression. Possibly the best candidate for such a discussion is Eratosthenes' treatise *On Means*.[51] His complete set of Aristoxenian conversions would have been a suitably impressive demonstration of the musical uses of the arithmetic mean, especially if the discussion was introduced by the description of the interval-halving procedure which survives as *Excerpta Neapolitana* 19. In such a context Eratosthenes may even have singled out three of these converted Aristoxenian tetrachords, the enharmonic, tonic chromatic and tense diatonic, to show that the mathematical attributes which connected them in Aristoxenus' system were still present in their rational conversion (see fig. 4.2).

Separating the sources of the scales among Eratosthenes' works is perhaps the most plausible solution to the whole dilemma, and on these arguments Schlesinger's conjecture may have some value. But we need not follow her in positing miscreant 'Pythagorean enthusiasts' (1933: 94) to explain his ditonic diatonic. There is no

[51] Περὶ μεσοτήτων (Papp. VII, 636.24–5).

reason to think that such a scale should not have appeared in the *Hermes* or the *Platonicus*. On this reading, Ptolemy's Eratosthenean tetrachords tell us more about the agenda of the compiler than of Eratosthenes himself, who could quite conceivably have pursued two different harmonic agendas at two different times, in two different contexts or frames of reference.

Where does this leave the monochord? As far as Eratosthenes' harmonic activities are concerned, the instrument is now completely off the radar. After its two appearances at the beginning of the third century BC, we have lost sight of it again, and it is far from clear whether Eratosthenes ever made use of it. There is no reason to suspect that he did, but equally there is no reason to suspect that he did not: it is quite possible that his 'division of the monochord' was just that, and that he used the instrument to demonstrate either the ratios of Aristoxenus' tetrachords or some version of the cosmic scale of the *Timaeus*. Given Nicomachus' criticism of Eratosthenes' canonic division alongside that of Thrasyllus, the latter seems more likely. But because Eratosthenes does not provide any concrete evidence for the instrument, its role in the harmonic science of the later third century remains obscure. Nothing indicates that it was put to any demonstrative or heuristic purpose more sophisticated than that of the *Sectio canonis*.

It is significant that we can produce no evidence that a harmonic theorist whose name sprang so readily to Nicomachus' mind as the author of a *kanonos katatomē* actually constructed this *katatomē* with the help of the *kanōn*, when the instrument was probably available to him. If the monochord had been thrust into the limelight in the apodeictic project of the *Sectio*, it still had some ground to cover before its use could become so commonplace among mathematical harmonicists as to allow Ptolemaïs to speak of a science called 'canonics', and before it could be put to work in demonstrations of greater rigour and complexity.

5

CANONIC THEORY

Insofar as harmonic science was treated in antiquity as a branch of mathematics, its pursuit in the Greek world was subject to demographics similar to those of mathematics more generally.[1] Those who could be given, even by the most generous reckoning, Aristotle's label 'mathematical harmonicists' (οἱ κατὰ τοὺς ἀριθμοὺς ἁρμονικοί, *Top.* 107a15–16) were certainly never in great abundance at any point, and from the time of Eratosthenes down to that of Ptolemy we can muster only a handful of authors and works (listed here with references in Porphyry's commentary on Ptolemy's *Harmonics* as applicable):[2]

Panaetius the Younger (not earlier than late second century BC), *Concerning the Ratios and Intervals in Geometry and Music* (Περὶ τῶν κατὰ γεωμετρίαν καὶ μουσικὴν λόγων καὶ διαστημάτων, 65.21–3)

Ptolemaïs of Cyrene (early first century AD?), *Pythagorean Elements of Music* (Πυθαγορικὴ τῆς μουσικῆς στοιχείωσις, 22.23–4)

Thrasyllus (died AD 36), *On the Heptachord* (Περὶ τοῦ ἑπταχόρδου, 91.14, 96.16)[3]

Heraclides [Ponticus the Younger?] (mid-first century AD), *Musical Introduction* (Μουσικὴ εἰσαγωγή, 30.1–2)

[1] Harmonics had been included among the *mathēmata* since the time of Archytas' predecessors (see the opening lines of Archyt. fr. 1), and the organisation of Plato, *Republic* VII was the source of many later catalogues of sciences. The place and nomenclature of harmonics in such lists are discussed in this chapter.

[2] I have attempted to order the list chronologically, but the result should not be taken as anything more than guesswork. Panaetius, for instance, may well be later than Ptolemaïs; it is possible that Porphyry's Heraclides is in fact the fourth-century BC Academic of the same name, not the first-century AD grammarian; and we know nothing more of Demetrius than the miserably few details provided by Porphyry, whose approximate death-date (c. AD 305) provides the latest possible *terminus ante quem* for him. I include Demetrius here only because he evidently wrote about ratios in music theory, and (for all we know) may have lived before Ptolemy.

[3] On Thrasyllus see Tarrant 1993, especially on Porphyry's relationship to Thrasyllan texts (110).

Didymus 'the musician' (fl. c. AD 60), *On the Difference Between the Aristoxenians and the Pythagoreans* (Περὶ τῆς διαφορᾶς τῶν Ἀριστοξενείων τε καὶ Πυθαγορείων, 25.5–6)

Nicomachus of Gerasa (fl. c. AD 100), *Handbook of Harmonics* (Ἁρμονικὸν ἐγχειρίδιον)

Adrastus 'the Peripatetic' (early second century AD), [*Commentary*] *on the Timaeus* (Εἰς τὸν Τίμαιον, 96.1–2)

Dionysius 'the musician' (early second century AD?), *On Likenesses* (Περὶ ὁμοιοτήτων, 37.15–16)

Demetrius (not later than AD 305), *On the Composition of Ratio* (Περὶ λόγου συναφῆς, 92.25–6)

To this group we ought perhaps to add Theon of Smyrna, whose *Mathematics Useful for Reading Plato* (written in the generation immediately before Ptolemy) is our main source for the extant musicological fragments of Thrasyllus and Adrastus. But neither his treatise nor any of the others in the list above except that of Nicomachus has survived complete. Both the brevity of the list and the high ratio of fragments to whole works give the impression of a bleak time in the history of musical science, but this may be somewhat misleading. For one thing, the place of harmonics among the *mathēmata* ensured that it continued to be studied.[4] That said, it must be admitted that harmonics, like the other mathematical sciences, was limited in the Hellenistic period to 'higher education' (or demonstrations of virtuosity) and was in practice hardly a popular school subject at any time.[5] Theon wrote his treatise for educated readers interested in studying Plato whose lack of childhood schooling in mathematics made this 'tough going';[6] Ptolemaïs' introductory text, on the other hand, shows that as late as the first century AD (perhaps) there was still a need for books on harmonics written specifically for pedagogical purposes. Some comprehension of the subject was necessary for an understanding of the vast quantities of material commonly gathered together in the tradition of commentary on Plato's *Timaeus*, and thus many authors of this period – Latin as well as Greek – were

[4] Barker 1994a: 60. [5] Marrou 1977: 183, Clarke 1971: 45–54, Netz 2002.
[6] οὐ μὴν εὔπορον οὐδὲ ῥᾴδιον, 1.9.

evidently sufficiently acquainted with mathematical harmonics to make intelligent sense of what they read.[7]

But making sense of the work of previous generations, or even applying this work to one's own projects, is very different from producing the kind of original contribution we encountered in authors of the fourth century BC, or even in Eratosthenes, for that matter, despite the absurdities of his Aristoxenian tetrachords. Of those who could be counted experts in musical theory, only Didymus (as far as we know) published new tetrachordal divisions, and others, such as Ptolemaïs, appear to have been more interested in pedagogical issues than in tackling serious scientific problems in harmonics.[8] The monochord had evidently found a firm place both among original theorists like Didymus as well as among harmonic educators, since several authors who display no interest or competence in the intricacies of harmonics mention the instrument and its science in passing. We shall come to these in due course.

The sparseness of our record for the period is shared by other sciences as well, and certain points between the careers of Eratosthenes and Ptolemy are poorly documented phases in Greek intellectual life as a whole.[9] But the work which was undertaken in harmonics at this time, while hardly substantial in many ways, was not altogether pedestrian, either. Two aspects of this work give us a sense of the ideological force of the monochord in the harmonics of the period. The first is the suddenly explicit use of the instrument, and the development of a new vocabulary both to inscribe it into the definition of Greek harmonics and to identify in particular that brand of harmonics which many of our authors termed 'Pythagorean' as opposed to 'Aristoxenian'.[10] The way in which this vocabulary – 'canonics' and 'canonicists' – was used

[7] Of the many examples one might cite, Cicero is as good as any; the cosmological and harmonic aspects of his *Somnium Scipionis* (*De re publica* VI) display a thorough acquaintance with the fundamental details of Platonic/Pythagorean harmonic science. See also Vitruvius I.1.8, V.4.

[8] Netz is more negative: 'the quadrivium is a myth. Very few bothered at all in antiquity with mathematics, let alone became creative mathematicians' (1999a: 289).

[9] In astronomy, for example, and in mathematics (Netz 1999a: 284; 2002).

[10] Until now I have avoided this terminology because it can be misleading: the general dichotomy between approaches to harmonics was being made already by Plato and Aristotle (i.e. before Aristoxenus), and one could apparently be a Pythagorean without necessarily accepting 'Pythagorean' tetrachords (e.g. Archytas); likewise, one could

by non-specialist authors is particularly revealing about how the science was perceived by those with no great interest in it, and the light this sheds on the prominence of the monochord is valuable indeed. The second aspect is the continuing attempts (which began, as we have seen, with Eratosthenes) at a rapprochement between 'Pythagorean' and 'Aristoxenian' accounts of pitch and interval, in which the debate becomes focussed specifically on the roles granted to reason and perception in each approach to the science. It will turn out that the *kanōn*, because it occupies a unique bridging position through its appeal both to reason and to perception, becomes an important tool in the debate. Significantly, however, the monochord proves to be as divisive as it is unifying; in some contexts, in fact, it appears to threaten the thesis that reason and perception can agree over such fundamental issues as concord and discord. In other authors still, we discover that 'canonic division' may not have meant precisely that, since carrying out a division as Nicomachus would have it done presents serious practical challenges to the instrument's capabilities.

The chapter is divided into two parts. The first deals with the appearance of the new science, 'canonics', which had worked its way into lists of mathematical sciences (*mathēmata*) by the first century BC. We shall look at the vocabulary of the science and its practitioners in order to assess the prevalence of the *kanōn* in the harmonics of the period and to determine how closely the instrument was linked to the science both in specialist and non-specialist discourse. In part 2 we shall turn to the question of what 'canonics' involved: what 'canonicists' did, and how the monochord fit into the theory and the practice of their science. The practice, we shall see, is primarily the 'division of the monochord', a procedure which differs according to the theoretical and argumentative concerns of the author who employs it. We shall consider four canonic divisions: those of Panaetius, Adrastus, Nicomachus and Thrasyllus.

The goal of the chapter as a whole is to account both for what is being said about the *kanōn* and for what is being done with

pursue mathematical harmonics without having particular views on, say, vegetarianism or the transmigration of souls.

it. Questions of the 'laboratory' methods of its use by harmonic theorists at this time – not just in 'thought experiments' but in actual procedures which might credibly have been carried out – will come to be of prime importance in the final chapter, when we come to assess the extent to which Ptolemy's procedures, as well as his instruments, have precedents among his forebears in the discipline.

1 The language of canonics

Language is only the instrument of science, and words are but the signs of ideas: I wish, however, that the instrument might be less apt to decay, and that signs might be permanent, like the things which they denote.[11]

The earliest datable use of the term 'canonic theory' (*kanonikē theōria*), a single occurrence, appears in a work by Philo of Alexandria (early first c. AD).[12] The most detailed account of the term and its meaning is provided by Ptolemaïs of Cyrene, who may be contemporary with Philo.[13] She would be unknown to us were it not for Porphyry, whose commentary on Ptolemy's *Harmonics* contains the only extant fragments of her work.[14] There are three fragments in total, none of them very long; Porphyry quotes them all in commenting on parts of Ptolemy, *Harmonics* I.2. There can be no guarantee that the passages are quoted either continuously or verbatim; in fact, Porphyry warns his reader before quoting the third fragment that he will proceed 'altering a few things for

[11] Samuel Johnson, *A Dictionary of the English Language* (1755), preface.

[12] *Opif.* 96. The only datable event in Philo's life was in his later years, an embassy to Caligula in AD 39/40.

[13] She may conceivably have lived at any time between the third century BC and Porphyry (AD 234–c. 305), but Düring conjectured that Didymus was Porphyry's immediate source for the portions of Ptolemaïs preserved in his commentary. Barker, who identified Didymus 'the Musician' (Ptol. *Harm.* 67.21) as the *Suda*'s Neronian Didymus, consequently suggested that Ptolemaïs could have been writing as late as the first century AD (Düring 1934; Barker 1989: 230, 239 n. 133). On this view her career may have been contemporaneous with that of Thrasyllus (fl. c. 5 BC–AD 36), but it is not impossible that it was closer to that of Eratosthenes (c. 285–194 BC).

[14] Porph. *In Harm.* 22.22–24.6, 25.3–26.5. The label 'Cyrenean' (Πτολεμαΐς ἡ Κυρηναία, 22.23) may tell no more about Ptolemaïs than it does about Eratosthenes, who appears to have spent most of his career away from his native city.

the sake of brevity',[15] and there is at least one place where Porphyry's editing appears to have confused the original sense of the passage.[16]

Ptolemaïs' treatise was an introductory textbook: Porphyry calls it an *eisagōgē* (23.24). Its catechetic style, which survives Porphyry's (possibly second-hand) quotation of it, is common enough among musical introductions: ready examples can be found in the treatises of Bacchius and Dionysius.[17] This feature serves to highlight one of the priorities of Ptolemaïs' book – it was written for memorisation and repetition, like all such textbooks. In content, too, it must have provided a quotable summary of musical basics, through which the student of harmonics graduated from simple toward more complex and challenging material.

Two things in particular make Ptolemaïs a valuable witness to the history of the monochord and its use. The first is that she mentions the monochord at all – in fact, she mentions it more frequently than any earlier author whose writings we possess. Duris gave us no more than a single-word reference; the last two chapters of the *Sectio canonis* offered only a brief set of instructions for monochord division; and the work of Eratosthenes, such as we have it, allows no ready answer to the question of whether he used the instrument at all. Now, quite suddenly, Ptolemaïs gives us not only the *kanōn*, but *kanonikē pragmateia*, 'canonic science'.[18] The fact that monochord use had become codified in this way represents a significant departure from its obscurity among her musicological predecessors. The second item of importance to the present discussion is her attempt to isolate the methodological characteristics that define a 'canonicist', as opposed to a 'Pythagorean' or an 'Aristoxenian'. Both of these aspects of Ptolemaïs' approach to the science distinguish her from her extant predecessors: the way in which she writes about the monochord and its role shows that the instrument and its place in harmonics had already begun

[15] ὀλίγα τῆς λέξεως συντομίας ἕνεκεν παρακινοῦντες, Porph. *In Harm.* 25.6–7, trans. Barker 1989: 241.

[16] See Barker 1989: 242 n. 148, commenting on 25.23–6.

[17] The style is also employed by Augustine in his *De musica*.

[18] Ptolemaïs ap. Porph. *In Harm.* 22.25. Formulations of both *pragmateia* and *theōria* with an adjective in *-ikē* (both nouns are feminine) are very common, and are used interchangeably in many contexts.

to change the way the science was taught and practised. The further developments made possible by monochord use formed one important part of the background to the work of Claudius Ptolemy.

The terminology and its usage

The first of these points, the mere mention of the monochord, comes as no surprise. But the fact that Ptolemaïs includes it in an elementary instruction manual is revealing: it is now a part of mathematical harmonics that every student needs to know about. The *kanōn* had evidently become so closely associated with its science as to have spawned the terms 'canonics' (*kanonikē* [*pragmateia*]) and 'canonicists' (*kanonikoi*);[19] the assumption that the science was named for its primary instrument was common enough that Ptolemaïs could attempt to refute it:

ἀπὸ τίνος κανονικὴν αὐτὴν λέγομεν; οὐχ ὡς ἔνιοι νομίζουσι ἀπὸ τοῦ κανόνος ὀργάνου παρονομασθεῖσαν, ἀλλ' ἀπὸ τῆς εὐθύτητος ὡς διὰ ταύτης τῆς πραγματείας τὸ ὀρθὸν τοῦ λόγου εὑρόντος καὶ τὰ τοῦ ἡρμοσμένου παραπήγματα.

From what do we derive the term '*kanonikē*'? Not, as some people think, by transference from the instrument called the *kanōn*, but from straightness, on the grounds that it is through this science that reason discovers what is correct, and discovers the *parapēgmata* of what is well attuned.[20]

On the contrary, she argues, 'it is rather that the instrument was named "*kanōn*" by derivation from the science of *kanonikē*'.[21] She also claims that the term *kanonikē* was the old word for *harmonikē* ('harmonic science') among the Pythagoreans: 'for what we now call *harmonikē* they used to name *kanonikē*'.[22] It was not applied rigidly to the monochord, either:

[19] *kanonikē pragmateia*: Porph. *In Harm.* 22.25, 23.4; otherwise simply *kanonikē*: 22.26, 27; 23.1, 21; *kanonikos* (sg.): 23.5; *kanonikoi* (pl.): 23.7, 8, 14, 18.

[20] Ptolemaïs ap. Porph. *In Harm.* 22.27–30, trans. Barker 1989: 239. I have discussed her use of the term *parapēgma* in ch. 1 part 2.

[21] μᾶλλον οὖν τὸ ὄργανον ἀπὸ τῆς κανονικῆς πραγματείας κανὼν ὠνομάσθη, 23.4–5, trans. Barker 1989: 239.

[22] ἣν γὰρ νῦν ἁρμονικὴν λέγομεν, ἐκεῖνοι κανονικὴν ὠνόμαζον, 22.26–7, trans. Barker 1989: 239.

κανονικήν γέ τοι καλοῦσι καὶ τὴν ἐπὶ συρίγγων καὶ αὐλῶν καὶ τῶν ἄλλων πραγματείαν, καίτοι τούτων μὴ κανονικῶν ὄντων, ἀλλ᾽ ἐπεὶ αὐτοῖς οἱ λόγοι καὶ τὰ θεωρήματα ἐφαρμόζουσι, κανονικὰ καὶ ταῦτα προσαγορεύουσι.

They also call '*kanonikē*' the investigation that employs *syringes* and auloi and the rest, though these are not strictly canonic; but they call them 'canonic' too, because the ratios and the theorems fit them.[23]

Ptolemaïs' contention about the origin and priority of the term *kanonikē* hardly warrants refutation,[24] but it is curious that she makes so much of it. Her concern seems to be at least partly to show that the science is about more than a single instrument. What does it mean for something to be 'canonic'? That 'the ratios and the theorems fit' it, according to Ptolemaïs. Thus a *kanonikos* is not exclusively a *harmonikos* who works with monochords, but one who practises *harmonikē* according to the rational procedures of the Pythagoreans and Platonists. 'A *kanonikos*, in general, is a harmonic theorist (*harmonikos*) who constructs ratios in connection with attunement.'[25]

Ptolemaïs also gives two further labels for groups of people who concern themselves with music. First are the *mousikoi*:

διαφέρουσι δὲ μουσικοὶ καὶ οἱ κανονικοί. μουσικοὶ μὲν γὰρ λέγονται οἱ ἀπὸ τῶν αἰσθήσεων ὁρμώμενοι ἁρμονικοί, κανονικοὶ δ᾽ οἱ Πυθαγορικοὶ οἱ ἁρμονικοί.

Mousikoi and *kanonikoi* are different; for '*mousikoi*' is the name given to the harmonic theorists (*harmonikoi*) who begin from perceptions, while '*kanonikoi*' is that given to the Pythagorean harmonic theorists (*harmonikoi*).[26]

For Ptolemaïs, *mousikoi* (which in some contexts indicates performing musicians as opposed to musical theorists) are to be distinguished from *kanonikoi* as perception-based (later to be identified as 'Aristoxenian') theorists; neither are performers. 'But each', she

[23] 23.1–4, trans. Barker 1989: 239.

[24] Origin: cf. Ptolemy's oblique statement at *Harm.* 5.12–13, and see Oppel 1937: 17–20. Priority: Ptolemaïs, if she predates Philo, would be the earliest extant author to use the term *kanonikē* (Barker 1989: 239 n. 134), whereas *harmonikē* was common in the fourth century (Arist. *An. post.* 79a1–2; *Metaph.* 997b21, 1078a14; *Ph.* 194a8; cf. Aristoxenus, who normally uses it as an adjective: 1.5, ⟨2.7,⟩ 39.5, 43.25 (*harmonikē pragmateia*); 8.4, 34.31 (*harmonikē epistēmē*); Plato uses *harmonia* and *harmonikos*, but never *harmonikē*).

[25] κανονικὸς δ᾽ ἐστὶ καθόλου ὁ ἁρμονικὸς ὁ περὶ τοῦ ἡρμοσμένου ποιούμενος τοὺς λόγους, 23.5–6, trans. Barker 1989: 239.

[26] 23.6–8, trans. Barker 1989: 239–40.

adds, 'are in the generic sense *mousikoi*'[27] – which must mean, as it did in Plato, something like 'people versed in things to do with music'.[28]

Later she introduces the *organikoi* (instrumentalists), who 'preferred perception', and 'gave no thought at all, or only feeble thought, to theory'.[29] Didymus, whose account closely parallels that of Ptolemaïs in many respects, pairs the *organikoi* with the *phōnaskikoi*, the 'vocal trainers', in his list of those whose procedure he characterises as 'non-rational' (*alogos*).[30] If we synthesise Ptolemaïs' and Didymus' accounts, the terminology for those who concern themselves with the study of music could be summarised in the following way: all are in the generic sense *mousikoi* (Porph. *In Harm.* 23.8–9), and those with some interest in theory are *harmonikoi* (23.5–8). The remaining terms could be grouped in a spectrum, from those most concerned with reason to those most focussed on perception, in the order *mathēmatikoi* (23.12), *kanonikoi*, *mousikoi*, *organikoi* and *phōnaskikoi*. There are subtleties and confusions in the classification which do not concern us here; some of these will become relevant in part 2 of this chapter.

For all Ptolemaïs' attempts to distance canonics from the *kanōn*, her own usage betrays the extent to which the former relied on the latter. The second part of the first excerpt quoted by Porphyry begins with a telling synonym for 'canonics':

ἡ κατὰ τὸν κανόνα θεωρία, ἐκ τίνων σύγκειται; ἐκ τῶν παρὰ τοῖς μουσικοῖς ὑποτιθεμένων καὶ ἐκ τῶν παρὰ τοῖς μαθηματικοῖς λαμβανομένων. ἔστι δὲ τὰ παρὰ τοῖς μουσικοῖς ὑποτιθέμενα, ὅσα παρὰ τῶν αἰσθήσεων λαμβάνουσιν οἱ κανονικοί, οἷον τὸ εἶναί τινα σύμφωνα καὶ διάφωνα διαστήματα καὶ τὸ εἶναι σύνθετον τὸ διὰ πασῶν ἔκ τε τοῦ διὰ τεσσάρων καὶ τοῦ διὰ πέντε καὶ τὸ εἶναι τόνον τὴν δ' ὑπεροχὴν τοῦ διὰ πέντε παρὰ τὸ διὰ τεσσάρων καὶ τὰ ὅμοια. τὰ δὲ παρὰ τοῖς μαθηματικοῖς λαμβανόμενα, ὅσα ἰδίως οἱ κανονικοὶ τῷ λόγῳ θεωροῦσιν ἐκ τῶν τῆς αἰσθήσεως ἀφορμῶν μόνον κινηθέντες, οἷον τὸ εἶναι ἐν ἀριθμῶν λόγοις τὰ διαστήματα καὶ τὸ εἶναι ἐξ ἀριθμῶν συγκρουστῶν τὸν φθόγγον καὶ τὰ παραπλήσια. τὰς ὑποθέσεις οὖν τῆς κανονικῆς διορίσειεν ἄν τις

[27] εἰσὶ δὲ καὶ ἑκάτεροι τῷ γένει μουσικοί, 23.8–9, trans. Barker 1989: 240.

[28] Cf. Pl. *Phdr.* 268d–e.

[29] τὴν δ' αἴσθησιν προέκριναν οἱ ὀργανοκοί, οἷς ἢ οὐδαμῶς ἔννοια θεωρίας ἐγένετο ἢ ἀσθενής, 25.14–16, trans. Barker 1989: 241.

[30] Didymus ap. Porph. *In Harm.* 26.13–15.

ὑπάρχειν τῇ τε περὶ τὴν μουσικὴν ἐπιστήμῃ καὶ τῇ περὶ τοὺς ἀριθμοὺς καὶ τὴν γεωμετρίαν.

The theory[31] that uses the *kanōn* – of what does it consist? Of the things postulated by the *mousikoi* and those adopted by the *mathēmatikoi*. The things postulated by the *mousikoi* are all those adopted by the *kanonikoi* on the basis of perceptions, for instance that there are concordant and discordant intervals, and that the octave is compounded from the fourth and the fifth, and that the excess of a fifth over a fourth is a tone, and similar things. Those adopted by the *mathēmatikoi* are all those which the *kanonikoi* study theoretically in their own special way, only beginning from the starting points given by perception, for instance that the intervals are in ratios of numbers, and that a note consists of numbers of collisions, and other things of the same sort. Hence one might define the postulates of *kanonikē* as lying both within the science concerned with music, and within that concerned with numbers and geometry.[32]

The last three sentences are of uncertain authorship,[33] but even on the basis of the first two we find Ptolemaïs describing canonics as a science defined in terms of the instrument on which its demonstrations were carried out, whose postulates originated among the *mousikoi* and the *mathēmatikoi*. A distinction can then be made between *mathēmatikoi* and *kanonikoi*: the former concern themselves with things like the first nine propositions of the *Sectio canonis*;[34] the latter are interested in the same things, but 'beginning from the starting points given by perception' (23.18–19). The definition of *kanonikē* (23.21–2) is that of a science whose postulates (*hypotheseis*) combine those of *mousikē*, *arithmētikē* and *geōmetrikē* – an unusual claim to which I shall return – but it is clear from Ptolemaïs' phrase ἡ κατὰ τὸν κανόνα θεωρία ('the theory that uses the *kanōn*', 23.10–11) that it is by virtue of the instrument itself, with its appeal to the primary elements of all three sciences, that this is so. For all her protests in Porphyry's first quotation (22.27–30), it appears that the *kanōn* has become intricately bound up in the definition of mathematical harmonic theory after all, and its practitioners and postulates along with it.

[31] *Theōria* does not mean simply 'theory' in our sense, but in various scientific and philosophical contexts 'investigation', 'consideration', 'contemplation', 'speculation' or even 'reckoning' (LSJ s.v.).

[32] 23.10–22, trans. Barker 1989: 240.

[33] Düring separates them as belonging to Porphyry; Barker translates the entire passage without a break, but admits that it may not all be Ptolemaïs' work (1989: 240 n. 138).

[34] Barker 1989: 240 n. 139.

'Canonics' may not have been limited to the monochord, but the instrument's influence had certainly come to be felt in the language in which the science was discussed.

Ptolemaïs' definition of canonics is confirmed in a more general statement by an author Porphyry calls Panaetius 'the Younger' (*In Harm.* 65.21).[35] Panaetius uses the expression 'canonic theory' (*kanonikē theōria*) twice in the course of the fragment Porphyry preserves, both times in the context of an argument about the incorrectness of the word 'semitone' (*hēmitonion*):

ἐπὶ δ᾽ ἐπογδόου διαστήματος μέσος ἀνάλογον οὐκ ἔστιν ἐν ἀριθμοῖς, οὐδὲ κατὰ τὴν κανονικὴν θεωρίαν τὸν τόνον φασὶ δίχα τέμνεσθαι. διόπερ οὔτ᾽ ἐπὶ τὰς ποιότητας ἀναφερόντων, οὔτ᾽ ἐπὶ τὴν κανονικὴν θεωρίαν τὸ ἡμιτόνιον ἥμισύ ἐστι τοῦ τόνου, ἀλλὰ καταχρήσει ὀνόματος λέγεται μόνον, καθάπερ ἡμίφωνον καὶ ἡμίονος. οὐδὲ γὰρ ἐν τούτοις ἔνεστι τὸ ἥμισυ τοῦ φωνήεντος ἢ τὸ ἥμισυ τοῦ ὄνου.

In the case of the epogdoic interval there is no mean proportional in numbers, and in canonic theory they deny that the tone can be divided in half. Hence neither when people refer the matter to qualities nor in canonic theory is the 'half-tone' half of the tone, but the term is merely a misuse of language, as in the case of the 'semivowel', and the 'half-ass' [mule]. For here too there is not half of a vowel or half of an ass.[36]

Like Ptolemaïs' canonic science, Panaetius' 'canonic theory' is mathematical harmonics, the kind of harmonics in which semitones cannot be divided equally.[37]

One of the most frequently recurring contexts for the use of the term *kanonikē* in this sense is in lists of mathematical sciences (*mathēmata*). Since the time of Archytas' predecessors, 'those concerned with the mathematical sciences' (τοὶ περὶ τὰ μαθήματα),[38] the list in its simplest form had only four items: arithmetic, geometry, astronomy and harmonics (the mediaeval quadrivium). For Archytas the last of these was simply the science 'about music'

[35] Younger, that is, than the more famous Panaetius (of Rhodes, c. 185–109 BC), the Stoic scholar associated with P. Cornelius Scipio Aemilianus. We cannot date him otherwise.

[36] Porph. *In Harm.* 67.3–8, trans. Barker 1989: 238–9. See also Philo, *Opif.* 126; Macrob. *In Somn.* II.1.21.

[37] The details of the argument will concern us more directly in part 2 of this chapter.

[38] Archyt. fr. 1 (Porph. *In Harm.* 56.5).

(περὶ μωσικᾶς);[39] Plato's usual word was *harmonia*.[40] The word *harmonikē* appears in such lists as early as Aristotle,[41] but from at least the time of Geminus (mid-first century AD), *kanonikē* often replaces *harmonikē*.[42] The relationships between sciences as expressed by Plato and Aristotle are frequently articulated as the lists expand, and Aristotle's notion of the subordination of some sciences to others is often apparent.[43] Canonics is thus normally paired in some way with arithmetic, as it is in Heron of Alexandria:

τίνα τίσι προσεγγίζει τῶν μαθημάτων; συνεγγίζει μᾶλλον τῇ μὲν ἀριθμητικῇ ἡ λογιστικὴ καὶ ἡ κανονική· καὶ γὰρ αὕτη ἐν ποσότητι λαβοῦσα κατὰ λόγους ἀριθμοὺς καὶ ἀναλογίας πρόεισι· τῇ δὲ γεωμετρίᾳ ἡ ὀπτικὴ καὶ ἡ γεωδαισία, ἀμφοτέραις δὲ καὶ ἐπὶ πλέον ἡ μηχανικὴ καὶ ἀστρολογική.

Which of the mathematical sciences are near to which? Calculation and canonics are closest to arithmetic; for arithmetic, too, proceeds by finding numbers and proportions in terms of quantity according to ratios. Optics and mensuration, on the other hand, are closer to geometry, and beyond these, mechanics and astronomy are related to both.[44]

For Heron, mathematical harmonics is called *kanonikē*, not *harmonikē*, and its closest scientific relative is arithmetic, because like arithmetic, it is concerned with numbers and proportions as

[39] Archyt. fr. 1 (Porph. *In Harm.* 56.10).

[40] E.g. *Resp.* 531a. From *Resp.* 530d it is clear that he understood the science (*epistēmē*) which studies *harmonia* as that which is concerned with harmonic motion (ἐναρμόνιος φορά), in parallel with the visible motion of its sister-science, astronomy.

[41] E.g. *Ph.* 194a8: 'the more physical of the mathematical sciences, such as optics, harmonics and astronomy' (τὰ φυσικώτερα τῶν μαθημάτων, οἷον ὀπτικὴ καὶ ἁρμονικὴ καὶ ἀστρολογία). Cf. also *An. post.* 79a1, *Metaph.* 997b21, 1078a14; Plut. *Non posse* 1093d, *Quaest. Plat.* 1001f.

[42] This new quadrivium – arithmetic, geometry, canonics and astronomy – is what we find in Plutarch, for instance, who derives it explicitly from book VII of Plato's *Republic* (ἀριθμητικὴν καὶ γεωμετρίαν καὶ κανονικὴν πᾶσαν καὶ ἀστρολογίαν, Plut. fr. 147 = Stob. IV.18a, 10). Geminus (ap. Procl. *In Euc.* 38.2–15) divided the sciences into those with intelligible as opposed to sensible objects, of which there were two (arithmetic and geometry), and those with sensible objects, of which there were six (mechanics, astronomy, optics, mensuration (*geōdaisia*), canonics and calculation).

[43] On harmonics as a subordinate science in Aristotle, see ch. 3 part 1.

[44] *Deff.* 138.7; his complete list of sciences is at 138.5. The text as transmitted is not unified, and it has been suggested (Knorr 1993) that the earliest components of it were written not by Heron (fl. AD 62) but by Diophantus (possibly mid-third century AD).

numerable quantities (as opposed to magnitudes, e.g. lengths or volumes), and expresses them by means of ratio.[45]

This is the orthodox account of the sciences, and many repeat it. But the process of classification sometimes places canonics in unusual positions. The fifth-century Neoplatonist Proclus, who makes his primary division of the sciences (following Geminus) according to whether they are concerned with intelligible or sensible objects, sees canonics as a science whose use of the monochord fixes its focus on audible phaenomena:

πάλιν ὀπτικὴ καὶ κανονικὴ γεωμετρίας εἰσὶ καὶ ἀριθμητικῆς ἔκγονοι . . . ἡ δ' αὖ κανονικὴ τοὺς φαινομένους λόγους τῶν ἁρμονιῶν σκοπεῖται, τὰς τῶν κανόνων κατατομὰς ἀνευρίσκουσα καὶ τῇ αἰσθήσει πανταχοῦ προσχρωμένη καὶ ὥς φησὶν ὁ Πλάτων ὦτα τοῦ νοῦ προστησαμένη.

Again optics and canonics are offshoots of geometry and arithmetic . . . The science of canonics deals with the perceptible ratios between notes of the musical scales and discovers the divisions of the monochord, everywhere relying on sense-perception and, as Plato says, 'putting the ear ahead of the mind' [*Resp.* 531a–b].[46]

Although Ptolemaïs had admitted that *kanonikē* adopted some of its postulates from sense-perception,[47] on balance her treatment of the science presents not the *kanonikoi* but the Aristoxenian *mousikoi* as 'everywhere relying on sense-perception'.[48] Proclus, by contrast, damns *kanonikē* with the very words Plato had used to criticise the minimal unit-finding procedures of Glaucon's harmonic theorists in *Republic* VII. As we saw in chapter 3, a monochord would be useless in such procedures (at least as Plato describes them); it seems then that for Proclus the mere fact that *kanonikē* is practised with the help of an audible instrument is what relegates it to the cave. What is canonics about? Canonic division, of course: one of the defining procedures of the science, for him, is 'discovering the divisions of the monochord' (τὰς τῶν κανόνων κατατομὰς ἀνευρίσκουσα, 40.23–4), an activity which he evidently sees as completely dependent on sense-perception. It

45 Nowhere in his extant works does Heron use the word *harmonikē* to indicate the science; his only use of the word (*Stereom.* I.29.1) is as an adjective in a context where *kanonikē* could not have been substituted for it (*harmonikē analogia*, 'harmonic proportion').

46 *In Euc.* 40.9–10, 40.22–41.2, trans. Morrow 1970: 33.

47 Porph. *In Harm.* 23.10–17.

48 With Proclus' statement cf. Ptolemaïs ap. Porph. *In Harm.* 24.1–6, 26.1–4.

attends to the 'perceptible ratios' of the *harmoniai*. Plato advocated a kind of 'inaudible harmony', but the division of the monochord, like the empirical string-torturing of Glaucon's theorists, pulls harmonics back into the realm of sense-perception. The *kanōn* is problematic, then, because it makes ratios audible. And 'canonics' is the only type of harmonics with which Proclus concerns himself.[49] Mathematical harmonics has come to be thought of as monochord-science, and this has philosophical consequences.

We cannot tell whether Geminus' account included an explicit reference to canonic division;[50] Heron's does not, but Panaetius' does.[51] Thus the association between canonic theory and canonic division in Proclus' account is not a fifth-century anachronism; it would not be unreasonable to suggest that it dates from the first century AD. The use of the term *kanonikē theōria* by Philo of Alexandria (*Opif.* 96) and his awareness of the use of the *kanōn* in setting out scales of different genera (*De posteritate Caini* 104) hint in this direction, but no first-century author makes the link as explicitly as Proclus does.[52] I shall return to the practice of canonics shortly, but first we must deal with one anomalous classification of the science.

Ptolemaïs defined *kanonikē* as a science whose postulates are drawn not only from *mousikē* and *arithmētikē*, but also from *geōmetrikē*.[53] In light of the fact that Heron, Proclus and (probably) Geminus all classified canonics as an arithmetical science, her

[49] *Harmonikē* occurs only once in Proclus' *Commentary* (34.16), in a discussion of precision (*akribeia*) and demonstration (*apodeixis*) in the sciences, where he observes that arithmetic is more precise than *harmonikē*. Compare the usage of Simplicius, who lists 'harmonics, otherwise known as canonics', among the *mathēmata* (*In Phys.* II.2, 293.9–10).

[50] What survives of Geminus' account of the mathematical sciences is preserved by Proclus (see n. 42 above).

[51] His brief canonic division (ap. Porph. *In Harm.* 66.20–30) comes immediately before his two references to *kanonikē theōria* (67.4, 5–6); it is discussed in part 2 of this chapter.

[52] Philo's single use of *kanonikē theōria* is problematic, because he invokes the science as an authority for the special harmonic status of the ratio 5:2. This is the ratio of the octave plus major third, a discord by all ancient definitions. It is hard to tell whether Philo is deliberately attempting to mislead, or simply out of his depth. In either case, however, he provides an example of the ways in which the language and practice of the monochord spilled over from harmonics into wider intellectual discourse in the century before Ptolemy.

[53] Ptolemaïs ap. Porph. *In Harm.* 23.21–2.

inclusion of the postulates of geometry seems odd.[54] I have argued already that although mathematical harmonic science adopted geometrical diagrams, the language which governed them constrained their use in the same way as we find geometrical diagrams constrained in Greek arithmetical proofs.[55] The result, I have suggested, is that although methods were available to harmonic theorists by which geometrical operations might have been performed with the monochord, the conception of the science as a species of arithmetic ruled them out.[56] Ptolemaïs' summary suggests that thinking of harmonics in terms of the *kanōn* introduced a complication: the instrument must be sounded by line division. Line division lies outside arithmetic; thus the act of canonic division involves first admitting the postulates of geometry into harmonics, before arithmetical entities (ratios) can be set out on the instrument. For to perform a canonic division one must first postulate the possibility of drawing a straight line from one point to another (Euc. *El.* I post. 1).

As we shall see in chapter 6, Ptolemy would later incorporate more of the methods of geometry into harmonics. Geometrical postulates never replace arithmetical ones, because the language of mathematical harmonics always belongs to arithmetic (irrational intervals, for instance, are always impossible). But geometry becomes a part of the science with the monochord, and Ptolemy would use it for tasks in which arithmetic was of no help.[57]

Ptolemaïs' inclusion of the postulates of geometry is thus not as surprising as it might first seem. She is also not the most extreme case. Ptolemy's contemporary Aulus Gellius distanced canonics from arithmetic, classifying it, together with optics, as a branch of geometry.

pars quaedam geometriae ὀπτική appellatur, quae ad oculos pertinet, pars altera, quae ad auris, κανονική uocatur, qua musici ut fundamento artis suae

[54] The arithmetic-harmonics connection derives, as we have seen, from statements by authoritative authors like Aristotle (e.g. *An. post.* 79a1); cf. *Suda* ε 2627.5 (s.v. ἐπιστήμη). But cf. Ptol. *Harm.* 94.16–20 for a more balanced approach.

[55] See ch. 1 part 1.

[56] One such method would have been the application of Eratosthenes' *mesolabon* to the tetrachords of Aristoxenus (see ch. 4).

[57] See e.g. *Harm.* III.2 (discussed in ch. 6).

utuntur. utraque harum spatiis et interuallis linearum et ratione numerorum constat.

One branch of geometry is called *optikē*, the one which belongs to the eyes; another branch, which belongs to the ears, is called *kanonikē*, and musical theorists employ it as the foundation of their science. Both of these rely on the spaces and intervals between lines and on ratios of numbers.[58]

Gellius is, so far as I can tell, unique in associating harmonics with geometry instead of arithmetic. This is clearly unorthodox, but two things about his account are worth noting. First, his classification is reasoned: canonics cannot be practised without expressing ratios as lines; this draws it into the realm of geometry, like optics. Second, his definition of *kanonikē* is itself unique: for him it is the science that measures not just the pitches, but also the durations, of sounds.[59] Rhythm is not included in any other ancient definition of canonics, but the *kanonikoi* are credited with statements about the similar (rational) nature of rhythm and melody.[60] Obviously the ratios of rhythm cannot be studied with the monochord, nor can they be thought of as a kind of quantity accessible to geometry. Gellius' account of canonics is therefore inconsistent, but it helps to show that the science was perceived to have some resemblance to geometry, and that its attention to ratios could be transferred to rhythm as well as to melody. Canonics seems to have been defined and classified not only by its use of the *kanōn* but by the instrument's governing principles, which included the basic postulates of geometry and arithmetic.

[58] Gell. *NA* XVI.18.1–2. On *kanonikē* in Aulus Gellius see Holford-Strevens 2003: 325–7.

[59] '*Kanonikē*, on the other hand, measures the durations and pitches of sounds. The measure of the duration of sounds is called *rhythmos* (rhythm), and the measure of their pitch is called *melos* (melody)' (κανονική *autem longitudines et altitudines uocis emetitur. longior mensura uocis* ῥυθμός *dicitur, altior* μέλος, Gell. *NA* XVI.18.4).

[60] Porphyry quotes Dionysius 'the musician' to this effect: 'according to the *kanonikoi* the nature of rhythm and melody is nearly one and the same; they also think that what is high-pitched is fast and what is low-pitched is slow. And in general they hold that attunement (*to hērmosmenon*) is a proportion between movements, and that well-tuned intervals are in ratios of numbers' (κατὰ μέν γε τοὺς κανονικοὺς μία σχεδὸν καὶ ἡ αὐτὴ οὐσία ἐστὶ ῥυθμοῦ τε καὶ μέλους, οἷς τό τε ὀξὺ ταχὺ δοκεῖ καὶ τὸ βαρὺ βραδύ. καὶ καθόλου δὴ τὸ ἡρμοσμένον κινήσεών τινων συμμετρία καὶ ἐν λόγοις ἀριθμῶν τὰ ἐμμελῆ διαστήματα, Porph. *In Harm.* 37.17–20). Porphyry expands on Dionysius' account and fleshes out the connection between the ratios of concord and those of rhythmic metres (37.25–38.2). Cf. also [Arist.] *Pr.* XIX.39.

2 The practice of canonics

> I certainly would not deny the role of the abacus for Greek arithmetical concept-formation. The question is different: whether any arithmetical proof, oral or written, was ever conducted with the aid of pebbles . . . [T]he special advantage of pebbles over other types of arithmetical representations is a result of their direct, physical manipulations, which are necessarily tied up with actual operations. It is not the mere passive looking at pebbles which our sources mention: they mention pebbles being moved and added.[61]

We are fortunate to have better evidence in harmonics than in arithmetic for the involvement of instruments in the presentation of proofs. One of the reasons for this, as we saw in chapter 1, is that the instruments of harmonics, unlike those of arithmetic, can be inscribed in the proof itself according to the conventions of the lettered diagram. What the monochord provides its science is in general similar to what the abacus offers arithmetic: a means of tying conceptual operations to physical manipulations. But the histories of the involvement of these instruments in their sciences are different; we can no longer entertain the possibility that the monochord played a role in the early formation of any of the fundamental concepts of mathematical harmonics. And yet concepts and activities are closely intertwined: scientific instruments, by the mere fact of their physicality, by their appeal to sense-perception, enable certainty; they help to convince. If seeing is believing, then seeing *and* hearing must produce an even more convinced belief; seeing and hearing as a result of one's own theoretically grounded manipulations is of a higher order again. Thus the role of apparatus in Greek science is closely connected with the process both of thinking through conceptions, and of acting them out: in the right circumstances, the division of the monochord can be the performance of the proof. It can be a way of simultaneously enacting a mathematical/harmonic proposition visibly and audibly, of converting concept to diagram, and of musically animating the diagram. Canonic division is thinking diagrams aloud.

The process also works in the opposite direction: in canonics, the demonstration that 'the octave is less than six tones' (*Sect.*

[61] Netz 1999a: 64.

can. prop. 14) comes as the final, perceptibly accessible point of an argument which, after the introduction of initial postulates from perception, could be followed entirely by calculation. But if one followed the argument by monochord all the way, then the agreement of perception with calculations based on initial perception-based postulates at each stage would serve to confirm – especially where the end results are theoretically contentious – that perception and reason work along parallel tracks. The monochord was the clearest (though not quite the only) means of direct two-way communication between the faculties of *aisthēsis* and *logos* available to the ancient Greek musical theorist.

The purpose of the remainder of this chapter is to examine the practice of canonics – that is, what precisely *kanonikoi* did, and how the *kanōn* was involved in that practice. Proclus' answer was 'canonic division', but this requires some fleshing out. What were the aims and objectives of canonic division? What theoretical purposes did it serve? What, in fact, were the physical manipulations conventionally designated by that term, and what theoretical operations did they animate? If the monochord was invented after (and as a result of) the formation of the basic postulates and theorems of mathematical harmonics, did its presence in the following centuries shape subsequent developments in Greek harmonics? Questions about the practice of *kanonikē* turn out to be questions about the role of the instrument in the connections between concept and demonstration in Greek harmonics.

Reason versus perception

The necessity of providing a framework for the relationship between the two indispensable faculties of reason and perception in musical theory had already been addressed by Plato and Aristoxenus.[62] Their arguments, however, only opened the debate. Plato had divided harmonic theorists into two groups (Socrates' and Glaucon's theorists, as we encountered them in chapter 3);[63] Aristotle recognised a similar distinction, of 'mathematical

[62] See especially Aristox. *El. harm.* 33.3–10, 34.26–30, 38.29–39.4.
[63] Pl. *Resp.* 531a–c.

harmonics' and 'harmonics by ear'.[64] Ptolemaïs expresses this dichotomy with the labels 'Pythagoreans' and 'Aristoxenians'. One of the most important and polarising points in the history of this opposition must have been the composition of the *Sectio canonis*, where the monochord had a central role to play in the apodeictic refutation of Aristoxenian harmonic arguments.

The *kanōn* had thus become involved, from the earliest point in its history, in what came to be known as the Pythagorean versus Aristoxenian debate. One of the most important consequences of the argument of the *Sectio* was that by it the mathematical harmonicist – the *kanonikos*, we may now call him – was able to use Aristoxenus' own prized criterion, perception, to prove him wrong. The Aristoxenian assertion that there are six tones in an octave, for example, can now be demolished by evidence presented simultaneously to reason and perception. As a straightener of the senses, the monochord speaks directly and incontrovertibly in the realms of music and mathematics, as Ptolemaïs was aware.

It would seem, then, that the monochord's arrival only widened the ideological rift between mathematical and Aristoxenian harmonics. Yet within a century of the *Sectio*, at least one scholar (Eratosthenes) had already begun to work toward some rapprochement between the diverging schools. Later, Adrastus would seek to bring them together by means of instruments[65] – a project whose methods were developed with even greater sophistication by Claudius Ptolemy. Although Ptolemy's aims were different from those of Adrastus, the unique brand of apodeictic empiricism which makes his *Harmonics* a pivotal contribution to ancient science owes its inception in part to the use Adrastus makes of demonstrative instruments, and in part to the way in which the opposition between 'Pythagorean' and 'Aristoxenian' harmonics was framed by certain earlier theorists.

Ptolemaïs was perhaps the earliest (though we cannot be certain) to focus the debate around the authority each camp gave either to reason or to perception. Her book was apparently entitled *Pythagorean Elements of Music*, but her allegiances are not so

[64] ἁρμονική ἥ τε μαθηματική καὶ ἡ κατὰ τὴν ἀκοήν, *An. post.* 79a1–2.

[65] Though his intention 'is not precisely to reconcile the two traditions' (Barker 1989: 210), and as we shall see, the monochord sits somewhat uncomfortably within his method.

plainly straightforward.[66] She was interested in the monochord and clearly considered its rational principles fundamental to harmonic investigation, but she was also well aware of the fact that mathematical harmonic theory cannot do without perception entirely. Her list of postulates culled from perception falls into two groups:[67]

1. those which the *kanonikoi* adopt from the *mousikoi*
   a. that there are concordant and discordant intervals
   b. that the octave is made up of fourth and fifth
   c. that the tone is the excess of a fifth over a fourth
   d. and similar things (καὶ τὰ ὅμοια)
2. those adopted by the *mathēmatikoi* and studied by the *kanonikoi*
   a. that intervals are in ratios of numbers
   b. that a note consists of numbers of collisions
   c. and other things of the same sort (καὶ τὰ παραπλήσια)

The first group contains items which are indisputably perception-based: there is no other faculty by which they can be independently ascertained. The second group, however, relies on a much less immediate appeal to perception; these postulates can only be inferred from perceptibles via arguments like those in the introduction of the *Sectio canonis*.[68] Neither of her examples (2a–b) would be endorsed by a perception-based Aristoxenian *mousikos* (to usc her term). In other words, we are dealing in group 2 with a set of postulates which rely ultimately on perception, but whose status as such would not be admitted by the harmonic theorists for whom the evidence of perception counted the most.

A word must be said here about Ptolemaïs' treatment of the Aristoxenians on this particular question. 'Aristoxenus puts no faith in reason, but much in the senses', wrote Boethius, apparently citing the opinion of Ptolemy.[69] Ptolemaïs' treatment of him is

[66] Title: Porph. *In Harm.* 22.23–4. Ptolemaïs' terminology for reason and perception is different from Aristoxenus'; her opposition is between *logos* and *aisthēsis*, whereas his is usually (though not always) between *dianoia* and *akoē*. Ptolemaïs' distinction is both more explicitly mathematical than Aristoxenus' (*logos* as opposed to *dianoia*) and not linguistically limited to harmonics ('sense-perception' as opposed to 'hearing').

[67] This scheme is a summary of Ptolemaïs ap. Porph. *In Harm.* 23.10–22 (quoted in part 1 of this chapter). It is possible that the sentences on which the scheme is based were written by Porphyry (see n. 33 above).

[68] *Sect. can.* props. 11–13 are founded on 2a.

[69] *Aristoxenus nihil rationi, sed tantum sensibus credit* (*Mus.* 355.13–14). The second clause has some foundation (cf. e.g. *El. harm.* 43.6–9, 56.31–3), but the first is an exaggeration of his considerably more subtle position.

more sensitive than this. She credits Aristoxenus with accepting reason and perception 'in the same way, as being of equal power' (ὁμοίως ἀμφότερα ἰσοδυναμοῦντα):[70]

οὔτε γὰρ αἰσθητὸν δύναται συστῆναι καθ' αὑτὸ δίχα λόγου, οὔτε λόγος ἰσχυρότερός ἐστι παραστῆσαί τι μὴ τὰς ἀρχὰς λαβὼν παρὰ τῆς αἰσθήσεως, καὶ τὸ τέλος τοῦ θεωρήματος ὁμολογούμενον πάλιν τῇ αἰσθήσει ἀποδιδούς.

For what is perceived cannot be constituted by itself apart from reason, and neither is reason strong enough to establish anything without taking its starting points from perception, and delivering the conclusion of its theorising in agreement with perception once again.[71]

Insofar as Aristoxenus treated the two faculties differently, she says, he took perception as leading reason 'in order, not in power' (τῇ τάξει, οὐ τῇ δυνάμει, 25.23). The two faculties are not so equally yoked on departure from Pythagorean stables, however:

βούλονται γὰρ αὐτοὶ τὴν μὲν αἴσθησιν ὡς ὁδηγὸν τοῦ λόγου ἐν ἀρχῇ παραλαμβάνειν πρὸς τὸ οἱονεὶ ζώπυρά τινα παραδιδόναι αὐτῷ, τὸν δὲ λόγον ἐκ τούτων ὁρμηθέντα καθ' ἑαυτὸν πραγματεύεσθαι ἀποστάντα τῆς αἰσθήσεως, ὅθεν κἂν τὸ σύστημα τὸ ὑπὸ τοῦ λόγου εὑρηθὲν τῆς πραγματείας μηκέτι συνᾴδῃ τῇ αἰσθήσει, οὐκ ἐπιστρέφονται, ἀλλ' ἐπεγκαλοῦσι λέγοντες τὴν μὲν αἴσθησιν πλανᾶσθαι, τὸν δὲ λόγον εὑρηκέναι τὸ ὀρθὸν καθ' ἑαυτὸν καὶ ἀπελέγχειν τὴν αἴσθησιν.

For they [the Pythagoreans] wish to accept perception as a guide for reason at the outset, to provide reason with a spark, as it were; but they treat reason, when it has set out from these beginnings, as working on its own in separation from perception. Hence if the *systēma* discovered by reason in its investigation no longer accords with perception, they do not retrace their steps, but level accusations, saying that perception is going astray, while reason by itself has discovered what is correct, and refutes perception.[72]

Ptolemaïs is realistic about the deficiencies of the Pythagorean approach to harmonics. Despite the title of her work, it was evidently not one-sidedly pro-Pythagorean in tone or argument.

70 25.17, trans. Barker 1989: 241. 71 25.19–22, trans. Barker 1989: 241.
72 25.26–26.1, trans. Barker 1989: 242. Compare Didymus' version ap. Porph. *In Harm.* 26.15–25. I have passed over the problematic pair of sentences at 25.25–6, which seem to signal a lapse of attention on Porphyry's part: 'Who are those who treat both together alike? Pythagoras and his successors' (τίνες τὸ συναμφότερον ὁμοίως; Πυθαγόρας καὶ οἱ διαδεξάμενοι). The lines which follow do not match the introductory question, but they are consistent with what Ptolemaïs says about the Pythagoreans otherwise. On this textual problem see Barker 1989: 241 n. 148.

The second part of the second extract from Ptolemaïs (24.1–6, which probably followed the last quotation immediately even though it is placed earlier in Porphyry's text) presents an antithesis to the Pythagorean approach:

ἐναντίως δὲ τούτοις ἔνιοι τῶν ἀπ᾽ Ἀριστοξένου μουσικῶν φέρονται, ὅσοι κατὰ μὲν τὴν ἔννοιαν θεωρίαν ἔλαβον, ἀπὸ δ᾽ ὀργανικῆς ἕξεως προκόψαντες. οὗτοι γὰρ τὴν μὲν αἴσθησιν ὡς κυρίαν ἐθεάσαντο, τὸν δὲ λόγον ὡς παρεπόμενον, πρὸς μόνον τὸ χρειῶδες. κατὰ δὴ τούτους εἰκότως οὐ πανταχῇ αἱ λογικαὶ ὑποθέσεις τοῦ κανόνος σύμφωνοι ταῖς αἰσθήσεσιν.

An opposite position to this is held by some of the *mousikoi* who follow Aristoxenus, those who applied themselves to a theoretical science based in thought, while nevertheless setting out from expertise on instruments. For they treated perception as authoritative, and reason as attending on it, for use only when needed. According to these people, to be sure, it is only to be expected that the rational postulates of the *kanōn* are not always concordant with the perceptions.[73]

From her account of Aristoxenus' even-handed treatment of reason and perception quoted above (25.17–22), it appears that Ptolemaïs is not talking here about precisely the same sort of 'Aristoxenians'; accordingly they have been associated with the 'instrumentalists' (*organikoi*) she mentions a little later (25.15).[74] But whereas she describes this particular group of *mousikoi* as theorists (24.2), she is very reluctant to grant the *organikoi* such a status.[75] At any rate, Ptolemaïs defines these Aristoxenian *mousikoi* by their treatment of perception as authoritative (24.3), and their preliminary experience with instruments (24.3–4). She presents their attitude to reason as thoroughly cavalier, and regards it as unsurprising that they cannot make intelligent use of the *kanōn* (24.5–6).[76]

This is where we come to the crux of the matter. Differing schools of harmonic theorists can be distinguished by the way in which they handle the faculties of reason and perception, but for Ptolemaïs this is not merely a convenient but otherwise

[73] Trans. Barker 1989: 241. [74] Barker 1989: 241 nn. 142, 145.

[75] The *organikoi* 'gave no thought at all, or only feeble thought, to theory' (25.15–16, quoted in n. 29 above); the admission of their theorising (at the end of the sentence in Greek) seems a grudging afterthought. Plutarch clearly means performers as opposed to theorists at *Quaest. conv.* 657e, where the opposition between *logikoi* (theorists) and *organikoi* (instrumental performers) is essential to a musical joke about the theory and practice of drinking.

[76] Again, cf. Didymus' account, 26.6–15.

unimportant criterion, a set of pegs on which to hang musicologists. The *kanōn* speaks with authority in the languages of both faculties; it *ought* therefore to occupy a bridging position between the two (23.21–2). Insofar as it does, it does only for the *kanonikoi*, whose science admits postulates derived from both (23.10–22). Her spectrum of *harmonikoi* is built around the reason/perception test, and the monochord determines the middle ground. The extreme Aristoxenians (24.1–6) and the extreme Pythagoreans (25.26–26.1) are on either end; in the middle are the *kanonikoi*, for whom the 'rational postulates of the *kanōn*' (which touch reason and perception equally) are the rule of their science. This summary must be tentative given the state of our text, but if it is correct, then it places the monochord in a position similar to the one it would later occupy in the work of Ptolemy – as mediator between *logos* and *aisthēsis*.

Much of Ptolemaïs' discussion has been lost in Porphyry's ragged cutting and pasting, but the testimony of Didymus tends to confirm this reconstruction. His book *On the Difference Between the Aristoxenians and the Pythagoreans* may have been Porphyry's source for the Ptolemaïs fragments, and in the portions of it which Porphyry preserves, Didymus outlines three approaches to the study of music which conform very closely to those of Ptolemaïs.[77] He, too, separates them according to their emphasis on either reason or perception as a criterion.[78] Unlike Ptolemaïs, he does not mention the monochord in this context, but that may be merely an accident of preservation: that he knew and used the instrument is clear from what Ptolemy writes about him (*Harm.* II.13).

What is important about the testimony of Ptolemaïs and Didymus, then, is that in their work the dichotomy between Aristoxenian and Pythagorean harmonics was concentrated on one issue, an issue in which, in Ptolemaïs' account, only the *kanōn* can be used as a reliable tool for judgement. Sometimes *syringes* and auloi are used in *kanonikē*, but these, she says, 'are not strictly canonic' (23.2). The monochord's place has been established within the ideological spectrum of Greek harmonics, but all the finer detail is

[77] Didymus ap. Porph. *In Harm.* 26.6–29. On the important difference of emphasis, see Barker 2000: 70.

[78] He discusses these criteria further at 27.17–28.26.

left blank. We do not hear from either Ptolemaïs or Didymus how the instrument was employed in canonics, nor how precisely reason comes to discover what is correctly tuned (22.29–30). Beyond the *Sectio*, our first hint of a more detailed integration of the monochord into the reason/perception argument comes from Panaetius.

Mules, semitones and lukewarm coffee

Throughout the brief fragment of his book *Concerning the Ratios and Intervals in Geometry and Music* preserved by Porphyry, Panaetius juggles what he refers to as 'qualitative' and 'quantitative' approaches to harmonic analysis.[79] These categories can be mapped in a fairly general way onto Ptolemaïs' perception- and reason-based approaches without much difficulty, although the parallel is nowhere made explicit in what we possess of Panaetius' work. (There is an implied connection between his qualitative category and the evidence of perception, to which he refers frequently, but no link is drawn between reason and quantitative analysis.)

Panaetius' argument is complicated, dense and in many respects unique, and centres largely on the problems of establishing a reliable connection between perception and the quantitative aspects of pitch and interval. His main objective in the passage is to demonstrate the invalidity of the term 'semitone' (*hēmitonion*), and he does this by appealing first to qualitative and then to quantitative approaches to harmonics. The first assumes that we apprehend pitches as differing from one another in quality – they 'feel' different, as it were, just as different colours and different temperatures are sensed as differences in quality. In modern terms, we do not hear two notes as 440 and 220 cycles per second, or even as 2:1 – we hear 'high note' and 'low note'; we hear them together as 'octave'; given another reference tone we might identify them as *a′* and *a*. In Greek terminology we could equally say that we might perceive the notes as *oxys* and *barys*, or as *nētē* and *hypatē*, but not as 2:1. The quantitative is removed from the level of perception; we hear notes, not numbers. The part of our science which involves

[79] *In Harm.* 65.21–67.10. On what little we know of him see n. 35 above; on the relationship between his writings and those of Adrastus, see n. 90 below.

observation of phaenomena is carried out, Panaetius says, at this qualitative level: 'for the investigation that deals with concordances is not based on the magnitudes of the notes, but on their qualities'.[80] But quantitative analysis is more exact, because of the weakness of sense-perception. Thus while he regards both as necessary, he is careful to deny the quantitative any logical priority: it is not the cause of the qualitative, merely an accompanying attribute:

διόπερ ἐπειδὰν λέγωσι τὸ διὰ πασῶν ἐν διπλασίονι λόγῳ, οὐ τοῦτο λέγουσιν, ὅτι ὁ φθόγγος τοῦ φθόγγου διπλάσιος, ἀλλ' ὅτι αἱ χορδαί, ἀφ' ὧν οἱ φθόγγοι οἱ ποιοῦντες τὸ διὰ πασῶν, τοῦτον ἔχουσι τὸν λόγον.

Hence when they say that the octave is in duple ratio, they do not mean that the one note is double the other, but that the strings from which there arise the notes making the octave have this ratio.[81]

It may seem an odd introduction to an argument in favour of the *kanōn*, perhaps, but this is precisely what comes next. The monochord does not provide us with a window on the numbers which *constitute* the notes, but it does provide harmonic investigation with a level of accuracy which surpasses the qualitative distinctions offered by perception. On both levels of analysis, he claims, the 'semitone' will be shown to be an invalid term of reference. If one could conceive of the difference between white and black, or hot and cold, as an interval (*diastēma*, 65.27), one could not then divide this difference in half. As we experience it through our senses, it is a qualitative difference, and locating the shade or the temperature which divides this difference exactly in half is beyond the powers of sense-perception. (To use a modern example: it is only by converting the sensations of scalding-hot and stone-cold coffee into quantities (degrees Celsius, for example) with a thermometer that one could do better than a rather vague lukewarm if asked to produce a cup which divided this difference in half. Unassisted perception stumbles around the quantitative world in the dark.) A further, punning example is brought out near the end: a half-tone (*hēmitonion*) is no more properly half of a tone than a mule (*hēmionos*) is literally half of an ass – what we *perceive* is a

[80] οὐ γὰρ παρὰ τὰ μεγέθη τῶν φθόγγων ἡ περὶ τὰ σύμφωνα πραγματεία, ἀλλὰ περὶ τὰς ποιότητας, 65.29–30, trans. Barker 1989: 237.
[81] 66.30–67.2, trans. Barker 1989: 238.

new beast, a blend (*krasis*), qualitatively different from both horse and donkey.[82]

Quantitative analysis is therefore an aid to perception:

ὅτι οὔτε τῆς ὄψεως ἰσχυούσης κρίνειν τὰ σύμμετρα τῶν μεγεθῶν ἀλλ' εὑρημένου μέτρου, ᾧ καταμετρούμενα τὰ σύμμετρα κρίνεσθαι πέφυκεν, οὔτε τῆς ἁφῆς ἰσχυούσης κρίνειν τὴν κατὰ τὰ βάρη σύγκρισιν, ἀλλ' εὑρημένου ζυγοῦ, ᾧ κρίνεται τὰ βάρη. ἄτοπον δὲ δοκεῖ τὴν ἀκοὴν πολὺ ἀσθενεστέραν ὑπάρχουσαν τῆς ὄψεως χωρὶς μέτρου τινος καὶ κανόνος κρίνειν τὰ σύμφωνα τῶν διαστημάτων. οἱ γὰρ αὖ τῇ αἰσθήσει προσέχοντες ὡς ἐκ γειτόνων φωνὴν ἀκούοντες, ὅμοιοι φαίνονται τοῖς χωρὶς μέτρου διὰ τῆς ὄψεως περὶ τῆς κατὰ τὰ μεγέθη συμμετρίας ἀποφαινομένοις, οἳ πολὺ ἀφαμαρτάνουσι τῆς ἀληθείας.

Since sight is not strong enough to assess the relative dimensions of magnitude unless some measure is found, by which the relative dimensions are naturally suited to being measured and assessed, and neither is the sense of touch strong enough to assess a comparison in respect of weights, unless a weighing device is found by which the weights are assessed, it seems absurd to suppose that hearing, which is much feebler than sight, can assess concordant intervals without some *kanōn* as a measure. For those who concentrate on perception, as though listening to a sound from next door, seem like people who make assertions about the relative dimensions of magnitudes without the aid of any measure, and they miss the truth by a long way.[83]

[82] 67.3–8, quoted in part I of this chapter. The word *krasis* was often used in definitions of concord (e.g. *Sect. can.* 149.19–22; cf. also Theophr. ap. Porph. *In Harm.* 63.17, Panaetius ap. Porph. *In Harm.* 66.18, Aelianus ap. Porph. *In Harm.* 35.27, Bacch. I.10, Gaud. 8). The concept of 'blending' is linked explicitly with musical pleasure in the Aristotelian *Problems*: 'we enjoy concord because it is a blend (*krasis*) of opposites which have a ratio to one another' (συμφωνίᾳ δὲ χαίρομεν, ὅτι κρᾶσίς ἐστι λόγον ἐχόντων ἐναντίων πρὸς ἄλληλα, XIX.38). Plutarch plays on this sense in a passage about mixing wine in musical proportions (*Quaest. conv.* III.9).

[83] 66.7–15, trans. Barker 1989: 237–8; note the overt reference to Glaucon's theorists (Pl. *Resp.* 531a). Panaetius' comparisons could be summarised as follows:

Non-rational faculty of perception (αἴσθησις) *which naturally judges its objects by their qualities*	*Object of sense-perception to be judged*	*Rational invention by which senses are assisted in judging quantities* (εὑρημένον κριτήριον)
ὄψις sight	τὰ σύμμετρα τῶν μεγεθῶν relative dimensions of magnitude	μέτρον measure
ἁφή touch	ἡ κατὰ τὰ βάρη σύγκρισις comparison in respect of weights	ζυγόν weighing device (balance)
ἀκοή hearing	τὰ σύμφωνα τῶν διαστημάτων concordant intervals	κανών monochord

Thus if any theoretical method were to provide an exactly equal division of a tone so as to produce two semitones, it would be one which provided a quantitative measure of sound, and did not attempt to assess such differences by means of unassisted perception. This leads Panaetius to the monochord:

διόπερ ἄλλων κατ᾽ ἄλλας ἐφόδους παρὰ τῶν πρότερον ζητούντων τὸ προκείμενον ἐπὶ τοῦ λεγομένου κανόνος – ὃν ἐγὼ καὶ τοὔνομα οἶμαι ἐσχηκέναι, ἐπεὶ κριτήριόν ἐστι τοῦ κατὰ τὴν ἀκοὴν ἐν τοῖς συμφώνοις γινομένου πλήθους – εὕρισκον χορδῆς τεταμένης καὶ τοῦ ὑπαγωγέως κατὰ τὴν διχοτομίαν ὑπαχθέντος τὴν ὅλην πρὸς τὴν ἡμισεῖαν συμφωνοῦσαν τὸ διὰ πασῶν, ὑπὸ δὲ τὸ τέταρτον ὑπαχθέντος τὴν ὅλην πρὸς τὰ τρία μέρη συμφωνοῦσαν τὴν διὰ τεσσάρων, πρὸς δὲ τὸ τέταρτον τὸ δὶς διὰ πασῶν· καὶ ὑπὸ τὸ τρίτον τὴν ὅλην πρὸς μὲν τὰ δύο μέρη συμφωνοῦσαν τὸ διὰ πέντε, πρὸς δὲ τὸ τρίτον τὸ διὰ πασῶν καὶ διὰ πέντε· τὸν δὲ τόνον ἐν ἐπογδόῳ, ὅτι ἡ ὅλη πρὸς τὰ ὀκτὼ διάστημα ποιεῖ τὸ ἴδιον.

Thus, with different people among our predecessors investigating the problem by different methods on what is called the *kanōn* – which I take to have acquired this name because it is the criterion for the hearing of the quantity involved in the concords – they discovered that when a string has been stretched and the bridge is placed under it at the halfway point, the whole is concordant with the half at the octave: when it is placed at a quarter of the string the whole is in concord with the three parts at the fourth, and with the quarter at the double octave: when it is placed at the third of the string the whole is in concord with the two parts at the fifth, and with the third at the octave and a fifth: and the tone is in epogdoic ratio [9:8], because the whole makes its special interval in relation to the eight parts.[84]

Here, then, is the first of the four canonic divisions to be considered in this chapter. Panaetius uses it not merely to illustrate the procedures of his predecessors, but to make the point that if this method will not generate an exact half-tone, none will. And of course the answer is that none will. 'In the case of the epogdoic interval there is no mean proportional in numbers, and in canonic theory [in other words, 'in the theory of the *kanōn*, which I have just illustrated'] they deny that the tone can be divided in half'.[85] The process of canonic division, as a physical manipulation of sensible measuring-gadgetry to assist in 'the hearing of the quantity involved in the concords', will demonstrate that the tone is epogdoic; mathematics can then be invoked to show that this ratio

[84] 66.20–30, trans. Barker 1989: 238.
[85] 67.3–5, trans. Barker 1989: 238–9 (quoted more fully in part 1 of this chapter).

cannot receive a geometric mean expressible in numbers – which is as much as to say, for the *kanonikos*, that such an interval cannot exist.[86] The monochord is for hearing numbers; if an interval cannot be found in numbers, it cannot be heard. 'Hence neither when people refer the matter to qualities nor in canonic theory is the "half-tone" half of the tone, but the term is merely a misuse of language' (67.5–7).

In Panaetius' scheme, then, the *kanōn* is indispensable. It is his only way of allowing the senses to assess the quantities in music accurately, and without it his argument would rest limply on assertions about the inadequacy of perception and a pair of analogies from school-room and stable.

Two further observations should be made here about the use of instruments in Panaetius' method. The first concerns the detail of his brief canonic division – and its very brevity is what helps to indicate that there is something unusual about it. In *Sectio canonis* propositions 19–20, fourteen bridge positions were established to generate fourteen notes beyond that of the whole string. Panaetius' concern is to exhibit interval-ratios rather than the notes of a scale, but the brevity of his division is the product of something more than his modest goals. He sets out six intervals in all, at least five of which are generated by only three bridge positions. It seems that a fourth bridge position is to be constructed for the remaining interval, the tone, but he does not spell this out; its epogdoic ratio could be found in the distance between the second and third bridge positions. Thus at best Panaetius' division is twice as efficient as that of the *Sectio*, simply by virtue of his using the portions of string on both sides of the bridge. It seems straightforward enough, but it deserves mention because it is precisely this innovation with which Ptolemy credits Didymus (*Harm.* II.13). Panaetius was either following Didymus' procedures here (and hence lived no earlier than the mid-first century AD), or Didymus was not the first to propose them.

The second curious feature of Panaetius' account also involves the use of stringed instruments, but in this case not the monochord.

[86] Compare e.g. Adrastus ap. Theo. Sm. 50.12–21, where the condition under which sounds may be called 'notes' is that they be in some ratio of numbers to one another.

In his description of the search for the ratios of the concords by the Pythagoreans and mathematicians, he defines concord by reference to the behaviour of strings.

ἣν δὴ πολλὴ ζήτησις ἄνωθεν ἀρξαμένοις τοῖς Πυθαγορείοις καὶ ἑξῆς τοῖς ἀπὸ τῶν μαθημάτων, κατὰ τίνας λόγους ἐπὶ τῶν συμφώνων διαστημάτων ἐκ τῶν διαφερόντων κατὰ ποιότητα φθόγγων μία γίνεται κρᾶσις καὶ τῆς ἑτέρας χορδῆς πληχθείσης, ᾗ[87] σύμφωνος συγκινεῖσθαι πέφυκεν. ἐζήτουν εἰ καὶ ταῦτα κατὰ λόγους ἐλαχίστους συμβαίνει.

There was a great deal of investigation, beginning initially with the Pythagoreans, and subsequently among specialists in mathematics, as to which ratios they are, in the concordant intervals, in which from notes differing in quality there arises a single blend (*krasis*) even when just one of the strings is plucked, by which one in concord with it is naturally set in motion along with it. They enquired whether these intervals too correspond to the smallest ratios.[88]

The phenomenon of sympathetic vibration of strings is mentioned by other ancient authors,[89] but nowhere else is it linked to the discovery of the concord-ratios. Panaetius assumes a connection between concord and sympathetic vibration, as though the latter were a physical manifestation of the former. The only other account of the phenomenon of concordance which invokes sympathetic vibration is that of Adrastus, whose treatment of the connection is more complex than Panaetius'.[90] For the moment, let us simply note that Panaetius evidently believed that the link between number-ratio and concord was no arbitrary human construct, but one which occurred at the level of nature itself.[91] Strings being

[87] ᾗ] ἡ MSS and Düring. It is very difficult to make good sense of the nominative here; I am grateful to Andrew Barker for his suggestion of the dative as a tentative emendation.

[88] 66.16–20; Barker's translation, slightly modified (see previous note).

[89] [Arist.] *Pr.* XIX.24, 42; Adrastus ap. Theo. Sm. 50.22–51.2; Aristid. Quint. 90.1–6. Cf. also Vitruvius V.5 (on tuned sounding vases for the theatre which functioned by sympathetic vibration).

[90] There are several other points at which Panaetius' argument includes things discussed by Adrastus, but none of them provide a clear indication of which account is prior to which. Adrastus' *Timaeus* commentary appeared early enough in the second century to be quoted by Theon, who was writing during the reign of Hadrian (AD 117–38).

[91] In this sense ratio as a quantitative measure of musical intervals is not like a measure of distance (e.g. *stadia*), but like the measure of the diameter of a circle by its circumference (π). The measures of human invention differ from the natural measures primarily by being constructs: degrees Celsius, for instance, are connected to the unchanging imperatives of nature (they are hundredths of the interval between the boiling and freezing points of pure water at standard temperature and pressure), but they are still a human construct and are not directly discoverable in nature as are the ratios of concord.

moved without being touched is a phenomenon which sparks human curiosity; investigation follows, and the monochord is employed to make the underlying principles plain. Again, we find the *kanōn* in a clarifying role; perception can indicate the existence of concords, but cannot measure them without the instrument's help.[92] Instruments, in general, like the balance for weights, are a necessary intermediary between brute sense-perception and rational, quantifying, thought. Without such instruments there can be no accurate assessment of quantities, and therefore no scientific investigation; without them we are stuck, as it were, in the stable.

The sore thumb: canonic division and the octave plus fourth

Adrastus' use of sympathetic vibration is more detailed than Panaetius', and carries greater implications for the debate about reason, perception and the monochord.[93] He defines 'concord' in this way:

[92] The untrustworthiness of perception on its own is so much a factor in the explanations of [Arist.] *Pr.* XIX.42 that the author seems to indicate that the phenomenon of sympathetic vibration may in fact be illusory.

[93] A synopsis of the excerpts on harmonics from Adrastus quoted by Theon may be helpful in advance (page and line references are to Hiller's edition):

49.6	The elementary constituents of musical sound
50.2	Pitch relations are determined by ratios
50.4	Pythagorean account of the connection between physical acoustics and ratio-types
50.22	Concord defined by the criterion of sympathetic vibration
51.4	The primary concords and the reasons for their names
52.1	The compound concords (including octave plus fourth)
52.9	The range (*topos*) of the voice and the genera of melody
53.1	The tone as a measure of this range
53.8	The semitone and why it is not really half of a tone
53.17	The tetrachord *systēma* and its three genera: diatonic (54.6), chromatic (54.16), enharmonic (55.8)
56.9	Pythagoras the first to identify the concord ratios; a list of these ratios (including 8:3) and those of the tone and 256:243 'semitone'
57.1	The various instruments on which Pythagoras investigated the ratios
57.11	Demonstration of the ratios by canonic division (excluding 8:3)
58.13	All the concords contained by the *tetraktys*
59.4	Investigations of the ratios with instruments by Lasus, Hippasus and others (lacunae at 59.10, 60.10, 60.12; other textual problems in this section)
60.13	Account of pitches by appeal to the acoustics of wind instruments (authorship uncertain)
61.11	Eudoxus and Archytas cited to support foregoing harmonic/acoustic arguments

συμφωνοῦσι δὲ φθόγγοι πρὸς ἀλλήλους, ὧν θατέρου κρουσθέντος ἐπί τινος ὀργάνου τῶν ἐντατῶν καὶ ὁ λοιπὸς κατά τινα οἰκειότητα καὶ συμπάθειαν συνηχεῖ· κατὰ ταὐτὸ δὲ ἀμφοῖν ἅμα κρουσθέντων ἡδεῖα καὶ προσηνὴς ἐκ τῆς κράσεως ἐξακούεται φωνή.

Notes are in concord with one another if, when one or the other is struck on a stringed instrument the other one also sounds with it, through some sort of kinship and sympathy: under the same conditions, if both are struck together, a sweet and agreeable sound arising from the mixture (*krasis*) is heard.[94]

The definition is revealing. Adrastus appeals to physical phenomena as empirical evidence: sympathetic vibration is the first criterion of concordance between notes. The test can be set up, and one can follow it to separate concordant from discordant intervals. If two notes are concordant, the second string will vibrate in sympathy with the first, and struck together they produce a 'sweet and agreeable sound' – two criteria, in fact, and both based in perception (visual and aural). It follows that if the criteria are not met, the interval is discordant, but Adrastus does not commit himself this far. We note also that his definition of concord bypasses the rational potential of the *kanōn* altogether: not only are ratios not in view, but a single-stringed *kanōn* will be insufficient to perform the test.

Adrastus then proceeds to list the concords, based on these criteria: fourth, fifth, octave (51.4–20); and after them:

προσανηύρηνται δὲ ταύταις ἕτεραι πλείους. τῇ γὰρ διὰ πασῶν πάσης ἄλλης προστιθεμένης, καὶ ἐλάττονος καὶ μείζονος καὶ ἴσης, ἐξ ἀμφοῖν ἑτέρα γίνεται συμφωνία, οἷον ἥ τε διὰ πασῶν καὶ διὰ τεσσάρων, καὶ διὰ πασῶν καὶ διὰ πέντε, καὶ δὶς διὰ πασῶν, ἔτι δὲ πάλιν τῇ διὰ πασῶν εἰ προστεθείη τούτων τις, οἷον ἡ δὶς διὰ πασῶν καὶ διὰ τεσσάρων, καὶ ἐπὶ τῶν ἄλλων ὁμοίως μέχρι τοῦ δύνασθαι φθέγγεσθαι ἢ κρίνειν ἀκούοντας.

61.18 Return to Adrastus' account of reason, perception and the role of instruments
62.4 The combinations and divisions of the concords (including 8:3)
63.25 The size (*megethos*) of the *systēma* according to Plato and Aristoxenus
65.10 The greater numbers go with the lower notes; instrumental examples of this
66.12 The fourth is the most important concord
66.19 The intervals that make up the fourth
67.13 The ratio of the *leimma*
70.7 The ratio of the tone, and its indivisibility into equal parts

[94] Adrastus ap. Theo. Sm. 50.22–51.4, trans. Barker 1989: 214.

Several other concords were discovered in addition to these. For when any other concord, smaller, greater or equal, is added to the octave, another concord arises from the two, as for instance the octave and a fourth, the octave and a fifth, and the double octave; and the same would hold if any of these were again added to the octave, as for instance the double octave and a fourth, and similarly for the others, to the limit of our capacity to make sounds or to judge them by ear.[95]

Now it is true that strings tuned a fourth, a fifth or an octave apart will resonate in sympathy with one another. The principle, which was not known in Adrastus' time, is that the fundamental or an excited partial of the sounded string excites an identically pitched partial (or the fundamental) of the untouched string, causing it to resonate. Thus if two strings are tuned an octave apart and the higher one is struck, the lower (whose first partial is of the same frequency as the fundamental of the higher string) will vibrate in sympathy. The same will occur if the procedure is reversed, and likewise for the fifth and fourth. The principle holds equally for the higher, compound concords: sympathetic vibrations at the octave plus fifth (3:1) and double octave (4:1) are particularly easily observed.

But this is where two difficulties enter the picture. First, the further removed from the fundamental the common frequency is, the fainter the sympathetic vibration. Second, the series of intervals which are *theoretically* included by the phenomenon is not limited to the three smaller concords and their compounds with the octave; it includes not only all intervals of multiple and epimoric ratio (not all of which are concords), but in fact all intervals which can be expressed in ratios of numbers – the series is infinite. *Practically*, of course, not all intervals will generate observable sympathetic vibrations: two strings tuned to the 256:243 *leimma*, for example, share their first common frequency at their 255th and 242nd partials – too far removed from their fundamentals for the phenomenon to be observable. But the main determining factor is proximity to the fundamental, not concordance, as Adrastus thought, and this turns out to be critical to his account.[96] For, as it

[95] 52.1–9, trans. Barker 1989: 215.

[96] Another factor is how strongly the partials of the sounded string are excited. Plucking (or striking with a plectrum) does not excite as many partials as strongly as does bowing or hammering, but neither of the latter are attested in Greek musical texts.

happens, sympathetic vibration is more easily observed between two strings tuned a double octave plus major third (5:1) apart than between two strings tuned an octave plus fourth (8:3) apart. The latter is a concord by Adrastus' definition (since it is a compound of a simple concord and an octave), and yet by his criterion of sympathetic vibration the former, a discord, would be more readily admitted as a concord.[97]

What, then, can we make of the place of strings – indeed, of empirical and demonstrative procedures in general – in Adrastus' (and Panaetius') harmonics? In terms of procedure, we should probably conclude that although some informal experimentation undoubtedly occurred at some point (enough to connect the most readily observed sympathetic vibrations with the first three multiple intervals, which are also concords), it was neither sufficiently careful nor sufficiently extensive to discover that the criterion was, like the category of multiple and epimoric ratios, too broad to exclude discords, and that (for practical instrumental reasons) it can also appear to exclude some concords. In terms of scientific method, we must conclude that whether or not he understood the implications of his approach, Adrastus was willing to include sensitivity of observation in his formal definition of concord. This of course is one of the prime dangers of admitting the evidence of sense-perception into a theoretical science – in the wrong method, the limit of perception can become the limit of thought. A definition of concord based on this criterion could lead to endless disputes about whether or not a sympathetic resonance was just barely perceptible, or perceptible on one instrument and not on another, and leaves itself open to the familiar criticism of Glaucon's unit-hunters, regardless of the rationality of

[97] See Taylor 1965, Fletcher and Rossing 1998. To be fair to Adrastus, it is possible that he means that the first three (simple) concords were discovered by observation of sympathetic vibration, and that the discovery of the compound concords was then merely a result of recognising the fourth or fifth in combination with one or more octaves; this would explain his emphasis on the judgement of the ear (52.8–9). But (a) he does not make such a distinction explicitly, and (b) if the discovery were really the result of observing sympathetic vibrations, the octave plus fifth (3:1), a compound concord, would be more readily discovered than the fourth (4:3), a simple concord.

the subsequent analysis: it amounts to putting the ears ahead of the mind.[98]

This is of some significance in our assessment of Adrastus' use of the monochord and canonic division. In the present argument, he is attempting to connect perception to reason, much as Panaetius did, by appeal to the natural acoustical properties of strings – he has, in the paragraph immediately before his definition of concord, given the 'Pythagorean' account of pitched sound, in words which have a distinctly Euclidean ring.[99] Here it is clear that notes, as melodic sounds, all exist through 'rational' relations (that is, the relative speeds of their movements can be expressed as whole-number ratios), and that 'irrational relations' do not correspond to intervals of notes, but to mere noises. Of the rational relations, he says, 'some are merely in attunement, but those in the primary and best known and most important ratios, both multiples and epimorics, are also concordant'.[100] The building-blocks of music are therefore constituted by the ratios of their impacts, and the special ratios of certain intervals are physically displayed to perception by the behaviour of strings.

After a careful and methodical discussion of scale-systems which begins with the tone and semitone,[101] and continues through the tetrachord (of which all three genera are named and detailed), Adrastus comes to the ratios themselves, which up to this point have not been specified. 'It seems that Pythagoras was the first to have identified the concordant notes in their ratios to one another,'

[98] Pl. *Resp.* 531a–b, applied to canonics by Proclus (*In Euc.* 41.2, quoted in part I of this chapter). I am grateful to Professor Colin Gough for discussing the physics of sympathetic vibration with me. It is his opinion that in the case of the 8:3 ratio, any sympathetic resonance would be very difficult to detect on an instrument constructed and played as Greek lyres were; it must occur, of course (the seventh partial of one string will excite the second of the other), and would be detectable under modern laboratory conditions if an experiment were set up with this objective.

[99] 50.4–21, with which cf. *Sect. can.* 148.3–149.24.

[100] οἱ μὲν ἄλλοι μόνον ἡρμοσμένοι, οἱ δὲ κατὰ τοὺς πρώτους καὶ γνωριμωτάτους καὶ κυριωτάτους λόγους πολλαπλασίους τε καὶ ἐπιμορίους ἤδη καὶ σύμφωνοι, 50.19–21, trans. Barker 1989: 214.

[101] Declared an invalid term here, as in Panaetius, but with a somewhat more careful version of the 'semivowel' comment, followed by a mathematically ridiculous explanation (53.8–16; cf. Panaetius ap. Porph. *In Harm.* 67.7, Philo *Opif.* 126, Macrob. *In Somn.* II.1.21).

he says,[102] and proceeds to match each concord with its ratio in a list which begins at the epitritic (4:3) fourth and ends at the 4:1 double octave, and includes the 8:3 octave plus fourth (56.11–57.1). In addition he gives the ratios of the epogdoic tone and the 256:243 interval, 'what is now called the semitone but was then [i.e. in Pythagoras' time] called the *diesis*'.[103]

Here again (as in his earlier list of concords) Adrastus seems to be exerting himself to include the disputed octave plus fourth interval in the list of concords.[104] He does not always agree with Aristoxenus (compare, for example, his direct refutation of Aristoxenus on the semitone, 53.8–11), but on this issue he does. His awareness of the fact that this interval was not universally accepted as a concord is indicated by a curious and prominent feature of his account of the ratios. His definition of the octave plus fourth is twice as long as that of any other interval in the list: its notes are 'those at an octave and a fourth in a ratio of 8 to 3, *which is multiple-epimeric, since it is two and two-thirds times as great*'.[105] The purpose of this unique expansion (the two clauses I have italicised) is to defend his inclusion of the interval with a supplementary definition based on mathematically determined ratio categories. What the extra clauses permit him to do is to slip this interval, with its mathematical defence, into an account of what the great sage Pythagoras 'seems to have been the first to discover' (πρῶτος ἀνευρηκέναι δοκεῖ, 56.10). Moreover, he goes on, Pythagoras discovered the intervals in this list by investigations with lengths, thicknesses and tensions of strings (some involving the attachment of weights), with widths of bores and breath-experiments in wind instruments, and with discs and vessels (57.1–7). 'For whichever of these is taken according to one of the ratios mentioned, other factors being equal, it will produce the concord that corresponds

[102] 56.9–10, trans. Barker 1989: 217.

[103] 56.18–19, trans. Barker 1989: 217. *Diesis* is the term used by Philolaus (fr. 6a), but most later writers call it *leimma*, as does Adrastus himself, including once when citing Plato (67.6, 13; 68.11; 70.3, 4; he also uses the adjective λειμματιαῖος at 69.3).

[104] He marks a clear separation between the ratios of the concords and those 'of the other attuned notes' (καὶ τῶν ἄλλων ἡρμοσμένων, 56.17).

[105] τοὺς μὲν διὰ πασῶν καὶ διὰ τεσσάρων ἐν λόγῳ τῶν η′ πρὸς γ′ ὅς ἐστι πολλαπλασιεπιμερής, διπλάσιος γὰρ καὶ δισεπίτριτός ἐστι, 56.12–15, trans. Barker 1989: 217, modified according to LSJ rev. suppl. (1996) s.v. δισεπίτριτος. On the category of 'multiple-epimeric' ratios, see n. 120 below.

to the ratio.'[106] If this was good enough for Pythagoras, Adrastus seems to imply, it should be good enough for us. The 8:3 ratio is multiple-epimeric, and when it is set out on an instrument, we hear the octave plus fourth.

The list of acoustic experiments which Adrastus attributes to Pythagoras is similar to those recorded in other late sources; I have discussed these in chapter 2 and have no more to say about them here. But we ought at least to register Adrastus' inclusion of several procedures which will not generate the ratios he says they will; only those involving string lengths and discs will support his argument. This observation is an important adjunct to what we have already learned about the place of strings in his definition of concord (50.22–51.4): clearly he did not try out for himself all of the procedures he reports. Perhaps he tested none; but he bases his argument on them nonetheless.

At this point, having – he thinks – established the ratios of the concords through several stages of argument and with the authority of acoustics, sympathetic vibration, mathematics and finally the reputed investigations of Pythagoras himself, Adrastus now brings in the monochord:

ἀρκείτω δ' ἡμῖν ἐν τῷ παρόντι διὰ τοῦ μήκους τῶν χορδῶν δηλῶσαι ἐπὶ τοῦ λεγομένου κανόνος. τῆς γὰρ ἐν τούτῳ μιᾶς χορδῆς καταμετρηθείσης εἰς τέσσαρα ἴσα ὁ ἀπὸ τῆς ὅλης φθόγγος τῷ μὲν ἀπὸ τῶν τριῶν μερῶν ἐν λόγῳ γενόμενος ἐπιτρίτῳ συμφωνήσει διὰ τεσσάρων, τῷ δὲ ἀπὸ τῶν δύο, τουτέστι τῷ ἀπὸ τῆς ἡμισείας, ἐν λόγῳ γενόμενος διπλασίῳ συμφωνήσει διὰ πασῶν, τῷ δὲ ἀπὸ τοῦ τετάρτου μέρους γενόμενος ἐν λόγῳ τετραπλασίῳ συμφωνήσει δὶς διὰ πασῶν. ὁ δὲ ἀπὸ τῶν τριῶν μερῶν φθόγγος πρὸς τὸν ἀπὸ τῶν δύο γενόμενος ἐν ἡμιολίῳ συμφωνήσει διὰ πέντε, πρὸς δὲ τὸν ἀπὸ τοῦ τετάρτου μέρους γενόμενος ἐν λόγῳ τριπλασίῳ συμφωνήσει διὰ πασῶν καὶ διὰ πέντε. ἐὰν δὲ εἰς ἐννέα διαμετρηθῇ ἡ χορδή, ὁ ἀπὸ τῆς ὅλης φθόγγος πρὸς τὸν ἀπὸ τῶν ὀκτὼ μερῶν ἐν λόγῳ ἐπογδόῳ τὸ τονιαῖον περιέξει διάστημα.

For the present let us be content to give a demonstration through the length of a string on what is called the *kanōn*. When the single string on the instrument is measured off into four equal parts, the note from the whole string, in relation to that from the three parts, will sound in concord at the fourth, being in epitritic ratio; in relation to that from the two parts, that is, from the half string, it will

[106] ὅ τι γὰρ ἂν ληφθῇ τούτων κατά τινα τῶν εἰρημένων λόγων, τῶν ἄλλων ⟨ἴσων⟩ ὄντων, τὴν κατὰ τὸν λόγον ἀπεργάσεται συμφωνίαν, 57.7–10, trans. Barker 1989: 217–18. Cf. also the list of apparatus (κατασκευαί) at 66.20–3.

sound in concord at the octave, being in duple ratio; in relation to that from the fourth part it will sound in concord at the double octave, being in quadruple ratio. The note from the three parts in relation to that from the two parts will sound in concord at the fifth, being in hemiolic ratio; and in relation to the note from the fourth part it will sound in concord at the octave and a fifth, being in triple ratio. If the string is divided by measure into nine parts, the note from the whole string, in relation to that from the eight parts, will bound the interval of a tone, in epogdoic ratio.[107]

Adrastus settles on the *kanōn* as the best way of illustrating the connection between numbers and concords, but it will be noted immediately that his canonic division omits the disputed octave plus fourth interval (8:3) he was at pains to include in his two previous concord lists. There are, however, procedural reasons for the omission which make it seem less arbitrary, and hence less pointed, than we might first suppose. The six intervals are given in the order fourth (4:3), octave (2:1), double octave (4:1), fifth (3:2), octave plus fifth (3:1), tone (9:8). This is not an immediately intuitive arrangement from the point of view of the intervals themselves: there is no attempt to keep epimorics or multiples together, nor is there any concern for listing the intervals in order of magnitude, as there was in both previous lists (51.4 and 56.9).

But what appears erratic on paper appears much less so if one reads the passage as a direct account of the process of canonic division, with the instrument in mind. If we do this, a pattern emerges. It is clear from the text that there are two stages to the division, each involving a separate marking out of the *kanōn*, after which the intervals from that marking are listed. The first stage marks off quarters (57.12–14), and the intervals are then extracted from this first measurement in what is most straightforwardly a linear sequence (it could be done otherwise, of course, but the simplest way to follow the directions is, in fact, linear).

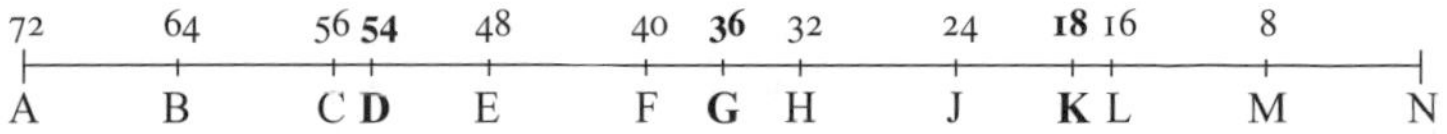

72 64 56 **54** 48 40 **36** 32 24 **18** 16 8
A B C **D** E F **G** H J **K** L M N

Fig. 5.1 Adrastus' canonic division (ap. Theo. Sm. 57.11–58.12) applied to a 72-unit *kanōn*. Figures in boldface indicate bridge positions for the first stage of the procedure, in which the string is divided into quarters.

107 57.11–58.12, trans. Barker 1989: 218.

The second stage of the division marks off ninths (58.9–12), and generates the epogdoic tone from comparison between eight and nine of these ninths. If we summarise Adrastus' division schematically (see fig. 5.1), we discover that the apparent irregularity of his procedure turns out to conceal a sequential arrangement after all. It is not a sequence by magnitude or by ratio form, but by spatial relationships on the instrument itself:

I	**i**	1.	NA:ND = 4:3
		2.	NA:NG = 2:1
		3.	NA:NK = 4:1
	ii	4.	ND:NG = 3:2
		5.	ND:NK = 3:1
II		6.	NA:NB = 9:8

In my scheme, the division is carried out from left to right (one could as easily perform it from right to left, of course, but the linear sequence on the instrument would remain): in stage **I** the movable bridge positions are D, G, K (intervals 1–3); the reference point then shifts from A to D (stage **I.ii**), and bridge positions G and K are used again (intervals 4–5). Finally in stage **II** the single bridge position B (for comparison with the fixed bridge A) generates the tone (interval 6). Stage **II** could, of course, be carried out at either end of the string (NA:NB = 9:8 or AN:AM = 9:8), but its position at the end of the division is explained by the necessity of marking off the *kanōn* in a greater number of units than was required to generate the other five intervals. One could easily complicate the procedure by arbitrarily changing directions over the course of the division, but if one proceeds in the simplest fashion, Adrastus' intervals come out in a spatially determined, linear order, either left to right (as in my scheme) or right to left.

Now let us imagine how Adrastus would have introduced his favoured 8:3 interval into this procedure, had this been his intention. He could not have introduced it at any point in stage **I**, because the string would not yet have been divided into more than four parts; a minimum of eight are required. He would, then, have had to bring it out in stage **II**. The marking of the instrument in ninths at this point results from the need to generate the epogdoic tone, and therefore it would have been disruptive to insert the 8:3 interval between intervals 5 and 6. Thus the only place for the construction of the octave plus fourth in Adrastus' division is at the end, as a seventh note (NB:NJ = 8:3 in my scheme).[108]

This presents an awkward situation: the intervals generated in stage **I** are all concords; stage **II** represents, in effect, the procedural separation between concord and discord in Adrastus' division. He is, at this point, caught in a dilemma: if he places the 8:3 interval at the end, separated from the concords and after the tone, it will appear to have been relegated to the world of discords; but we know from Adrastus' determined inclusion of the interval in his two previous concord lists that this is not how he thought of it.[109] Nevertheless, there is no way, short of overhauling the division entirely, to include the octave plus fourth in stage **I** with its fellow concords. Such an overhaul would involve marking off the *kanōn* into eighths (at least) at stage **I**; but in this case the octave plus fourth would still stand out as the only interval in stage **I** whose ratio comes out in lowest terms – all the other concords would have their terms doubled.[110] (He would then have to begin: 'When the single string on the instrument is measured off into *eight* equal parts, the note from the whole string, in relation to that from the *six* parts, will sound in concord at the fourth, being in epitritic ratio', and so on.) There is, in fact, no way of including the interval in a canonic division without drawing attention to the fact that it does not fall into the same mathematical category as the other concords.[111]

108 NE:NK is also 8:3; constructing the interval this way has the advantage of using a bridge position determined in stage **I**.

109 It is particularly curious that the diagram accompanying Adrastus' division in Hiller's edition (printed between 58.2–3) differs from the text by including the octave plus fourth and excluding the tone.

110 Unless, of course, he used only half of the string to construct the other concords. But in this case the octave plus fourth would stand out as the only concord whose construction required the entire string. Furthermore, the difficulty of obtaining a good sound from one eighth of any monochord string (as required for the multiples in such an arrangement) makes the procedure less desirable from a practical point of view. This was a problem noted by Ptolemy (*Harm.* 83.9–11); he solved it, in a different context, by introducing the fifteen-stringed *kanōn*, on which all the notes of his two-octave system could be found without compromising the sonority of the highest pitches.

111 Ptolemy's solution (*Harm.* 18.22–19.15) is probably the best possible one. It is the result of an ingenious pairing of the linear approach of the present division with the use

There are three possible conclusions we might draw from this. The first is that the division did not come to Theon from Adrastus' book.[112] The second is that it was from Adrastus' pen, but that Adrastus was himself copying from some source whose harmonic goals or priorities were inconsistent with his own, and that he made no attempt at integrating them. Thirdly, we might argue that Adrastus is indeed the source of the division, but that either he simply forgot to include the 8:3 interval, or faced with the impossibility of including it in such a way as to conceal its anomalous character, he omitted it: it was a sore thumb which could not be bandaged; it could either stick out or be amputated, and faced with the choice Adrastus decided on the latter.[113]

All three conclusions present difficulties of one sort or another. The third involves Adrastus either being remarkably forgetful or knowingly losing face over the interval. This is possible, of course, but not all of Theon's quotations from Adrastus suggest such a poor scholar, and to admit our third conclusion may well be to do him a disservice. That said, however, it is clear from his account of Pythagoras' investigations immediately before the division that Adrastus was willing to report procedures gleaned from earlier sources without testing them himself. This weighs in favour of the second conclusion, which would only necessitate a view of Adrastus as intellectually lazy: he cribbed the division from somewhere else and did not think through its implications for his own argument.

The first conclusion is the most charitable of the three. It has not met with universal approval; Barker includes the entire disputed passage (57.11–61.17), but does allow that portions of it may not

of the segments of string on both sides of the bridge (cf. Panaetius' division, discussed above), along with the omission of the tone. It also avoids the necessity of sounding any length less than a fifth of the whole string (see ch. 6).

[112] Burkert entertained this possibility: the quotation from Adrastus 'apparently extends at least as far as 57.10 ... and possibly to 59.3, and is then taken up again at 61.17' (1972: 377 n. 36).

[113] The way the division is introduced (57.11–12) probably rules out a possible fourth conclusion, that the source of the entire division is the later interpolation of marginal material. The same cannot be said for the paragraph which follows it (58.13–59.3), but this is a dangerous game to play with Theon's text, where separating Theon's voice from the many others which appear in it is complicated by many factors, not least of which the document's textual history (on which see Hiller's preface).

have been taken from Adrastus' work.[114] His argument in favour of inclusion is based on the fact that although Theon writes, 'So much for the discovery of the concords: let us return to the things handed down by Adrastus' at 61.18–19,[115] which would appear to indicate a break in the quotation, the content of the following sentence is in line with that of the disputed passage:

φησὶ γὰρ ὅτι τούτοις τοῖς εἰς τὴν ἀνεύρεσιν τῶν συμφωνιῶν ὀργάνοις κατὰ μὲν τοὺς λόγους προπαρασκευασθεῖσιν ἡ αἴσθησις ἐπιμαρτυρεῖ, τῇ δὲ αἰσθήσει προσληφθείσῃ ὁ λόγος ἐφαρμόζει. πῶς δὲ καὶ οἱ τὸ λεγόμενον ἡμιτόνιον περιέχοντες φθόγγοι πρὸς ἀλλήλους εἰσὶν ἐν λόγῳ τῷ τῶν σνϛʹ πρὸς σμγʹ, μικρὸν ὕστερον ἔσται φανερόν. δῆλον δὲ ὅτι καὶ αἱ συνθέσεις καὶ αἱ διαιρέσεις τῶν συμφωνιῶν ὁμόλογοι καὶ συνῳδοὶ θεωροῦνται ταῖς τῶν κατὰ ταύτας λόγων συνθέσεσί τε καὶ διαιρέσεσιν ἃς πρόσθεν ἐμηνύσαμεν.

He [Adrastus] says that when these instruments designed for the discovery of the concords have been previously prepared in accordance with the ratios, perception agrees with their testimony, while when perception has been taken first, reason is in attunement with it. How it is that the notes bounding the so called semitone are in the ratio of 256:243 to one another will become clear a little later. But it is plain that the combinations and divisions of the concords are found to be in agreement and in consonance with the combinations and divisions of the ratios that correspond to them, the combinations and divisions that we mentioned previously.[116]

Barker's solution is straightforward: 'Perhaps the sense is only that Theon is returning from what Adrastus reports of others to what he says in his own right' (1989: 220 n. 46). This is certainly possible: there is no reason why Adrastus should not have quoted the intervening material, which would have complemented his general argument well enough. The material is rather poorly placed, however: since it constitutes further examples of the very simplest instrumental demonstrations of only the first three concords, one would expect it to have appeared earlier, between Pythagoras' demonstrations and the canonic division. But it does not; instead, it rather interrupts the flow of Adrastus' argument. We have gone from a canonic division constructing six intervals to a set of instrumentally inferior procedures displaying only three

[114] Barker 1989: 220 n. 46.

[115] ταυτὶ μὲν περὶ τῆς εὑρέσεως τῶν συμφωνιῶν· ἐπανέλθωμεν δὲ ἐπὶ τὰ ὑπὸ τοῦ Ἀδράστου παραδεδομένα, trans. Barker 1989: 220.

[116] 61.20–62.4, trans. Barker 1989: 220.

intervals at most. Then Adrastus speaks of 'instruments designed for the discovery of the concords'. Not all of the instruments in the disputed passage could properly be described in these terms: the *syrinx* (60.7, 20) and auloi (61.2) could be 'previously prepared in accordance with the ratios', perhaps, but could hardly be said to have been designed for the purpose of either discovering or demonstrating them. Furthermore, the passage contains several lacunae, and in some other places the text as we have it is certainly not sound.[117]

Let us imagine for a moment that Adrastus' original text, or something like it, ran from 58.12 (the end of the canonic division) directly into 61.20 without any of the intervening material.[118] If this were the case we would interpret 'these instruments designed for the discovery of the concords' as those mentioned briefly in connection with Pythagoras, and especially the *kanōn*. The *kanōn* would then be very much an instrument designed for the discovery of the concords – designed, that is, to elucidate the numbers behind the intervals identified through the observation of sympathetic vibrations (50.22–51.4). Furthermore, the *kanōn* has indeed been 'previously prepared in accordance with the ratios' in Adrastus' division: it has been marked off in quarters before any intervals are constructed. But here we encounter a difficulty. Adrastus' argument in the passage just quoted – which Barker underlines as the very pith of Adrastus' harmonic project (1989: 210) – is that reason (*logos*) and perception (*aisthēsis*) are unified when allowed to present their evidence to one another on such instruments. If reason sets out the division, perception agrees that it is correct; if perception sets it out, reason accords with it (61.23). Canonic division, then, is the process by which harmonic science can be demonstrated as unified.

The difficulty arises when we recollect the content of the division itself. On the evidence of perception (sympathetic vibration in

117 Hiller marks lacunae at 59.10, 60.10 and 60.12; see his apparatus for a full account of other problems and proposed emendations.

118 See the synopsis in n. 93 above. This would exclude the two sentences about the concords being contained by the *tetraktys* (58.13–59.3), which while consistent with the content (though not with the procedure) of the preceding division, seem more appropriately linked to the reports that follow.

strings), Adrastus has included the octave plus fourth interval in two previous concord lists (at 52.4 and 56.13); then he offers a division of the *kanōn* which includes every other concord in those lists except this one (57.11–58.12); then he says that it is in instruments such as the *kanōn* that reason and perception can meet on common ground. He uses canonic division to establish the unity of his science in the link between reason and perception, but omits from his division the one interval which is difficult to accommodate in a consistent rational definition of concord.

The difficulty is real, but it may not pose as serious a challenge to Adrastus' project as it seems to do. Canonic division only becomes a problem if reason is taken to dictate either that all concordant intervals are either multiple or epimoric, or that they can all be illustrated by the same simple division of the monochord. Adrastus maintains neither of these assumptions consistently or explicitly. His statement on the first is that while some notes are merely in attunement, 'those in the primary and best known and most important ratios, both multiples and epimorics, are also concordant'.[119] Notably absent is a statement to the effect that all concords are in multiple or epimoric ratios, or that intervals which are not in such ratios are discordant.[120] On the second Adrastus makes no comment at all. Thus while the apparent difficulty raised by his division is not wholly illusory, it is perhaps a kind of problem he would not have considered important. His own assumptions about harmonics gave him no reason not to include the octave plus fourth in his division, and had he done so, it would have demonstrated simply that reason and perception agree on its 8:3 ratio, no matter how its construction was carried out.

If Adrastus had no strong reason to omit the octave plus fourth from his canonic division, then it remains a mystery why he did. To explain the anomaly we might reasonably suspect either that (1) the division was Adrastus', but that he meant it to confirm

[119] 50.19–21, quoted above (n. 100).

[120] Adrastus' interest in the categories of ratio is apparent from other quotations in Theon's treatise: see Adrastus' list of ratio-types at 76.1–7, which includes the category of multiple-epimerics (οἱ πολλαπλασιεπιμερεῖς). When this category is later defined, 8:3 is the first example given (79.15–19). Similarly, when the 8:3 ratio was first applied to the octave plus fourth, it was defined as multiple-epimeric (see n. 105 above).

only that certain ratios correspond to certain intervals; or that the division was not Adrastus', either because (2) he quoted it from some other author, or because (3) Theon interpolated it. There is little to decide between these possibilities, and I confess that I cannot say with any confidence which of them is most likely. But I shall close this discussion by citing one small piece of evidence which might support suggestions (2) or (3).

In the final sentence of the passage quoted above (62.1–4), Adrastus makes clear that he is not talking merely about the three fundamental concords (the only ones generated in the disputed intervening reports of 59.4–61.17), but about their 'combinations and divisions' (αἱ συνθέσεις καὶ αἱ διαιρέσεις) as well. Through the agency of the instruments, he says, these will be seen to correspond to the combinations and divisions of the ratios: that is, once reason and perception have agreed on the basic correspondence between certain intervals and their rational counterparts, parallel manipulations can be performed on each with parallel results. To which combinations and divisions is he referring? To those 'which we mentioned previously' (ἃς πρόσθεν ἐμηνύσαμεν, 62.4). If we assume that he is referring back to the intervals of the canonic division, these would be the double octave (4:1) and octave plus fifth (3:1), the only two compounds constructed there; the octave plus fourth, as we noted, is conspicuously absent. And yet in his expansion and clarification of 'the combinations and divisions we mentioned previously', the octave plus fourth reappears, with its ratio, just before the octave plus fifth (63.2–5). Evidently, then, he cannot have been referring to the canonic division, but to one (or both) of the preceding concord discussions, where the interval was included (52.4 and 56.13).

I have discussed this anomaly at some length not only because it bears on the question of Adrastus' authorship of one of the four canonic divisions considered in this chapter, but also because it highlights the very different methodological issues which can be at stake when canonic division is employed in a harmonic argument. The fact that one such methodological conflict seems not to have perturbed Adrastus does not mean that he was unaware of the theoretical problems introduced along with the monochord, however. On the contrary, as we shall see, he seems to have recognised (imperfectly, it must be admitted) one of the most serious of these.

The problems of canonic division

When Adrastus comes to discuss the impossibility of dividing the epogdoic tone equally (70.19–72.20), he refers to canonic division again. Here too, secure attribution of the passage to Adrastus is frustrated by doubts about the text in the immediately preceding pages. But the moderately greater intellectual sophistication of the passage in question compared with the confused thinking of what comes before it weighs in favour of Adrastus.[121]

The tone, he says, is conceived of in four different ways. There is the 'ideal' tone (ὁ μέν τις νοήσει λαμβάνεται, 71.1), there is the one found in numbers, the one in intervals – which he soon specifies as *visible* intervals (ὁρατοῖς διαστήμασιν, 71.4) – and the one which is audible (ὁ δὲ δι᾽ ἀκοῆς ἐν φωναῖς, 71.2). Throughout the passage he concerns himself chiefly with the last two, the perceptible tones. (He returns to the ideal tone briefly at 72.19, to declare it equally divisible (the only one of the four); the impossibility of dividing the numerically apprehended tone into equal parts has already received his attention at 53.13–16 and perhaps at 69.12–70.19, but both passages offer seriously flawed explanations.) The monochord, significantly, is brought into the discussion to argue against the equal division of the visible, not the audible, tone.[122]

ἐπὶ γὰρ τοῦ κανόνος αἰσθητὸς ὢν ὁ ὑποβολεὺς πάντως ἕξει τι πλάτος καὶ οὐκ ἔσται οὕτως ἀπλατής, ὡς μὴ πάντως τι ἐπιλαβεῖν ἐν τῇ διαιρέσει τοῦ τόνου καὶ τοῦ πέρατος τοῦ πρώτου μέρους καὶ τῆς πρώτης ἀρχῆς τοῦ δευτέρου, καὶ διὰ

[121] See Barker (1989: 223 nn. 59–61, 224 n. 64), who excludes 69.12–70.19 on the grounds that 'Adrastus was a competent mathematician, as many passages in Theon testify' (223 n. 60), whereas the author of the paragraphs in question shows no such competence.

[122] Canonic division plays a role in assessing visible quantities in the accounts of other authors as well. Compare, for example, Nicom. *Ar.* II.27.1, where the division of the monochord and of an aulos is used as a way of visualising the procedure of finding arithmetic, geometric and harmonic means. In the case of the *kanōn*, the bridge indicates the mean. The subsequent elaboration of the three means (with examples) has no further recourse to music or to the sound of the ratios thereby generated, and the reference to the monochord (μουσικὸς κανών) seems therefore to be entirely for the purpose of visualising the calculations. For a more integrated view of sight and hearing cf. Panaetius ap. Porph. *In Harm.* 66.7–12, Ptol. *Harm.* 93.11–94.20.

τοῦτο ἀπαναλωθήσεταί τι τοῦ τόνου. ἔτι ἐν ταῖς διαιρέσεσι τρία ἐστί, δύο μὲν τὰ διαιρούμενα, τρίτον δὲ τὸ ἐξαιρούμενον. τῶν δὲ διαιρουμένων ἀπ' αὐτῆς τῆς διαιρέσεως ὡς ἐπὶ πρίονος ἐν τῇ τομῇ ἀναλοῦταί τι τὸ ἐξαιρούμενον ὑπ' αὐτῆς τῆς τομῆς. ὡς οὖν ἐπ' ἐνίων αἰσθητῶν ἐξαιρεῖταί τι, οὕτω καὶ ἐπὶ πάντων κἂν ἐκφεύγῃ τὴν αἴσθησιν πάντως ἀναλωθήσεταί τι ἐν τῇ τομῇ. δόρυ γοῦν ἢ κάλαμον ἢ ἄλλο ὁτιοῦν αἰσθητὸν μῆκος ἂν πρὶν ἢ διελεῖν μετρήσῃς, ἔπειτα διέλῃς εἰς πολλὰ μέρη, εὑρήσεις τὸ τῶν διαιρουμένων πάντων κοινὸν μέτρον ἔλαττον ὂν τοῦ ὅλου πρὶν ἢ διῃρῆσθαι. ἔτι χορδὴν ἂν διέλῃς, εἶτα διακόψῃς, ἡ ἔκτασις μετὰ τὴν διακοπὴν ἀνέδραμε, κἂν πάλιν τὰ διακοπέντα τείνῃς, ἀνάγκη ἀφῃρῆσθαί τι τοῦ μεγέθους εἰς τὰς ἐξάψεις τῶν ἑκατέρωθεν ἁφῶν τοῦ τεινομένου. καὶ διὰ τοῦτο οὐκ ἔσται τέλεια δύο ἡμιτόνια.

For on the *kanōn* the bridge, being perceptible, will always have some width, and in the division it will not, by being without width, avoid occupying some part of the tone – both a part of the last boundary of the first section and a part of the initial origin of the second – and hence some part of the tone will have been used up. Again, in the divisions there are three things, the two things divided off and thirdly the thing taken out; and some part of the things divided off is expended in the division itself, just as when one uses a saw something is expended in the cutting, that which is taken out by the cutting itself. Thus just as in various perceptible cases something is taken out, so it is in all cases that even if it escapes perception still something is always expended in cutting. For instance, if you measure a stick or a reed or any other perceptible length before dividing it, and then divide it into many parts, you will find that the joint measurement of all the parts divided off is smaller than that of the whole before it was divided. Again, if you divide a string and then cut it up, after the cutting the extension has contracted again; and if you again stretch the pieces that were cut up, it is necessary to take away something of the magnitude in order to fasten the stretched piece at its points of contact at each end. Hence there will not be two complete half-tones.[123]

As Barker has pointed out, 'this last point is a little naive' (1989: 225 n. 67). Not only that, but the entire problem has been formulated in such a way as to represent the visible tone as a single length of string, and its indivisibility into equal parts merely a result of the fact that the bridge itself has width and so takes away from the total length in an otherwise straightforward halving procedure. This, of course, is ridiculous: a distance on a string is a note, whereas a tone is a relationship between two distances. But the general point is an important one, and would need to be addressed even if Adrastus' example were reformulated to treat the visible tone as a ratio of lengths. In that case it would run

[123] 71.4–72.3, trans. Barker 1989: 224–5.

something like this (see fig. 5.1):[124] the interval between lengths NA and NB is a tone (NA:NB = 9:8). A point P on the *kanōn* is found such that NA:NP::NP:NB. When point P is found it will also be the case that AP + PB = AB. The ratio of NA:NP and NP:NB will not be a ratio of numbers.[125] A movable bridge, Q, is then placed under the string at P. Measurement will then reveal that because the width of the bridge takes up some of AB, AQ + QB < AB, even though NA:NQ = NQ:NB and Q has been placed directly on top of P. In other words, even if two exactly equal intervals can be found to divide the tone, when constructed they will no longer add up to the tone, because the physical existence of Q intrudes on the procedure in a way that the purely ideal existence of P does not.

The problem with this summary, of course, is that the difficulties raised at the end must then be applied to the beginning, and the whole procedure then collapses. The problem has nothing to do with whether or not an interval can be expressed in ratios of numbers. As far as canonic division is concerned, at least, one is forced to admit that if the P/Q discrepancy disallows equal division of the tone, then a similar difficulty will disallow construction of the tone in the first place, since this involves placing the movable bridge at B in order to claim that NA:NB = 9:8. We can imagine 9:8, but we cannot construct it either visibly or audibly. The same holds for any other interval. Canonic division is therefore impossible: it does not do what it claims to do. The *kanonikos* is caught in the crawl-space between the ideal and the sensible worlds.

That this is an argument about reason and perception is clear even from Adrastus' version of it. He seems to anticipate the counter-argument that the bridge takes up so little of the string's length that its interference is effectively negligible, i.e. that the P/Q discrepancy is imperceptible and can therefore be discounted as theoretically insignificant. He emphasises the point that the discrepancy exists even if it cannot be perceived: 'Thus just as in various perceptible cases something is taken out, so it is in all

[124] Barker gives a similar example (1989: 225 nn. 67, 69).

[125] That is, it will not be a ratio of positive integers. It can be found in geometry (see ch. 1 part 1).

cases that *even if it escapes perception* still something is always expended in cutting' (71.14–16). His point that perceptibility and reality are different, and that reality is what is at stake, is paralleled in Nicomachus' account of volume. Just as there are 'bodies which reveal no weight on a balance – bits of chaff or bran or other such things',[126] but which when added together in sufficient quantity tip the scales, so too there must be sounds too faint to hear, but which are no less real than apparently weightless chaff, even if they are inaudibly quiet. One is tempted to respond that if the point of canonic division is to render the ratios of certain intervals visible and audible – that is, to display things which are ideal and eternal temporarily in ephemeral and perceptible guise[127] – then any imperceptible discrepancy between the two is, in fact, unimportant: the procedure has still performed its required function.

In chapter 1 we saw that the way the *kanōn* was used in Greek mathematical harmonics is similar to the way lettered diagrams were used in Greek mathematics. Now we see that the criticisms which, by Adrastus' argument, could be levelled at the use of the monochord could also be applied to the use of diagrams. If the bridge has width, so too does the line of ink (as opposed to the ideal line) which bisects another line, and whether or not it is immediately evident to unassisted perception, the two line segments when measured will be found to fall short of the original line by an amount equal to the width of the bisecting line. One might apply the same counter-argument here as in the case of the monochord: if a diagram is so well executed that the eye does not perceive any inequality between two line segments which the proof declares to be equal, then the diagram may be deemed effective in communicating the proof. But mathematical diagrams do not regularly meet such high standards – certainly not in the manuscripts, at any rate. Mathematical proofs do not rely logically on the precision of their diagrams.[128] The monochord, by contrast,

[126] *Harm.* 240.9–10, trans. Barker 1989: 250. Nicomachus seems to be confusing pitch and volume here (see Barker's n. 11 ad loc.), but this is irrelevant in the present context.

[127] This definition is one with which Nicomachus might have agreed; see below.

[128] See Netz 1999a: 54–5 for an extreme example: a case where a diagram must *mis*represent its object in order to illustrate a *reductio ad absurdum* argument (Euc. *El.* III.10).

must present its divisions to a consistently higher standard of accuracy, or its results will be rejected by the senses, and by the sense of hearing in particular.

The argumentative roles of monochord and diagram turn out, then, not to be quite the same after all. Even Aristoxenus, who gave no hint of the monochord's existence, was aware of the different roles of perception and reason in mathematics and harmonics:

ἔστι δὴ τὸ μὲν ὅλον ἡμῖν ⟨ἡ⟩ θεωρία περὶ μέλους παντὸς μουσικοῦ τοῦ γιγνομένου ἐν φωνῇ τε καὶ ὀργάνοις. ἀνάγεται δ' ἡ πραγματεία εἰς δύο, εἴς τε τὴν ἀκοὴν καὶ εἰς τὴν διάνοιαν. τῇ μὲν γὰρ ἀκοῇ κρίνομεν τὰ τῶν διαστημάτων μεγέθη, τῇ δὲ διανοίᾳ θεωροῦμεν τὰς τούτων δυνάμεις. δεῖ οὖν ἐθισθῆναι ἕκαστα ἀκριβῶς κρίνειν· οὐ γάρ ἐστιν ὥσπερ ἐπὶ τῶν διαγραμμάτων εἴθισται λέγεσθαι· ἔστω τοῦτο εὐθεῖα γραμμή, – οὕτω καὶ ἐπὶ τῶν διαστημάτων εἰπόντα ἀπηλλάχθαι δεῖ. ὁ μὲν γὰρ γεωμέτρης οὐδὲν χρῆται τῇ τῆς αἰσθήσεως δυνάμει, οὐ γὰρ ἐθίζει τὴν ὄψιν οὔτε τὸ εὐθὺ οὔτε τὸ περιφερὲς οὔτ' ἄλλο οὐδὲν τῶν τοιούτων οὔτε φαύλως οὔτε εὖ κρίνειν, ἀλλὰ μᾶλλον ὁ τέκτων καὶ ὁ τορνευτὴς καὶ ἕτεραί τινες τῶν τεχνῶν περὶ ταῦτα πραγματεύονται· τῷ δὲ μουσικῷ σχεδόν ἐστιν ἀρχῆς ἔχουσα τάξιν ἡ τῆς αἰσθήσεως ἀκρίβεια, οὐ γὰρ ἐνδέχεται φαύλως αἰσθανόμενον εὖ λέγειν περὶ τούτων ὧν μηδένα τρόπον αἰσθάνεται.

Taken as a whole, our science[129] is concerned with all musical melody, both vocal and instrumental. Its pursuit depends ultimately on two things, hearing and reason. Through hearing we assess the magnitudes of intervals, and through reason we apprehend their functions. We must therefore become practised in assessing particulars accurately. While it is usual in dealing with geometrical diagrams to say 'Let this be a straight line', we must not be satisfied with similar remarks in relation to intervals. The geometer makes no use of the faculty of perception: he does not train his eyesight to assess the straight or the circular or anything else of that kind either well or badly: it is rather the carpenter, the wood-turner, and some of the other crafts that concern themselves with this. But for the student of music accuracy of perception stands just about first in order of importance, since if he perceives badly it is impossible for him to give a good account of the things which he does not perceive at all.[130]

Didymus, working from this text in his summary of Aristoxenian harmonics, offers a similar but slightly expanded version:

οὐ γὰρ εἶναι λογικὸν μάθημα μόνον τὴν μουσικήν, ἀλλ' ἅμα αἰσθητὸν καὶ λογικόν, ὅθεν ἀναγκαῖον εἶναι μὴ ἀπολείπεσθαι θατέρου τὸν γνησίως πραγματευόμενον, καὶ προηγούμενον τιθέναι τὸ τῇ αἰσθήσει φαινόμενον, εἴπερ ἐντεῦθέν

[129] The word is *theōria*, on which see n. 31 above.
[130] Aristox. *El. harm.* 33.2–26, trans. Barker 1989: 150–1.

ἐστιν ἀρκτέον τῷ λόγῳ. γεωμέτρῃ μὲν γὰρ ἐνέσται ἐπὶ τοῦ ἄβακος τὸ κυκλοτερὲς ὑποθεμένῳ ὡς εὐθὺ διανύειν τὸ θεώρημα ἀνεμποδίστως διὰ τὸ ἀφροντιστεῖν τοῦ πεῖσαι τὴν ὄψιν περὶ τοῦ εὐθέος λογικὴν ὕλην διεξάγοντι. μουσικῷ δ' οὐκ ἔσται ὑποθεμένῳ τὸ μὴ διὰ τεσσάρων ὡς διὰ τεσσάρων θεωρῆσαί τι δεόντως, ὅτι προσομολογηθῆναι τοῦτο δέον ἐστὶ τῇ αἰσθήσει καὶ τὸν λόγον τὸ ἀκόλουθον τῷ φανέντι ἐπισυνάπτειν, ὅθεν μὴ κατ' ὀρθὸν τούτου συναφθέντος τῇ αἰσθήσει καὶ τὸν λόγον διαμαρτήσεσθαι τἀληθοῦς.

For [Aristoxenus says that] music is not only a rational branch of learning, but is perceptual and rational at the same time, and hence that it is necessary for the genuine student not to neglect either of the two, while putting what is evident to perception in first place, since it is from there that reason must begin. For a geometer can take a curved line on his drawing-board and postulate that it is straight, and can complete his theorem without any hindrance, since he is not concerned to persuade the eye about what is straight, since the subject matter which he investigates is in the domain of reason. But a student of music who postulates that something which is not a fourth is a fourth cannot consider anything correctly, since it [i.e., the supposed fourth] ought to be made to agree with perception too, and reason should attach, in addition, that which follows in accordance with what is perceptually evident; and hence when this [the supposed fourth] is incorrectly viewed by perception, reason too will go astray from the truth.[131]

The difference, then, is that geometry has no need to judge its figures by perception: the diagrams are there to help us form the ideal figure in our mind's eye. 'The geometer makes no use of the faculty of perception' is Aristoxenus' blunt version of the statement. The harmonic theorist, by contrast, is likely to be misled if he does not concern himself with the accuracy of his perceptible constructions.

If the diagram is exempted in this way from the difficulty raised by Adrastus' account of the semitone, so too is the abacus, but for different reasons. The monochord is in some respects the abacus of harmonic science: like the abacus, it presents quantities to sense-perception, and sensory evidence to reason. The abacus, however, deals only in discrete quantities, not in the quantities of position.[132]

[131] Didymus ap. Porph. *In Harm.* 28.9–19, trans. Barker 1989: 244.

[132] Pebble-figures (squares of four, triangles of three, etc.) use the composite positions of shape to assist the eye in discriminating odd from even, and other such things; but regardless of their relative positions on the ground there can be no debate about how many pebbles there are. Position is not fundamental to the operations of the abacus in the same way as it is to those of the monochord. (On the abacus and the monochord, see ch. 1 part 2.)

Nine pebbles and eight pebbles are a relationship between discrete quantities, one which can be felt with the fingers or seen with the eye, but not heard musically. Nine units of string and eight units of string are a relationship of quantities which relies, for its musical significance, on exact physical dimensions, on quantities of distance between three points. The abacus therefore escapes the dilemma which vexes the monochord, in that there cannot be any discrepancy between its manipulable physical objects (pebbles) and the ideal quantities they represent (positive integers). Because of the requirement of hearing, the monochord is not strictly arithmetical, and this is the source of the problem. Canonics is about connecting quantities to sounds in the medium of linear distance – physical, not just ideal, distance. It is therefore *practised* as a kind of geometry as much as it *is* a kind of arithmetic, because its numbers are conceived and represented spatially as well as abstractly. According to geometrical practice, it requires (audible) diagrams to explicate its spatial relationships. But in geometry, the ideal spaces represented in the diagrams are themselves the manipulable objects of the science; in canonics, on the other hand, spaces are merely the visible, tangible intermediary between pure number and pure sound. Eye and hand are placed between reason and the ear. The monochord falls prey to the P/Q dilemma partly because its task of representation is neither precisely that of the geometrical diagram nor that of the arithmetical abacus.

The wider significance of Adrastus' flawed 'cutting' argument probably escaped him, but I have tried to show that it was important nonetheless because it contained the seeds of a very real objection that can be raised against the claim that canonic division may be a scientifically meaningful activity. The objection cannot be written off completely; it is in fact the opposite side of the same coin which established the instrument's position as a criterion in the first place. Because the *kanōn* is physically constructed and physically divided according to rational principles, it both appeals to and 'straightens' our senses.[133] But because of its very physicality, it cannot be constructed or divided exactly according to rational principles, which are never perfectly instantiated in material objects.

[133] Cf. Ptol. *Harm.* 5.11–13.

The objection is an important one, insofar as it highlights the need for rigorous checks and tests on the instrument itself before its evidence can be deemed trustworthy. Such tests are not a solution, but a practical response to a theoretical dilemma which affects every science that relies on instruments, one that can be minimised but never fully resolved. Adrastus' account merely shows the vague beginnings of this sort of thinking, and to see further progress on this front we must wait for the advances of Ptolemy (perhaps only a generation later). The way in which Adrastus employs the notion is simplistic and misguided, but even so his observations constitute an important step in the history of argument by which apparatus and method were integrated in Greek harmonic science.

Canonic(al) divisions

Beyond the issue of the legitimacy of doing canonic division in the first place, questions may be raised about how it should be done. The answers to such questions depend largely on what the harmonicist's objectives are. In the examples we have studied so far, the goals have been simple: to demonstrate that certain ratios of numbers correspond to certain musical intervals (so far, only the tone and five concords). The objectives of the *Sectio canonis* are somewhat more sophisticated, but the basic scheme of the project, along with the standard set of intervals it treats, seems to have passed from the third century BC to the second century AD without substantial change. In chapter 6 we shall see that in at least two cases (Didymus and Ptolemy) the scientific programme in which the monochord was involved became significantly more complex, but these will appear to be the exception, rather than the rule. In general the project and its building-blocks were simple.

For this reason it is not surprising to find another example of the simplest sort of canonic division in the only treatise on mathematical harmonics written between Euclid and Ptolemy which survives complete, Nicomachus' *Handbook of Harmonics*. Nicomachus lived in roughly the same generation as Theon and Adrastus, and in many ways his approach to harmonics is similar to theirs.

After establishing the acoustical foundations of his discussion (ch. 4), Nicomachus digresses with a historical account of the major discoveries and developments in harmonic science (chs. 5–9). He then returns to the point where he left off in chapter 4, and continues with 'what comes next' (τὰ ἑξῆς) at the beginning of chapter 10. This, he says, is to show that although the quantity of the note according to its tension (*tasis*, 254.7) is higher with higher pitch and lower with lower pitch (a thesis worked out in ch. 4),[134] the inverse is shown in demonstrations with strings or auloi.

ἐν τούτοις γὰρ ἀνάπαλιν αἱ μὲν βραχύτητες ὀξυτονοῦσιν, αἱ δὲ πλειονότητες βαρυτονοῦσιν. εἰ γάρ τις χορδῆς μακρᾶς ὑπὸ μίαν καὶ τὴν αὐτὴν τάσιν κειμένης ἐπί τινι κανόνι, ἐξηρμένης δ' αὐτοῦ ἐφ' ὅσον μὴ ψαύειν, τὸν ἀπὸ τῆς ὅλης κρουσθείσης φθόγγον συγκρίνοι πρὸς τὸν ἀπὸ τῆς ἡμισείας, ἀποληφθείσης τῆς χορδῆς ὑπαγωγεῖ ἢ τοιούτῳ τινὶ ἐκ τοῦ μεσαιτάτου, ἵνα μὴ περαιτέρω τοῦ ἡμίσους ὁ τῆς κρούσεως κραδασμὸς χωρήσῃ, διὰ πασῶν εὑρήσει τὸν ἀπὸ τῆς ἡμισείας πρὸς τὸν ἀπὸ τῆς ὅλης ψόφον μείζονα, ὅπερ ἐστὶ διπλάσιον, ἐναντιοπαθῶς ταῖς τοῦ μήκους ἀνταποδόσεσιν. εἰ δὲ κατὰ τὸ τρίτον μέρος ἀκριβῶς μετρηθὲν κατάσχοι τὸν κραδασμὸν, τὸ ἀπὸ τοῦ διμοίρου φθέγμα ἡμιόλιον ἀναγκαίως ἔσται πρὸς τὸ ἀπὸ τῆς ὅλης, ἀντιστρόφως τῷ μήκει. εἰ δὲ κατὰ τὸ τέταρτον μέρος τῆς χορδῆς ἐγκόψεις τῇ κρούσει, περαιτέρω προχωρεῖν οὐκ ἐῶν τὸν κραδασμὸν, ἐπίτριτον ἂν πρὸς τὸ ἀπὸ τῆς ὅλης ἠχήσειε τὸ ἀπὸ τῶν τριῶν μερῶν, ἐναντίως τῇ ἐν τοῦ μήκους σχέσει.

For in these cases it is the other way round, small dimensions yielding high pitch and greater ones low pitch. For if you take a long string at a single uniform tension lying above a measuring rod (*kanōn*), and fixed away from the rod so as not to touch it, and if you compare the note from the whole string when it is plucked with that from half of it, the string being divided off exactly in the middle by a bridge or something of the sort, so that the vibration from the blow does not extend further than half-way, you will find that the sound from the half string stands at an octave to the larger sound from the whole – that is, it is double that sound, being qualified in the opposite way to the relationships of the lengths. And if you exclude the vibration from a third part of the length, accurately measured, the sound uttered by two thirds of the string will necessarily stand in hemiolic ratio to that from the whole, the opposite way round to the length. And if you cut off a fourth part of the string from the blow, not allowing the vibration to extend further, the sound from three parts of the string will stand in epitritic ratio to that from the whole, the opposite way round to the relationship of length.[135]

[134] On this notion see also Thrasyllus ap. Theo. Sm. 87.9–89.8 and Boeth. *Mus.* IV.5; Adrastus devotes some energy to a complicated argument refuting it (Theo. Sm. 65.10–66.11). See also Barker 2000: 49 n. 12.

[135] *Harm.* 254.11–255.3, trans. Barker 1989: 262.

Nicomachus goes on to add parallels from auloi and *syringes*, following the same steps. It is the same sort of division as that found in several of the dubiously attributed demonstrations quoted by Theon (59.4–61.17), simpler even than that of Panaetius. Three intervals are constructed and quantified, in the order octave, fifth, fourth. If it were not for Nicomachus' development of the inversion argument the passage would be thoroughly unoriginal – not, it must be said, an undesirable quality in an introductory handbook. The practical details of the division, on the other hand, are just what such a book requires; this is a description designed for the beginner.

With the division performed, Nicomachus can now move on to a more detailed account of harmonic structures made up of these intervals. In chapter 11 he sets out the fifteen notes of the Greater Perfect System in the diatonic genus. But then, after this, he makes a brief and puzzling reference to canonic division. It is the only place in the *Handbook* where Nicomachus actually uses the term *kanonos katatomē*; in chapter 10 he did not put a name to his demonstration on the *kanōn*. The reference comes amid a list of things Nicomachus has decided to set aside for inclusion in a larger work to be written at a later date (which, if he wrote it, has not survived).[136] The last item on the list is a canonic division – not, he makes clear, of any current two-octave system such as the one presented earlier in the same chapter, but of the entire four-octaves-plus-sixth division of *Timaeus* 35b–36b:

καὶ προσεκθησόμεθα τὴν τοῦ Πυθαγορικοῦ λεγομένου κανόνος κατατομὴν ἀκριβῶς καὶ κατὰ τὸ βούλημα τοῦδε τοῦ διδασκάλου συντετελεσμένην, οὐχ ὡς Ἐρατοσθένης παρήκουσεν ἢ Θράσυλλος, ἀλλ' ὡς ὁ Λοκρὸς Τίμαιος, ᾧ καὶ Πλάτων παρηκολούθησεν, ἕως τοῦ ἑπτακαιεικοσιπλασίου.

In addition we shall set out the division of the so-called Pythagorean *kanōn*, worked out accurately and completely according to Pythagoras' own intention, not in the manner of Eratosthenes or Thrasyllus, who misunderstood it, but in that

[136] It is uncertain whether the fragments collected by von Jan under the title *Excerpta ex Nicomacho* derive from the missing work promised in *Harm.* 11. It is also possible, as Bower has argued (1989: xxiv–xxix, xxxiv, 126), that the canonic division given in Boeth. *Mus.* IV.5–13 is that of Nicomachus' lost book. It is true that the 'inversion' argument at the opening of IV.5 echoes that of Nicom. *Harm.* 10.

of Timaeus of Locri, whom Plato also followed, right up to the twenty-seven-fold ratio.[137]

We have encountered this passage once already, in considering the possibility that Eratosthenes may have employed the monochord (ch. 4). There we were forced to conclude that in the case of Eratosthenes at least, a canonic division could be produced without using the *kanōn*. What precisely Nicomachus meant by the term *kanonos katatomē* is then as much in question here as determining his views on its purpose.

One disturbing possibility is that he did not count the division of chapter 10 as a real *kanonos katatomē* at all. If this is so, there may be several reasons for it: differences between the division of chapter 10 and that hinted at in chapter 11 are primarily those of range, completeness and purpose. The brevity of Nicomachus' remarks on the subject denies us a clear picture of the criteria by which he judged the practice of canonic division, but we may nevertheless gain a sense of what the issues were through a closer look at the passage and those he mentions in it.

Which Timaeus?

Our confidence in using Nicomachus as a source, either for his report that Eratosthenes produced a division of the monochord, or for the judgement that, along with Thrasyllus, he got it wrong, is justifiably shaken by Nicomachus' apparent confusion about his Platonic source. It is not altogether clear what text Nicomachus was intending to use as the basis for his *kanonos katatomē*.

The most likely candidate is a work supposedly written by Timaeus of Locri that was composed in such a way as to appear a credible source for Plato's *Timaeus*.[138] This book survives in its

[137] Nicom. *Harm.* 260.12–17, trans. Barker 1989: 266, slightly modified. Cf. Adrastus ap. Theo. Sm. 63.25–65.9.

[138] Thesleff offers a résumé of its contents (1961: 23) and an edition (1965: 205–25); for a more recent text and translation see Tobin 1985. The document is about one-fifth the length of the authentic Platonic dialogue on which it is based. Its full title varies from Τιμαίω Λοκρῶ Περὶ ψυχᾶς κόσμω καὶ φύσιος in some MSS to Τιμαίω Λοκρῶ Περὶ φύσιος κόσμω καὶ ψυχᾶς in others. It is sometimes cited by its Latin title *De universi natura*; I shall refer to it as the *Timaeus Locrus* and abbreviate it as *Ti. Locr.* (following LSJ).

entirety; it is a short and derivative document which was known to Iamblichus and others.[139] It cannot have been written long before the date of Nicomachus' *Handbook*,[140] and because of its content, Nicomachus' reference to 'Timaeus of Locri' has been taken as a reference to this text.[141]

The numerical harmony of the creation of the world soul is, not surprisingly, one of the topics addressed in the *Timaeus Locrus*. The author makes the observation that the entire scale, including the movable notes, can be rendered in whole numbers if one begins at 384; the highest number will then be 10,368 (*Ti. Locr.* 21–2; see fig. 5.2a).[142] This was by no means an original contribution to the tradition of comment on the passage.[143] Plato's division contained seven terms, the last of which was twenty-seven times the first.[144] The project of the *Timaeus Locrus* is the same, except that all the gaps between Plato's terms are filled with epogdoic tones, *leimmata* (256:243) and *apotomai* (2187:2048), a procedure which brings the total number of terms to thirty-six. After an extensive elaboration of the resulting numbers (ch. 22), the author summarises the operation he has just completed: 'The entire scale is four octaves and a fifth and a tone, [extending] up to the twenty-seven-fold ratio' (τὸ πᾶν τετράκις διὰ πασῶν καὶ διὰ πέντε καὶ τόνος, μέχρι τοῦ ἑπτὰ καὶ εἰκοσαπλασίου, 23, 212.19–20). Then, in a deliberate clarification of the numbers of the preceding chapter so as to link them with those of Plato's *Timaeus*, he continues: 'for 10,368 is the twenty-seventh multiple of 384' (τῶν γὰρ τπδ′ ὁ ͵ατξη′ ἑπτὰ καὶ εἰκοσαπλάσιος, 23, 212.24; see fig. 5.2a–b).

[139] Iambl. *In Nic.* 105.11, Procl. *In Ti.* I.1.

[140] Taylor (1928: 656–7) was inclined to date the work to the first century AD; Tobin (1985: 3–7) comes to a similar conclusion.

[141] Taylor 1928: 656, Tobin 1985: 3. Nicomachus would then be the first ancient author to mention the *Timaeus Locrus*; Calvenus Taurus (fl. AD 145) is the next.

[142] Passages in the *Timaeus Locrus* are cited by chapter and, where necessary, by page and line number in Thesleff's edition (1965). Chapters 21–3 correspond to 96b–c.

[143] Taylor identifies Eudorus (fl. c. 25 BC) as a possible source (1928: 658); Thesleff (1965: 209) suggests Crantor (c. 335–275 BC), because Plutarch says that Eudorus followed Crantor in taking 384 as the first number of the Platonic series (*De an. procr.* 1020c). Crantor lends his name to the so-called 'Crantor diagrams', or 'lambda diagrams', by which the two geometric series of *Ti.* 35b–36b were separated so as to present the series of doubles (1, 2, 4, 8) along one arm of a large lambda, and the series of triples (1, 3, 9, 27) along the other, with the unit as the common point at the top (see fig. 5.2c).

[144] *Ti.* 35b–36b; the entire passage is quoted in ch. 3 part 2.

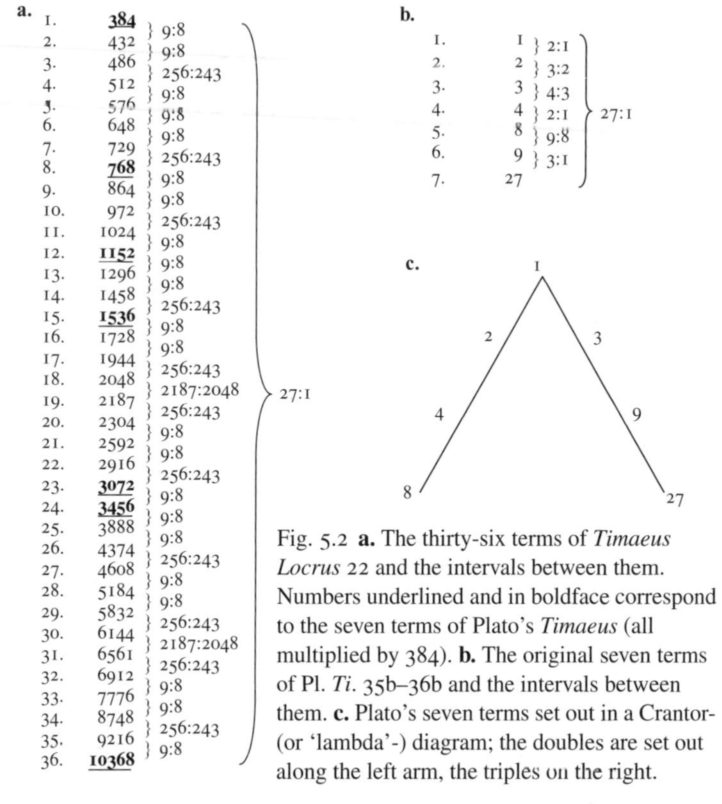

Fig. 5.2 **a.** The thirty-six terms of *Timaeus Locrus* 22 and the intervals between them. Numbers underlined and in boldface correspond to the seven terms of Plato's *Timaeus* (all multiplied by 384). **b.** The original seven terms of Pl. *Ti.* 35b–36b and the intervals between them. **c.** Plato's seven terms set out in a Crantor- (or 'lambda'-) diagram; the doubles are set out along the left arm, the triples on the right.

Matthias Baltes has argued that chapters 22 and 23 as we receive them are the product of later interpolations.[145] Yet he acknowledges that some sort of division must have occupied this place in the text, and even what he offers, on a sceptical reading, maintains all the essentials: thirty-six terms, beginning at 384 and extending to 10,368.[146] Nicomachus' aim of following Timaeus of Locri 'right up to the twenty-seven-fold ratio' (ἕως τοῦ ἑπτακαιεικοσιπλασίου) thus appears to be a direct reference to the scale of the

[145] Baltes 1972: 79–82. The main grounds for doubt are stylistic and dialectal.

[146] Baltes 1972: 81. Taylor noted that Proclus (*In Ti.* II.188) was critical of the *Timaeus Locrus* for positing thirty-six terms because these included the *apotomē*, an interval not mentioned by Plato (1928: 142–3). Thus either *Ti. Locr.* 22 had already become corrupt by the time of Proclus, or the original division also included thirty-six terms, including *apotomai*. Baltes favours the latter interpretation. Tobin (1985: 21–2) follows Baltes' reading, and prints only Baltes' table (1972: 43) between the first sentence of ch. 22 and the last sentence of ch. 23.

Timaeus Locrus, which also extended 'up to the twenty-seven-fold ratio' (μέχρι τοῦ ἑπτὰ καὶ εἰκοσαπλασίου). It seems likely, then, that Nicomachus intended to base his extended canonic division on the numbers of the *Timaeus Locrus*, not on those of Plato's *Timaeus*,[147] and that his criticisms of Eratosthenes and Thrasyllus should be interpreted in this light.

It will be evident that the *Timaeus Locrus* would have provided ample material for the sort of *kanonos katatomē* which Nicomachus promises his patroness. He may not have had a copy of the book in his possession when he wrote the *Handbook*; he was apparently travelling at the time.[148] By the same token he may not have had copies of Eratosthenes' and Thrasyllus' canonic divisions with him either. This makes his brief and uninformative remark that they both misunderstood the project even less useful. Because it is the first statement of its kind in the literature of the monochord, however, some attempt at an interpretation is warranted. The one I suggest here is necessarily speculative.

Misunderstanding canonic division

Perhaps the most puzzling aspect of the last sentence of chapter 11 of Nicomachus' *Handbook* is that he gives no reason for his dissatisfaction with the canonic divisions of either Eratosthenes or

[147] Barker (1989: 266 n. 88) raises the objection that 'in its scalar constructions . . . it [the *Timaeus Locrus*] does not even use the number 27'. But, strictly speaking, Nicomachus does not mention the number 27; what both texts refer to is the twenty-seven-fold ratio. Levin (1975: 92), to whom Barker refers, puts her objection differently: 'In fact, the treatise *Timaeus Locrus* does not even adopt the integer 27 in the series of terms, but 36 or the square of the first perfect number, 6.' This not only misses the point, but misreads the text: the integer 27 does not appear in the series of terms because the series of terms begins at 384; for the same reason, the integer 36 does not appear either. But there is, of course, a twenty-seventh term (assigned the integer 4608, *Ti. Locr.* 212.13) and a thirty-sixth term (10,368). Not even Plato adopted 'the integer 27 in the series of terms' (*Ti.* 35c).

[148] *Harm.* 237.10–11. Nicomachus seems not to have had a copy of Plato's *Timaeus* to hand either: this is suggested by the mess he makes of Pl. *Ti.* 36a–b when he quotes it at 250.6–11. His misquotation of this important passage changes its sense so as to imply that the range of the *Timaeus* scale was smaller than it really was. It seems most likely that quotation from memory is to blame, but it is just possible that Nicomachus was deliberately distorting Plato's text in order to make him appear the 'follower' of Timaeus of Locri (see Barker 1989: 259 n. 60 for a fuller discussion). Already by the end of the fourth century BC rumours of Plato's plagiarism of Pythagorean sources in the writing of his *Timaeus* had achieved a certain popularity (Diog. Laert. VIII.84–5, Gell. *NA* III.17); cf. Nicom. *Harm.* 260.17.

Thrasyllus. He does not even say whether they made the same error; they are simply lumped together as those who 'misunderstood'. But the fact that his intention seems to be to carry out his own division according to what he saw as its original Pythagorean 'purpose' (*boulēma*, 260.14) suggests at least that he thought Eratosthenes and Thrasyllus put the instrument to some other purpose, or used it in some way Pythagoras had not intended.

A second possible reason for Nicomachus' disapproval is that the divisions of Thrasyllus and Eratosthenes may both have been smaller in range than that of the *Timaeus Locrus* (and, for that matter, of the *Timaeus* itself).[149] All he writes about Thrasyllus and Eratosthenes is that their procedures involved a misunderstanding; all he writes about 'Timaeus of Locri' is that Plato followed him 'right up to the twenty-seven-fold ratio'; linking the two statements is tempting indeed. In the same sentence Nicomachus appears to emphasise three characteristics of the well-constructed *kanonos katatomē* (260.14–15): accuracy (the adverb ἀκριβῶς), completeness (the verb συντελέω) and adherence to Pythagorean dogma (the phrase κατὰ τὸ βούλημα τοῦδε τοῦ διδασκάλου). The four-octaves-plus-sixth range of the *Timaeus Locrus* division could be seen as satisfying the second and third of these criteria. One might object that in the very chapter in which we find the reference to Eratosthenes and Thrasyllus, Nicomachus himself has been setting about the construction of a two-octave system:

τόδε τοίνυν τὸ τοῦ διαγράμματος κύτος κατὰ τὸ διατονικὸν γένος ἐστὶν δὶς διὰ πασῶν τετραπλασίου πλάτους. τοσοῦτον γὰρ ἡ ἐναγώνιος φωνὴ διανύει χωρὶς κινδύνου τινὸς ἢ ὀλισθήματος, ἐφ' ἑκάτερον ἄκρον δυσέμπτωτος γινομένη, εἰς μὲν κοκκυσμὸν κατὰ τὸ νητοειδές, εἰς δὲ βηχίαν κατὰ τὸ βομβυκέστερον τῶν ὑπατῶν.

Now the scope of the diagram in the diatonic genus is a double octave of quadruple breadth. And this is the maximum that a voice trained for competition can traverse without some kind of risk or slippage, since it becomes difficult to pitch at each of the extremes, tending towards a mere squeak in the area of *nētē* and to a hoarse whisper around the deeper of the *hypatai*.[150]

[149] Thrasyllus' two-octave range was narrower than that of Eratosthenes, which apparently extended, like Aristoxenus', to two octaves and a fifth (see chapter 4).

[150] Nicom. *Harm.* 11, 255.23–256.4 trans. Barker (1989: 263); 'the deeper of the *hypatai* is *hypatē hypatōn*' (n. 78). With the last sentence cf. *Exc. ex Nicom.* 4, 274.11–20.

Evidently Nicomachus was not averse to two-octave scale-systems, nor did he disapprove of the consideration of practical concerns in the study of harmonics, as factors taken in isolation. But the canonic division projected at the end of chapter 11 was presumably going to go beyond anything Nicomachus had included in the *Handbook*, and thus we cannot fairly guess at his criteria for judging canonic divisions on the basis of his presentation of the *kanōn* in chapter 10 or the two-octave scale-system in chapter 11. Range, then, may well have been an important part of the project of 'the division of the so-called Pythagorean *kanōn*' as Nicomachus thought it should be done.

Of Eratosthenes' division we know no more than the intervals preserved by Ptolemy (on which see chapter 4), but Thrasyllus, for his part, set out a division with a total range of only two octaves, from *nētē hyperbolaiōn*, the highest note of the Greater Perfect System, to *proslambanomenos*, the lowest.[151] It is only in light of the four-octaves-plus-sixth scale of the *Timaeus* that Thrasyllus' canonic division could be considered small; those of the *Sectio*, Panaetius and Adrastus were all of two-octave range despite other differences between them. Thrasyllus begins by dividing the string into first two, then three, and finally four parts, and naming the notes which come out of each division (Theo. Sm. 87.9–89.8). At this point he introduces unit-numbers by way of clarification:

δῆλον δ᾽ ἂν γένοιτο τὸ λεγόμενον ἐπὶ τῶν ἀριθμῶν. εἰ γὰρ τὸ τοῦ κανόνος μέγεθος ιβʹ μέτρων ὁποιωνοῦν, ἔσται μὲν μέση δίχα διαιρεθείσης ⟨τῆς ὅλης χορδῆς, καὶ ἀφέξει⟩ ϛʹ ἑκατέρωθεν [διαιρουμένη]· ἡ δὲ ὑπάτη τῶν μέσων ἀπὸ τῆς ἀρχῆς δʹ· ἡ δὲ νήτη διεζευγμένων ἀπὸ τῆς τελευτῆς δʹ· καὶ τὸ μεταξὺ αὐτῶν δʹ. ἡ δὲ ὑπερυπάτη ἀπὸ τῆς ἀρχῆς τρία ἀφέξει μεγέθη, ἀπὸ δὲ τῆς ὑπάτης ἕν· ἡ δὲ ὑπερβολαία ἀπὸ μὲν τῆς τελευτῆς γʹ, ἀπὸ δὲ τῆς διεζευγμένης ἕν. μεταξὺ δὲ αὐτῶν ϛʹ, ὥστε ἀπὸ τῆς μέσης ἑκατέρα γʹ, καὶ γίνεται ἡ ὅλη διαίρεσις ἀπὸ μὲν τῆς ἀρχῆς ἐπὶ ὑπερυπάτην γʹ, ἐντεῦθεν δὲ ἐπὶ ὑπάτην ἕν, ἐντεῦθεν δὲ ἐπὶ μέσην δύο, εἶτ᾽ ἀπὸ μέσης ἐπὶ τὴν διεζευγμένην βʹ, ἐντεῦθεν δὲ εἰς τὴν ὑπερβολαίαν ἕν, ἀπὸ δὲ ταύτης εἰς τὴν τελευτὴν γʹ. γίνεται πάντα ιβʹ.

What has been said can be made clear by means of numbers. For if the *kanōn* has a magnitude of 12 units of whatever sort, *mesē* will arise when the whole string is

[151] Thrasyllus ap. Theo. Sm. 88.17–25. The scale he presents is in fact a combination of the Greater and Lesser Perfect Systems (Barker 1989: 229 n. 90), but its overall range is no different from that of the Greater Perfect System. A description of these systems is given in the Introduction.

divided in half, and will be at a distance of 6 units from each end. *Hypatē mesōn* will be 4 units from the origin, and *nētē diezeugmenōn* 4 units from the end: the distance between them is 4. *Hyperhypatē* will be at a distance of 3 magnitudes from the origin and 1 from *hypatē*. *Hyperbolaia* will be 3 units from the end, and 1 from *diezeugmenē*. Between them will be 6, so that each is 3 from *mesē*. The whole division is then as follows: from the origin to *hyperhypatē*, 3; then 1 to *hypatē*; then 2 to *mesē*; then from *mesē* to *diezeugmenē* 2; then 1 to *hyperbolaia*, and from it to the end, 3: in all there are 12.[152]

Thrasyllus then proceeds to specify in systematic fashion all the ratios which fall between the notes, reducing each to lowest terms as he goes (89.23–90.21):

4:3, 6:3, ⟨8:3,⟩[153] 9:3, 12:3 = 4:3, 2:1, ⟨8:3,⟩ 3:1, 4:1
6:4, 8:4, 9:4, 12:4 = 3:2, 2:1, 9:4, 3:1
8:6, 9:6, 12:6 = 4:3, 3:2, 2:1
9:8, 12:8 = 9:8, 3:2
12:9 = 4:3

He then places the remaining notes of the diatonic and chromatic genera, and makes a half-hearted attempt to include the enharmonic as well.[154] Altogether it constitutes much more than what was offered either in the *Sectio* or by any subsequent author in the context of a canonic division. It is detailed and systematic, both practically and mathematically.

Furthermore, it may owe something to the *Timaeus* tradition, not only because of Thrasyllus' substantial knowledge of Plato, but for a more specific reason as well. At the conclusion of his division Thrasyllus remarks that the entire division could be set out in numbers, rather than in ratios as he has just described it. The importance of this final paragraph is in the way he sets up the numerical division:

[152] Thrasyllus ap. Theo. Sm. 89.9–23, trans. Barker 1989: 227. *Hyperbolaia* and *diezeugmenē* are shortened forms for *nētē hyperbolaiōn* and *nētē diezeugmenōn* respectively.

[153] The ratio of the octave plus fourth was supplied from an apograph to fill a lacuna at 90.2–3; the restoration is secure: cf. 88.22–3. Of the four canonic divisions considered in this chapter, only that of Thrasyllus includes the octave plus fourth. Significantly, Thrasyllus specifies its ratio, but not its concordant quality. Thrasyllus' division is also unique among the four in being concerned with setting out the notes of a *systēma*, rather than a set of intervals. These two differences are closely connected.

[154] 90.22–93.2. For the enharmonic Thrasyllus gives no more than a brief sentence telling which notes to remove from each tetrachord; he does not offer notes to divide the enharmonic *pyknon*, which in his (ditonic) scheme is the 256:243 *leimma*.

εὕροιμεν δ' ἂν ταῦτα καὶ ἐν ἀριθμοῖς ἀπὸ τῆς νήτης τῶν ὑπερβολαίων ἀρχόμενοι, ὑποτεθείσης αὐτῆς μυρίων τξη'· οἱ ἐφεξῆς ἐπόγδοοί τε καὶ οἱ λοιποὶ κατὰ τοὺς προειρημένους λόγους λαμβάνονται, οὓς περίεργον ἐκτιθέναι· ῥᾴδιον δὲ τῷ παρηκολουθηκότι τοῖς προειρημένοις.

We could also find these things in numbers, beginning from *nētē hyperbolaiōn*, if we assume that it is 10,368. The successive epogdoics and the remainders are taken according to the ratios previously stated, which it is unnecessary to set out: it is an easy task for anyone who has followed what we have said.[155]

It is possible that on account of their position at the end of the quotation, these sentences were written not by Thrasyllus but by Theon. As we have seen in the case of Adrastus, Theon frequently offers little indication of the precise beginnings and endings of his many quotations from the works of other authors. But in this case, the next sentence offers some reassurance: 'That, then,' Theon sums up, 'is the method of dividing the *kanōn* which has been handed down by Thrasyllus.'[156] It is as firm an end-quotation mark as we can hope for.

Having allowed for some uncertainty about the authorship of the concluding lines of Thrasyllus' canonic division, we may note that they contain two important features. The first is that the now familiar number 10,368 has reappeared. Thrasyllus cannot have found this number in the *Timaeus*, since it does not occur there, and consequently any view of Nicomachus' criticism of Thrasyllus and Eratosthenes as a straightforward bias in favour of 'Pythagorean' versus 'Platonic' canonic divisions (however Nicomachus may have understood such a distinction) is not entirely satisfactory. Where did the number 10,368 come from? If Theon's hand is to blame, we might suspect that he got the number from the *Timaeus Locrus*; but if the paragraph belongs to Thrasyllus, then a look at earlier sources is probably in order. The *Timaeus Locrus* may just have been in circulation by Thrasyllus' time, but we cannot assume that he knew it. It is evident from remarks made by Plutarch in his treatise *On the Procreation of the Soul in the Timaeus* (1020c) that there had already been widespread and detailed manipulation of the numbers in the *Timaeus* by commentators as early as Crantor

[155] 93.2–7, trans. Barker 1989: 229.

[156] καὶ ἡ μὲν ὑπὸ Θρασύλλου παραδεδομένη κατατομὴ τοῦ κανόνος ὧδε ἔχει, 93.8–9, trans. Barker 1989: 229.

(c. 335–275 BC) and as late as Eudorus (fl. c. 25 BC). Both Crantor and Eudorus wrote *Timaeus* commentaries (Crantor's was especially influential in antiquity), and both apparently began the Platonic twenty-seven-fold division at 384, as did the author of the *Timaeus Locrus*. As noted above, this yields a final number of 10,368.

It is evident, then, that Thrasyllus had a wide pool of numerical analysis to draw from for his own work; by his lifetime the *Timaeus* had spawned a tradition of comment which went far beyond the original scope of the Platonic text. What is particularly strange about the appearance of the number 10,368 in this context is not the fact that it appears in the first place, but that while Thrasyllus' sources must have applied it to the twenty-seven-fold, four-octaves-plus-sixth division of the *Timaeus*, he applies it, perversely, to his diminutive and mundane double-octave 'immutable scale-system' (*ametabolon systēma*, 92.26–7). This may have been at least one of the spurs to Nicomachus' annoyance.

The second thing of note in the concluding lines of Thrasyllus' division concerns the fact that the number 10,368 is made to parallel not, as we might expect, the bottom note (*proslambanomenos*) of the two-octave scale, but the top note (*nētē hyperbolaiōn*). In the procedures he has described at 89.9–90.21 the division numbers, envisaged as lengths of the same string, were higher according to lower pitch, and lower according to higher pitch: hence '⟨the ratio⟩ of *nētē diezeugmenōn* to *hyperbolaia* will be 4:3, the epitritic ratio of the fourth',[157] not the other way around. In the 'numerical' division sketched at the end of the quotation, we must either suppose that *nētē hyperbolaiōn* (which Thrasyllus also calls *hyperbolaia*, as above) still corresponds, at 10,368, to the shortest length of string – giving *proslambanomenos* the extraordinarily high value of 41,472 (= 10,368 × 4:1) – or, on the other hand, that the numbers are no longer being conceived as string lengths and that Thrasyllus intends his reader to locate *proslambanomenos* at 2592 (= 10,368 ÷ 4:1). It is true that other authors, such as Plutarch (at least some of the time), carried out their divisions in this way.[158]

[157] ἔσται οὖν πρὸς μὲν τὴν ὑπερβολαίαν ⟨ὁ λόγος⟩ τῆς μὲν νήτης διεζευγμένων δ′ πρὸς γ′ ἐπίτριτος ὁ τοῦ διὰ τεσσάρων, 89.23–90.1.

[158] *De an. procr.* 1018d. But cf. 1020c–d, where he writes that he prefers to use a base number of 192 for the *Timaeus* division because the first tetrachord thereby generated

But in fact Thrasyllus' division can only be carried out in one direction: the requirement, stated at 93.3, was that the entire division be found 'in numbers' (ἐν ἀριθμοῖς), and if 10,368 is taken as the highest term, then one note cannot be assigned a number (in the sense meant by *arithmos*). Thrasyllus locates this note, *tritē synēmmenōn*, with two reference points (note 10 in fig. 5.3b): it is a tone below *tritē diezeugmenōn* and a *leimma* above *mesē* (92.2–6).[159] But if 10,368 is taken as the lowest term, *tritē synēmmenōn* will be found at 19,683. In fact, the highest note of the scale cannot be assigned a value less than 10,368. Thus there is only one way to read Thrasyllus' division: numbers refer to string lengths, and he is consistent throughout in attaching the higher numbers to the lower notes.

Now it is possible that the number 10,368 occurs in Thrasyllus' division merely because it happens to be the lowest value that can be given to his *nētē hyperbolaiōn*: he did the reckoning, and out it came. But he may well have had some help. He could hardly have been unfamiliar with the Platonic commentators who had applied it to the larger scale of the *Timaeus*, and since the basic intervals of his scale are all derived from the same ingredients as that of the *Timaeus*, the number 10,368 would have come as a helpful tip for his own project. Thrasyllus' adoption of this number for his divison of the smaller *ametabolon systēma* may therefore have seemed to him not merely a fortuitous numerical coincidence, but an opportunity to allude to the tradition of *Timaeus* commentary, and to illustrate the fact that the mathematical language of this tradition could be brought to bear on the analysis of other scale-systems in order to provide the most economical expression of their terms. We may even speculate that he thought of 10,368 as the meeting point between the structures of eternal and ephemeral music, and as a sort of proof that the two could be thought of as mathematically contiguous.

(192, 216, 243, 256) is in lowest terms, and the numbers of the *leimma* mentioned by Plato (256:243) are readily seen. From this it is clear that in Plutarch's conception 192 represents the highest note of the scale, since the *leimma* is never the highest interval in a tetrachord.

[159] Thus if 10,368 is the highest number in the system and is assigned, as Thrasyllus tells us, to *nētē hyperbolaiōn*, then *mesē* will be 5184 (= 10,368 ÷ 2:1), and 5184 × 256:243 is not an *arithmos* (a positive integer greater than one).

Nicomachus, as we have seen, made much of the necessary reversal of the numbers when applying them to the monochord in chapter 10 of his *Handbook*. His argument was that the causal constituents of pitch, impacts, have higher quantities according to higher pitch, and that in strings this is evident in the correlation between higher tension and higher pitch (ch. 4). For Nicomachus the note *as such* is properly quantified according to this arrangement, and in chapter 10 much of his energy is devoted to showing that the relationship of high pitch to *low* number displayed in the lengths of strings or pipes is the *opposite* of that previously explained.[160] That he takes the high number, high pitch correlation to be the 'real' relationship is clear from the way he sets out the ratios in his brief canonic division (if that is what he would have called it): 'you will find that the sound from the half string stands at an octave to the larger sound from the whole – that is, it is double that sound, being qualified in the opposite way to the relationships of the lengths',[161] and so on for the rest. In at least one other detailed discussion of the subject (*Exc. ex Nicom.* 2) Nicomachus also made the higher numbers correspond to the higher notes, as did Plutarch.[162]

Yet his objection cannot have been that Thrasyllus' division was 'backwards', for Thrasyllus' division is itself framed in similar terms. 'When they divide the magnitude in half', he says, 'the octave in duple ratio makes *mesē*, which in its movements is qualified in the converse way, having double the pitch upwards.'[163] Throughout his division Thrasyllus reminds his reader, as Nicomachus did, that 'the numbers of the movements are inversely proportional to the division of the magnitudes'.[164]

[160] Note especially his repetitive use of the vocabulary of oppositeness throughout the chapter: ἀντίστροφον, 254.10; ἀνάπαλιν, 254.11; ἐναντιοπαθῶς, 254.21; ἀντιστρόφως, 254.25; ἐναντίως, 255.3; ἀντιπαθῶς, 255.16.

[161] 254.19–22, quoted (with the Greek) above. [162] See n. 158 above.

[163] δίχα μὲν διελοῦσι τὸ μέγεθος μέσην ποιεῖ τὸ διὰ πασῶν ἐν τῷ διπλασίῳ λόγῳ, ἀντιπεπονθότως ἐν ταῖς κινήσεσι διπλασίαν ἔχουσαν τάσιν ἐπὶ τὸ ὀξύ, Thrasyllus ap. Theo. Sm. 87.9–11, trans. Barker 1989: 226.

[164] ἀντιπεπόνθασιν οἱ ἀριθμοὶ τῶν κινήσεων τῇ διαιρέσει τῶν μεγεθῶν, Thrasyllus ap. Theo. Sm. 88.15–16; repeated almost verbatim at 89.7–8. The principle is invoked again at 90.22. Compare the very similar statements in Nicom. *Harm.* 10 (254.21–2, 25; 255.3, 16–17).

Likewise, Nicomachus could not have objected to Thrasyllus' application of the higher numbers to the lower notes by taking the *Timaeus Locrus* as his model for canonic division. That text does not refer to any of the notes in the Greek scale by name, but it is clear from the arrangement of the intervals laid out in chapter 22 and from the way their combinations are highlighted that the higher numbers, as in Thrasyllus' division, correspond to the lower notes, and vice versa; were it not so, each tetrachord would be upside down, with the *leimma* at the top and the two tones below it. The author of the *Timaeus Locrus* proceeds sequentially, from the first term (the highest note), 384, to the last (the lowest note), 10,368, rather than in such a manner as to facilitate the construction of the intervals on a monochord by the reader, as in *Sectio canonis* propositions 19–20, or in Thrasyllus' division (see fig. 5.3).

The divisions of Thrasyllus and the *Timaeus Locrus* bear comparison in other ways as well. Proclus was uncomfortable with the two *apotomai* (2187:2048) which the author of the *Timaeus Locrus* placed between terms 18–19 and 30–1, because this interval is not mentioned in Plato's original scheme.[165] But the *apotomai* fall in musically useful places: by lying just above the *leimma* at the bottom of the tetrachord, they each supply a chromatic note, equivalent to Thrasyllus' *chrōmatikē mesōn* (note 13 in fig. 5.3b). An identical placement of the *apotomē* is specified in Thrasyllus' division. In fact, the entire lower octave of Thrasyllus' division (*mesē* to *proslambanomenos*, notes 11–19 in fig. 5.3b) is placed twice, note for note, between terms 16–24 and 28–36 of the *Timaeus Locrus* division (fig. 5.3a). Between these octaves lies a single tetrachord and a tone of disjunction. The two upper octaves (terms 1–8 and 8–15) are identical to one another, but different from those of Thrasyllus' division. Each replicates the central octave of the Greater Perfect System (that is, the tetrachords *diezeugmenōn* and *mesōn* separated by a tone of disjunction); they are conjoined at term 8, and separated from the lower octaves by the tone between terms 15 and 16.

The two tetrachords of the *Timaeus Locrus* division which contain *apotomai* (16–20 and 28–32) are constructed, like their

[165] See n. 146 above.

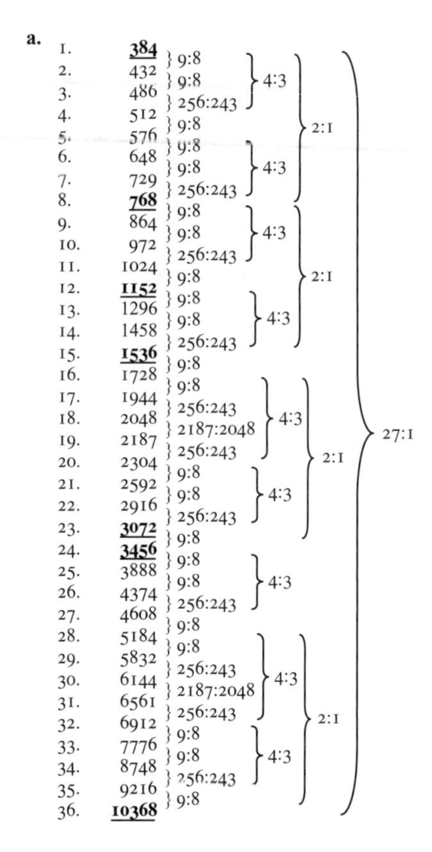

Fig. 5.3 **a.** The division of *Ti. Locr.* 22. The numbers are original; ratios between terms, and their groupings by tetrachord and octave, are my additions. **b**. Thrasyllus' division. Note-names and tetrachord groupings are as he specifies them, but all numbers except 10,368 have been added. Synonymous note-names are given in parenthesis. When two different notes occupy the same pitch they are separated by an oblique stroke. In both (**a**) and (**b**) high-pitched notes are at the top, and low-pitched notes at the bottom.

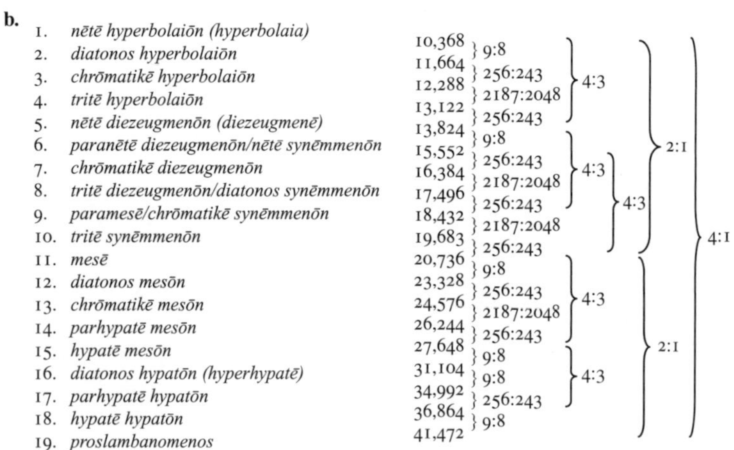

Thrasyllan counterparts, so as to allow dual use; if terms 17 and 29 are omitted, the tetrachords become chromatic, but if notes 18 and 30 are omitted they become diatonic. This version of the chromatic arises from a straightforward application of the ditonic scale: the *apotomē* is, after all, only the interval 'cut off' from the tone when a *leimma* has been taken away from it. There is no thorough-going reform of the standard ditonic tetrachord here; these tetrachords are nothing like those produced by Archytas, Eratosthenes and Didymus. Every step in the scale is either a tone or one of two fractions of a tone.

It seems surprising, then, that if Nicomachus took the division of the *Timaeus Locrus* as the standard by which to judge others, he would reject that of Thrasyllus, with which it shares so many essential features. Nicomachus cannot have objected to the lower octave of Thrasyllus' division, since it occurs twice in that of the *Timaeus Locrus*. Furthermore, the structures in Thrasyllus' upper octave merely involve more of the same procedures already applied in the *Timaeus Locrus* division, not the introduction of a new principle.[166] It seems, then, that formal structures were not at the root of Nicomachus' criticism of Thrasyllus.

Performing the division

This brings us back to the most significant difference between the divisions of Thrasyllus and the *Timaeus Locrus*: magnitude. This difference could be counted theoretically or even philosophically important, but if we are to take Nicomachus literally when he talks of a 'division of the monochord', it is first and foremost a practical difference. The four-octaves-plus-sixth range of the *Timaeus Locrus* division makes its construction on the monochord quite problematic. It is not merely the difficulty of dividing the *kanōn* into 10,368 units that causes the problem, though this is certainly the first hurdle; the units would need to be very small indeed. Yet we have seen already that Ptolemy (whose concern for the practicalities of canonic division is manifest)

[166] Thrasyllus' addition of the tetrachord *synēmmenōn*, which overlaps the tetrachord *diezeugmenōn*, is also paralleled by Nicomachus (*Harm.* 259.6–13), though this cannot be taken to dictate his preferences in the construction of a canonic division.

expressed the numbers in his division-tables in sexagesimals, thus allowing for 7200 theoretically possible bridge positions.[167] And if the *Timaeus Locrus* division can be criticised on this front, so can that of Thrasyllus, which requires 41,472 units. Some approximation is therefore necessary in both cases.[168]

The second problem is that the notes from the smaller lengths of string will be difficult to produce accurately and sonorously; very short end-lengths give out a poor sound, and their pitches are distorted by the height of the bridge. Whereas Thrasyllus' *ametabolon systēma* requires only a quarter of the string for its shortest length, the *Timaeus* scale (whether in this version or not) requires one twenty-seventh. On a Ptolemy-style 120-unit *kanōn* this comes out as a mere 4 27′ units.[169] The smallest distance between bridge positions used in the division (the interval between terms 3–4 in fig. 5.3a) will then be only 18 sixtieths of a unit. In Thrasyllus' division, by contrast, the shortest length will be 30 units, and the smallest step in the scale (the interval between notes 2–3 in fig. 5.3b) a distance of 1 48′ units. Thrasyllus' division is not beyond the capabilities of a carefully built instrument; the same cannot be said for the one Nicomachus favoured. A single attempt at reconstructing the highest octave of the *Timaeus Locrus* division on a monochord will suffice to convince the reader of the point: this is not a division meant to be demonstrated by *kanōn*.[170]

There are ways around the difficulty, of course, but they are disruptive to the procedure. One could transpose the upper octaves to a lower pitch and demonstrate them there; or one could set up a

[167] See ch. 4. Only rarely is less than half the string sounded in the scales of Ptolemy's tables, however; the true number of bridge positions that can be indicated by means of his numerical notation is thus reduced to about 3250.

[168] Here we discover just how relevant the P/Q discrepancy of Adrastus turns out to be once larger structures containing smaller intervals are at issue. We can never be sure that the bridge has been placed precisely at 36,864, and not (say) 36,867; even if it has, the bridge itself must take up several of of these units; and even if it has, we cannot hear the difference. The question becomes whether or not monochord division can still be a theoretically meaningful activity under such conditions.

[169] For the sexagesimal notation, see table 4.1.

[170] And yet Proclus also resorts to the monochord in order to explain the numbers of Pl. *Ti.* 35b–36b, which pose some of the same problems (*In Ti.* II.237.15–20). Significantly, however, the exercise he recommends to his reader is a sort of 'armchair' canonic division, not a practical one: νενοήσθωσαν οὖν οἱ ἀριθμοὶ πάντες ἐφ' ἑνὸς γεγραμμένοι κανόνος, ὥσπερ δὴ καὶ ποιεῖν εἰώθασιν οἱ ἁρμονικοί, 'Let all the numbers be *imagined* as being marked on a single monochord, as is the custom of the harmonicists.'

polychord *kanōn* with the strings tuned in octaves to one another. It is not inconceivable that monochords had been used by Pythagoreanising Platonists to explicate versions of the *Timaeus* scale.[171] Or perhaps the lack of unity and continuity in remedial polychord solutions could even have been turned to philosophical use: the scale cannot be contained or reproduced by earthly instruments; only the world soul is large enough to be harmonised by it. The insufficiency of the *kanōn* could itself be used to communicate the magnitude of the cosmic harmony.

Comparisons between the divisions of Thrasyllus and the *Timaeus Locrus* therefore help to reinforce the point that for Nicomachus, setting out 'the division of the so-called Pythagorean *kanōn* . . . accurately and completely according to Pythagoras' own intention' (*Harm.* 260.13–14) did not apparently mean including only the intervals of the ditonic diatonic. It seems rather to have meant setting out a scale that was not particularly monochord-friendly. We may quite reasonably wonder whether Nicomachus envisaged a role for the instrument in the procedure at all.

The charges against Eratosthenes and Thrasyllus

I have suggested that Nicomachus' objections to the divisions of his two predecessors may have been partly to do with range. The two remaining areas of conflict (which I have signalled already) are as follows. Thrasyllus devotes a considerable amount of attention to describing how to go about dividing the monochord, but only two brief and almost dismissive concluding sentences outlining the way to find the numbers corresponding to all the notes. His priorities are clear, and they are the exact opposite of those found in the *Timaeus Locrus*, despite the numerical similarities between the two divisions. In the latter, there is no practical instruction whatsoever, nor are any note-names supplied; all we are given is a list of terms from 384 to 10,368. Eratosthenes, for his part,

[171] Perhaps this is what Proclus meant by ὥσπερ δὴ καὶ ποιεῖν εἰώθασιν οἱ ἁρμονικοί (see previous note): not simply 'as is the custom of the harmonicists' – i.e. their custom of marking division numbers on monochords – but more specifically 'just the way the harmonicists customarily do it' – i.e. set out the *Timaeus* scale on a single-stringed *kanōn*.

offered tetrachords divided in ratios which bore no mathematical relation to the ditonic division which underlies all the harmonic analyses of Nicomachus, Thrasyllus and the *Timaeus Locrus*. What the approaches of Thrasyllus and Eratosthenes have in common is that both could be seen as involving concessions, whether practical or mathematical, which compromised the 'purity' either of the intervals themselves or of the process of their explication.

It is curious that while in chapter 11 of his *Handbook* Nicomachus treats the double-octave scale from a historical and practical point of view, his promise to set out a canonic division in a later treatise comes into the discussion only as a related, but separate, procedure. It appears that for him there are on the one hand the scales of human music, which are limited in range for vocal reasons, and whose development can be traced via the history of Greek musical instruments, and on the other there is the scale by which the world soul is harmonised, whose range is eternally fixed (as described by 'Timaeus of Locri'). It can be illustrated by (perhaps imaginary) canonic division, and although it admits none of the limitations of human musical capabilities, it bears startling resemblances to the general shape not only of the Perfect System outlined in chapter 11 of Nicomachus' *Handbook*, but of the Thrasyllan division which he dismissed as misguided.

We may wonder why Nicomachus settles on these two theorists as his target. If my summary of his views is correct, then any number of other Greek musicologists could equally have been in the line of fire – Didymus, for example, or the author of the *Sectio canonis*, or even a relatively early Pythagorean like Archytas, all of whom either diverged from the received intervals of Pythagorean harmonics, or allowed practical concerns to influence the formulation or presentation of the ratios of their scale-systems in some way. In this case, why should Nicomachus have singled out Eratosthenes and Thrasyllus? It is not a question which admits any easy answer, and perhaps we should conclude simply that these were the two names which sprang most readily to mind as he neared the end of his short and hastily written handbook.[172]

[172] There are only twelve chapters in the *Handbook*; cf. Nicomachus' apologies to his patroness for 'the hasty nature of this essay' (265.1, trans. Barker 1989: 269) in the final paragraph.

Summary

After Eratosthenes the monochord became more widespread among music theorists, perhaps largely as a pedagogical instrument. The catechetic manual of Ptolemaïs of Cyrene makes it clear that *kanonikē* had become a clearly delineated branch of harmonics, one which adopted both the rational postulates of mathematics and the perception-based postulates of the Aristoxenians. Although Ptolemaïs' definition of canonics included the use of other instruments (particularly auloi and *syringes*), the monochord is given special place as being 'strictly canonic'. The term 'canonics' was used by others as well, although not always with the monochord directly in mind. By some accounts (like that of Aulus Gellius), canonics also included the study of the ratios of rhythm. Many ancient authors list *kanonikē* in place of *harmonikē* among the mathematical sciences, usually categorising it as a branch of arithmetic because it treats ratios of numbers. Notably, however, canonics is also presented as a geometrical science, because its ratios must be constructed geometrically if they are to be heard.

The activity which defined the *kanonikoi* and their science, in Proclus' account, was canonic division. Although Ptolemaïs does not mention canonic division, I have argued that her discussion of the criteria of reason and perception seems to have given the monochord a mediating role between the two. Panaetius took the process a step further by using canonic division to confirm that equal division of the tone is impossible both in qualitative and quantitative harmonic analysis. Through the writings of Adrastus, however, two important problems with canonic division are raised. The first is that posed by the octave plus fourth, whose status as a concord cannot be confirmed by canonic division without drawing attention to the fact that its ratio is unlike those of the other concords. This difficulty could be taken to interfere with Adrastus' thesis that reason and perception reach fundamental agreement through instruments, but Adrastus himself does not draw this conclusion explicitly even though the octave plus fourth is the only concord omitted from the canonic division Theon apparently quotes from his book. The second problem (imperfectly conceived by Adrastus himself) is that the discrepancy between the ideal division and its physical construction on the instrument means that no interval viewed (or sounded) on the monochord is ever exactly the one reason has established. The importance of the second problem is that it provides a spur for the essential project of minimising error within the apparatus, something with which Ptolemy was greatly concerned.

The most complex division to survive from the period is that of Thrasyllus. In his work we can see the influence of previous analyses of the numbers harmonising the world soul in Plato's *Timaeus*, although in his version they are applied to a much smaller gamut. Nicomachus, who criticised Thrasyllus along with Eratosthenes for misunderstanding canonic division, appears to have favoured a kind of division exemplified by the *Timaeus Locrus*, in which the entire four-octaves-plus-sixth scale of Plato's *Timaeus* was worked out in detail, and for which the monochord is for practical reasons an inadequate demonstrative tool.

From these accounts of canonic theory a rather heterogeneous picture emerges. On the one hand the instrument seems to be thriving; it gives its name to its own brand of harmonics and those who practise this science. Procedures of canonic division are recorded, and at the most basic level the instrument has evidently become a key in the growing attempts to present harmonics as a unified science despite the rupture between Aristoxenian and Pythagorean approaches to it. On the other hand, some of those most engaged in the finer details of harmonic analysis begin to discover that at this level the instrument only forces the two approaches further apart. Likewise, those who model their divisions on the 'timeless' scale of the *Timaeus* run into practical difficulties which cannot be satisfactorily resolved on a single-stringed *kanōn*. The instrument is now at least notionally well established in Greek harmonic science, but evidently not always used in so-called 'canonic division', and no scholar had yet mounted a detailed and systematic attempt to integrate it into a unified harmonic project any more complex than that of the *Sectio canonis*. For this we must wait for Claudius Ptolemy.

6

PTOLEMY'S CANONICS

It was J. F. Mountford's opinion that 'no complete study of ancient Greek music can be undertaken without continual recourse to Ptolemy'.[1] His statement is even more applicable to a study of the early history of the *kanōn*, for no ancient harmonicist provided the instrument with a role more wide-reaching, more scientifically grounded and more conceptually indispensable than the one it occupies in Ptolemy's *Harmonics*.

Ptolemy has indeed made many appearances over the preceding pages. In chapter 1 we compared his introduction of the *kanōn* in the *Harmonics* with his introduction of two astronomical instruments in the *Almagest* in order to isolate the role of the *kanōn* between those of diagrams and tables; in chapter 2 he was our only source for the divisions attributed to Archytas; and in chapter 4 he provided our only indication of Eratosthenes' tetrachords. Other vital clues and fragments have been gleaned from Ptolemy's earliest and most extensive ancient commentator, Porphyry. At many other points along the way Ptolemy's work has served as a convenient benchmark by which to contextualise the struggles, errors and advances of his predecessors. References to Ptolemy so far have largely had the effect of moderating both the achievements and the mistakes of previous scholars: if they made some significant step forward, 'Ptolemy would do this much better'; or, on the other hand, if their work was hampered by some fundamental misunderstanding, '*even* Ptolemy did not satisfactorily resolve this issue'.

To move from Ptolemy's predecessors in the field of harmonics to Ptolemy himself presents a new interpretative challenge, one

[1] Mountford 1926: 74. 'Recourse to Ptolemy' among students of ancient Greek music has picked up in recent years: since the work of Düring (1930, 1932, 1934) the treatise has been translated with commentary by Barker (1989), Solomon (2000) and Raffa (2002), and is treated at length by Mathiesen (1999) and Barker (2000).

which we have not yet had to face. In most of the cases examined since chapter 2, we have been dealing with brief – often frustratingly brief – fragments and testimonia, some of which mention the *kanōn*, and some of which do not. At times it has been a hunt for a rare quarry through a thick wood. Now, with Ptolemy, we are overwhelmed with references to the *kanōn*. And not merely references: Ptolemy gives careful descriptions of a handful of harmonic instruments, sometimes including fine details of their parts and construction, and almost always with painstakingly clear instructions for their use. The statistics alone are impressive: the word *kanōn* appears twenty-five times in an instrumental context;[2] *monochordos* five times; *kanonion* (usually meaning 'measuring rod') nineteen times; *magas*, *magadion*, *hypagōgeus*, *hypagōgion* and *hypagōgidion* (words for bridges) thirty-three times jointly; *organon* ('instrument') fifteen times; and other more obscure instrumental vocabulary sometimes occurs as well.[3] The *kanōn* has gone from the sooty hearth to the prince's palace.

Yet it was not as sudden a translation as it might seem from the stark differences between our surviving texts. As we have seen already, there have been several coaches (and several pumpkins) along the way. Porphyry expends some energy on the allegation (poorly substantiated, in the end) that Ptolemy had taken 'the greater part, if not more or less all' (τὰ μὲν πλεῖστα, εἰ καὶ μὴ σχεδὸν πάντα) of his material from Didymus' book *On the Difference Between Pythagorean and Aristoxenian Music Theory*, without acknowledging the fact.[4] If Ptolemy's work were really this unoriginal, then the many new instruments and instrumental procedures he describes might also be assumed to predate him. But in two cases where Ptolemy has inherited an instrument from his predecessors (the *kanōn* and the *helikōn*), he is candid about the fact, as he is also when he discusses a technique pioneered by Didymus. There is no evidence (not even in Porphyry's commentary) to

[2] Usually it indicates an instrument; occasionally it signifies a ruler to be placed temporarily alongside the strings (e.g. 84.10). Ptolemy also refers to his numerical tables (II.15) as *kanones* (these uses are excluded from the count).

[3] E.g. *pelekēsis* (81.9), for a type of movable device to which tuning pegs could be attached (discussed below).

[4] Porph. *In Harm.* 5.7–15, on which see also Barker 1994a: 64–5.

suggest that the more complex instruments in the *Harmonics* were not original to Ptolemy. For these and other reasons, Porphyry's charges have not been taken seriously by modern scholars, and the reassessments of Ptolemy's originality in other fields have left his *Harmonics* relatively untouched.[5]

On the other hand, no arguments for Ptolemy's originality as a harmonic theorist have attempted to deny his dependence on ideas worked out by those who came before him. It is fairly clear, too, that Didymus was one of his main sources on the writings (and probably the divisions) of Archytas and Eratosthenes. Furthermore, in a passage where Ptolemy describes (and criticises) the canonic procedures and divisions of Didymus, it appears that the monochord had been employed in demonstrations of some complexity a century or less before the composition of Ptolemy's *Harmonics*.

This chapter is about Ptolemy's canonics: that is, about the place of the *kanōn* and related instruments in Ptolemy's conception and elaboration of harmonic science. I shall argue that although he never describes his harmonics as *kanonikē*, the instrument both defines and governs his exposition of the science, from its basic postulates to its wider conclusions. I have begun with the question of his originality because it is connected to the definition of his canonics as opposed to that of his predecessors. In general, Ptolemy's treatment of harmonic instruments can be distinguished from others we have examined in five respects. The first is that he confronts the issue of the accuracy and trustworthiness of the results they produce squarely, and repeatedly, throughout his treatise. The second is that he is explicit about his use of the *kanōn* in conjunction with tetrachordal divisions which deviate from the standard ditonic model. Thirdly, his canonics is not limited to the monochord: he not only describes polychord *kanones* but insists on their importance to his project; he also introduces other instruments built on more complex geometrical principles than the *kanōn*, and incorporates some of their features into his improved *kanones*. Fourthly, these instruments are employed not just apodeictically,

5 No modern scholarship on Ptolemy's *Harmonics* has judged him as harshly as, for example, Newton 1985 on his astronomical work.

but also empirically: they are involved in procedures where not only the answers to immediate questions, but also by extension Ptolemy's basic theses, are turned over to the judgement of perception. Finally, his canonics is more geometrical than any we have yet seen: the *kanōn* is introduced geometrically, as we saw in chapter 1, and the more elaborate instruments rely heavily on geometry to display their ratios to the eye.

Some of these aspects of Ptolemy's work evidently had precedents in earlier writers; Didymus seems to have used the monochord to set out non-ditonic tetrachordal divisions, and at least one of Ptolemy's more curious instruments (the *helikōn*) was one which by his own admission he had inherited from 'the mathematicians'. But in combination they add up to a very different brand of canonics than those we have considered so far. One of Ptolemy's starting points, as we shall see, is to define the roles of reason and perception in harmonic investigation. This was nothing new: Plato and Aristoxenus had opened the discussion in the fourth century, and Ptolemaïs and Didymus had both contributed to it in contrasting Aristoxenus' approach with those of other theorists. But Ptolemy goes beyond Ptolemaïs and Didymus in making a more forthright case for the indispensability of the *kanōn* in harmonics. If reason and perception are going to be shown to agree about what is beautiful in the realm of musical sounds (and one of Ptolemy's central concerns is to show that they do), then an instrument is required which speaks in the languages of both faculties. Because he puts so much argumentative weight on the thesis that the two faculties agree, the reliability of the instruments is therefore an essential concern.

This 'reliability anxiety' gives rise to Ptolemy's increased geometricising of the instruments, in the following way. We have seen already that the monochord was introduced in the *Sectio* as an extension of the lettered diagram. To allow reason and perception to contemplate the same objects simultaneously, it had to be diagrammatic in its presentation as much as it had to be physical in its construction. In the *Sectio*, this meant that canonic division was an arithmetical activity carried out harmonically through the medium of an audible diagram; its diagrammatic aspects were arithmetically limited, like the diagrams of the arithmetical books of Euclid's *Elements*. The geometry of sound was an arithmetical geometry.[6] But a point raised by Adrastus in his argument against the equal division of the tone turned out to have wider implications for this picture. Canonic division differs from line division because the bridge must have some width, and this must take up some of the string in a way that a point which divides a line does not. The monochord may be represented diagrammatically, but the instrument and the diagram will never be exactly identical, and therefore the parallel operations performed on each turn out not to be parallel after all.[7]

Ptolemy's reliability anxiety is more complex than the one I extrapolated from Adrastus' problematic objection to the semitone. Not only does the bridge have width, it also has height: it cannot divide the string without stretching it slightly. The *kanōn* therefore fails to conform to its diagrammatic representation (that of the *Sectio*, at any rate) in more than one way. As we shall see, Ptolemy's approach to this is two-pronged: first, he attempts to restrict his instruments to the realm of the diagram by appeal to geometry; second, he defines the relevance of the issue by a test of perceptibility. Both parts of the approach are a form of damage control: his instruments have already broken free of the conventions of the geometrical diagrams used in arithmetic, and his geometricising exercises are aimed at preventing them from escaping the realm of geometry altogether. The horse is already out of the barn; perhaps it can be kept in the paddock. Questions of the differential tension that arise from the stretching effect of a higher bridge are therefore postulated away, and the geometry of circles and triangles is invoked to show that there are no holes in the fence. The second part of Ptolemy's response is to devise a test for the relevance of the horse's location, as it were: if perception cannot tell the difference, does it matter that he is not in the barn? When lengths of string are made equal, to the greatest degree of precision possible with tools accessible to sight (the measuring-rod), and they give out pitches which sound identical to the musical ear, the theorist has reached the limit of the discrepancy's relevance within his project.

[6] See ch. 1 part 1. [7] See ch. 5 part 2.

One of the central features of Ptolemy's canonics, then, is that by placing his instruments firmly within the grasp of both reason and perception, he is forced to consider the questions of their reliability. His answers to these questions both geometricise canonics to a far greater degree than the diagrammatic presentation of the monochord in the *Sectio* had done, and also place perception in a very important discriminating role.

There are several aspects of Ptolemy's canonics which need to be considered before this summary can be fleshed out. I shall begin with his treatment of Didymus' contributions to the science, because they offer several points of comparison on methodological issues of primary relevance to the treatise as a whole.[8]

Debts to Didymus

We have encountered Didymus already in chapter 5, primarily on account of his discussions about the criteria of reason and perception. But he provides a useful starting point here too, not only because Porphyry quotes him when he comes to comment on the chapter in which Ptolemy first mentions the *kanōn* (*Harm.* I.2), but because Ptolemy himself refers to Didymus by name. In this respect he is part of a rather élite club, whose only other members are Pythagoras, Archytas, Aristoxenus and Eratosthenes. Besides the brief references in II.14 and the accompanying tables, Ptolemy devotes much of a chapter (II.13) to a critique of two of Didymus' contributions to harmonics: a new method of using the monochord, and three new tetrachordal divisions. This was certainly not the sum total of Didymus' work in the field; it has recently been argued that his harmonic output was substantial, compendious and of significant quality.[9]

[8] The subject of Ptolemy's instruments and the procedures for which he uses them has been treated in detail by Barker (2000, especially chs. 10–11). In this chapter I cover some of the same ground, and my arguments are indebted to his at many points. Other useful commentary is provided by Raffa (2002: 235–481) and by Solomon (2000), who frequently cites the scholia and other mediaeval commentators. Mathiesen treats Ptolemy in some detail (1999, ch. 5), including discussions of the manuscript tradition of the *Harmonics*, and on the later chapters of book III.

[9] Barker 1994a and 2000.

The two items which Ptolemy selects for discussion were both immediately relevant to his own method as well. Let us consider Didymus' divisions first. As they appear in the tables of the following chapter (on the textual problems of which see chapter 4), they include all three genera, even though Ptolemy complains in II.13 that Didymus posited three but only produced divisions for two.[10]

enharmonic	*chromatic*	*diatonic*
5:4	6:5	9:8
31:30	25:24	10:9
32:31	16:15[11]	16:15

Despite Ptolemy's criticisms about the smaller magnitude of the middle ratios of the chromatic and diatonic,[12] all three divisions exhibit two key characteristics shared by Ptolemy's own. The first is that they are non-ditonic. This is, of course, nothing new: the same was true of the divisions of Archytas and Eratosthenes (both of whose ratios are preserved in the same tables). But the observation is useful in that it helps to position Didymus in relation to the other theorists whose work we have examined. This particular issue, what we might call the 'purity' question, divides them into two groups: Philolaus (possibly), Plato, [Euclid], the author of the *Timaeus Locrus*, Plutarch, Thrasyllus, Adrastus, Theon and Nicomachus all worked with strictly ditonic divisions, while Archytas, Eratosthenes, Didymus and Ptolemy himself diverged from this pattern. The purity question is one which, for Ptolemy, evidently separated the more interesting from the less interesting accounts of tetrachordal division.

Secondly, Didymus' tetrachords employ exclusively epimoric intervals. Although Archytas' divisions revealed an inclination

[10] *Harm.* 68.17–18. The ratios which follow are found in the text at 68.20–6 (chromatic and diatonic) and in the tables at 70–3 (enharmonic, chromatic and diatonic).

[11] The lower ratio in the chromatic is indeed identical to that of the diatonic, as Barker notes (1989: 347 n. 123), despite the fact that both Barker's and Düring's tables indicate a different value for the chromatic *parhypatē*. The error appears to have been Düring's, since he prints the correct Greek numeral (ριβ′ λ′) and the incorrect Arabic one ($112^1/_9$) side by side in his table (1930: 72).

[12] Specifically, that the lowest ratio in the chromatic is greater than the middle ratio, and the highest ratio in the diatonic is greater than the middle ratio. As Barker points out (1989: 343 n. 105), two of Ptolemy's own diatonics violate the principle on which he here criticises that of Didymus.

toward epimorics, the highest and middle intervals of his chromatic (32:27 and 243:224) were exceptions to the rule. Eratosthenes, likewise, used epimoric intervals, but evidently did not consider them the only candidates for inclusion in tetrachordal divisions.[13] But Didymus' are uniformly epimoric, a feature shared only by those of Ptolemy; the only non-epimoric interval Ptolemy includes is the *leimma* in the ditonic diatonic, a tetrachord he is forced to admit since it played such a critical role in the attunement of Greek stringed instruments, and because he could not deny that it was used in practice.[14]

Ptolemy also made use of Didymus' new monochord techniques. He tells us two things about these: first, that Didymus used both sides of the bridge in his demonstrations, something which Ptolemy also does in various procedures of his own.[15] Secondly, he makes clear that Didymus' method was aimed at making it easier not simply to demonstrate scale-systems, but to *play melodies* with the monochord (68.11). Barker has argued, on the basis of a comparison between Didymus' tetrachords and the theoretical principles of Aristoxenus, that Didymus' project was something like that of Eratosthenes: he was seeking to provide a 'rationalised' set of Aristoxenian tetrachords which he could then demonstrate in a melodic context with the monochord (1994a: 67–71). Using both sides of the bridge enabled him to gain an extra note for every bridge position, thus potentially reducing the number of ungainly shifts, slides and pauses between notes. As Ptolemy remarks, however, this is only an advantage when the two notes generated by the bridge position happen to follow one another in the melody; he notes furthermore that some of the other more serious drawbacks of the instrument in this capacity remain unimproved by Didymus' advance (68.6–15).

There is a connection between the two aspects of Didymus' harmonic work which Ptolemy cites. Ptolemy's criticisms of Didymus' chromatic and diatonic (*Harm.* 68.15–69.12) are based not on rational principles, but on those of perception: his judgement of

[13] Eratosthenes' tetrachords are discussed in ch. 4.
[14] See Ptol. *Harm.* 39.16–19, 40.2–20, 43.19–44.12, 80.17.
[15] See especially 18.9–19.15, 46.7–47.17 and cf. Ptolemy's development of the idea in the instrumental instructions of III.2.

the middle intervals in the chromatic and diatonic is that they have been set out 'contrary to what appears to the senses' (παρὰ τὸ ταῖς αἰσθήσεσι φαινόμενον, 68.26), producing an effect which is 'in no way melodic' (μηδαμῶς ἐμμελοῦς, 68.28); in sum, he says, Didymus 'takes no account of the consequences of what is perceived' (οὐδέν τι προσποιεῖ τῶν φαινομένων ἐχόμενον, 68.16).[16] He may have been on the right track in attempting to facilitate the hearing of the tetrachords in their melodic context, Ptolemy argues, but he went astray for two reasons: firstly because he did not listen to the intervals of musical practice and allow these to influence his theoretical decisions, and secondly because his apparatus was not effective enough to allow him to hear his divisions in context.

γέγονεν οὖν αἴτιον ἅπασι τοῦ μὴ δεδοκιμασμένως προσεληλυθέναι τῇ τῶν λόγων ὑποθέσει τῷ μὴ πρότερον ἐπεσκέφθαι τὴν δι' αὐτῶν χρῆσιν, ἀφ' ἧς μόνης ἠδύναντο παραβάλλεσθαι ταῖς τῆς αἰσθήσεως καταλήψεσι, καὶ διὰ τοῦτο τοὺς μὲν τῶν συμφωνιῶν λόγους καὶ διὰ μιᾶς χορδῆς ἐξετάζεσθαι δυναμένους κατὰ τὸν εἰς δύο μερισμὸν ἐξειληφότες φαίνονται, τοὺς δὲ τῶν ἐμμελειῶν τῇ συνθέσει τοῦ δι' ὅλου συστήματος μόνως ἂν θεωρηθέντας, ὅπερ οὐκ ἐνῆν ἐπὶ μιᾶς χορδῆς ἀκριβῶς ἰδεῖν, καὶ πάνυ διεψευσμένως. ἐλεγχθεῖεν γὰρ ἂν ἐναργῶς, εἴ τις κατ' αὐτοὺς ποιοῖτο τὰς κατατομὰς ἐπὶ τῶν ἐκτεθειμένων ἡμῖν ἰσοτόνων ὀκτὼ χορδῶν, ἱκανῶν οὐσῶν ἤδη τὸν εἱρμὸν τοῦ μέλους ἐπιδεικνύναι ταῖς ἀκοαῖς, ἵνα καταμάθωσι τό τε γνήσιον καὶ τὸ μή.

The reason for all these things was his failure to embark on the imposition of the ratios with sufficient circumspection, having failed to consider in advance the way in which they are used in practice: only this makes it possible for them to be brought into conformity with the impressions of the senses. This is why they seem to have constructed the ratios of the concords, which are capable of being tested on a single string on the basis of its division into two, while those of the melodics can be understood only through the composition of the whole *systēma*, which could not be seen accurately on a single string, and their treatment of these is utterly mistaken. For they would be plainly refuted if one were to construct the divisions they propose on the eight strings of equal pitch that we have discussed, these being adequate to display to the hearing the sequence belonging to a melody, so that the genuine and the spurious can be distinguished.[17]

It appears that Didymus' use of the monochord involved *epideixis* without empiricism, in contrast to that of Ptolemy.[18] He and his predecessors used only a single-stringed *kanōn*, which Ptolemy

[16] The translations are those of Barker (1989: 343).

[17] Ptol. *Harm.* 68.32–69.8, trans. Barker 1989: 344.

[18] Barker 1994a: 71–2.

says is sufficient for judging the ratios of the concords (as we have seen it used by many of our authors so far), but which is unreliable with melodic intervals. Had they developed a more complex instrument such as the one Ptolemy recommends, they would have been 'refuted' (ἐλεγχθεῖεν . . . ἄν, 69.5) – by their senses, that is – on the basis of the results.

I shall return to the adaptations and additional instruments with which Ptolemy supplemented the monochord. For the moment let me summarise Ptolemy's debts to Didymus as follows: like Didymus, he insisted on the primacy of epimorics to the exclusion of other intervals (with the one exception noted above), and like Didymus, he used the medium of melody to give musical context to his theoretical constructions. He went a step further, however, in insisting that the ear must be allowed to judge the constructions of theory on its own criteria, and that the necessary conditions for a fair appraisal cannot be met with the monochord alone, despite Didymus' improved bridge-handling method. This very method would be taken up in Ptolemy's own many-stringed demonstrations. We may reasonably surmise that Didymus' activities played some role in Ptolemy's conception of larger experimental instruments: once the necessity of judging scales through melodies has been adopted (itself a combination of Didymus' and Ptolemy's priorities), the larger instruments must follow; they are the only satisfactory way of surmounting the difficulties posed by the monochord. If the tetrachords proposed by earlier theorists are to be improved upon in a way which both reason and perception will approve, advances in both method and apparatus will be necessary. It is clear from the outset of his treatise that Ptolemy's argument must involve such advances, and so we may turn now to his introduction.

Preparing the ground: Ptolemy's canonics

In the *Almagest*, Ptolemy introduces his first scientific instruments (those used to determine the obliquity of the ecliptic) in chapter 12 of book I; we have seen already that the armillary sphere (*astrolabon organon*) does not make its appearance until the first chapter of book V (though it is mentioned in I.6), and the parallactic instrument (*organon parallaktikon*) until V.12.[19] In each instance, the instrument is brought out to deal with a specific astronomical question.

In the *Harmonics*, Ptolemy also introduces his various instruments on a problem-by-problem basis, but the stage is prepared for their entrance much earlier than in the *Almagest*. In the *Harmonics*, it is clear from the outset that the successful pursuit of the science will depend ultimately on agreement between the twin criteria of hearing and reason (*akoē* and *logos*, 3.4). Accordingly, the entire first chapter is preparatory to the introduction of the *kanōn* at the beginning of I.2. Ptolemy does not leave the instrument until it is needed; his harmonics cannot be practised without it, so it must be involved from the start.

I have discussed already the affinities between the geometrical diagram and the monochord.[20] In his first chapter, Ptolemy uses (like Aristoxenus and Didymus) the example of the drawing of geometrical figures to make the point that what is generated by unassisted perception is never as accurate as it seems to be. Perception is inconsistent, and needs the didactic correction (παραπαιδαγώγησις) of reason, which is like a teacher's cane (βακτηρία, 3.19–20).

ὥσπερ οὖν ὁ μόνῃ τῇ ὄψει περιενεχθεὶς κύκλος ἀκριβῶς ἔχειν ἔδοξε πολλάκις, ἕως ἂν ὁ τῷ λόγῳ ποιηθεὶς εἰς ἐπίγνωσιν αὐτὴν μεταγάγοι τοῦ τῷ ὄντι ἀκριβοῦς, οὕτω κἂν μόνῃ τῇ ἀκοῇ ληφθῇ τις ὡρισμένη διαφορὰ ψόφων, δόξει μὲν εὐθὺς ἐνίοτε μήτε ἐνδεῖν τοῦ μετρίου, μήτε ὑπερβάλλειν, ἐφαρμοσθείσης δὲ τῆς κατὰ τὸν οἰκεῖον λόγον ἐκλαμβανομένης ἀπελεγχθήσεται πολλάκις οὐχ οὕτως ἔχουσα, τῆς ἀκοῆς ἐπιγινωσκούσης τῇ παραθέσει τὴν ἀκριβεστέραν ὡσανεὶ γνησίαν τινὰ παρ᾽ ἐκείνην νόθον.

Thus just as a circle constructed by eye alone often appears to be accurate, until the circle formed by means of reason brings the eye to a recognition of the one that is really accurate, so if some specified difference between sounds is constructed by hearing alone, it will commonly seem at first to be neither less nor more than what is proper: but when there is tuned against it the one that is constructed according to its proper ratio, it will often be proved not to be so, when the hearing,

[19] Ptolemy's handling of these instruments is discussed in ch. 1 part 2.
[20] See ch. 1, ch. 3 part 4 and ch. 5 part 2.

through the comparison, recognises the more accurate as legitimate, as it were, beside the bastardy of the other.[21]

Sense-perception is inadequate to construct harmonic systems on its own, but it is through this very faculty that errors are detected when the illegitimate (νόθος) is compared to the legitimate. The errors may be only slight, but in combination they add up, like the bits of chaff on Nicomachus' scales, until the difference is both considerable and evident to perception.[22]

The comparison can only be made, however, if there is an instrument whose evidence will be acceptable both to reason and to perception. So Ptolemy is able to end his first chapter with this appeal:

τὸν αὐτὸν τρόπον καὶ ταῖς ἀκοαῖς διακόνοις οὔσαις μάλιστα μετὰ τῶν ὄψεων τοῦ θεωρητικοῦ καὶ λόγον ἔχοντος μέρους τῆς ψυχῆς, δεῖ τινος ἀπὸ τοῦ λόγου, πρὸς ἃ μὴ πεφύκασι κρίνειν ἀκριβῶς, ἐφόδου, πρὸς ἣν οὐκ ἀντιμαρτυρήσουσιν ἀλλ' ὁμολογήσουσιν οὕτως ἔχειν.

For the ears, similarly, which with the eyes are most especially the servants of the theoretical and rational part of the soul, there is needed some method derived from reason, to deal with the things that they are not naturally capable of judging accurately, a method against which they will not bear witness, but which they will agree is correct.[23]

For Ptolemy, then, the *kanōn* resembles not only the geometrical diagram but also the compasses used to draw it, and the figure drawn is like the interval to be judged. Thus the *kanōn* is (as its name suggests) a tool – but unlike the compasses, it is a tool which must be contemplated even when its constructive tasks are complete. To use the categories of chapter 1, the *kanōn* functions in modes (b) and (c) – representation and analogy – as well as in modes (a) and (d) – generation and direct physical manipulation. The ruler and compasses, once the diagram has been drawn with them, can be returned to the toolbox; the *kanōn*, on the other hand, has become the diagram, and must remain on the table.

[21] 3.20–4.7, trans. Barker 1989: 277. See also the further parallel of drawing straight lines, 4.19–5.2, where length (rather than evenness) is at issue.

[22] Nicom. *Harm.* 240.9–10.

[23] 5.6–10, trans. Barker 1989: 278; cf. also Ptol. *Judic.* 1.6–7, 10.1–6.

Having set out the instrumental requirements of his science, Ptolemy can now introduce the *kanōn* at the opening of I.2:

τὸ μὲν οὖν ὄργανον τῆς τοιαύτης ἐφόδου καλεῖται κανὼν ἁρμονικός, ἀπὸ τῆς κοινῆς κατηγορίας καὶ τοῦ κανονίζειν τὰ ταῖς αἰσθήσεσιν ἐνδέοντα πρὸς τὴν ἀλήθειαν παρειλημμένος. ἁρμονικοῦ δ' ἂν εἴη πρόθεσις τὸ διασῶσαι πανταχῇ τὰς λογικὰς ὑποθέσεις τοῦ κανόνος μηδαμῇ μηδαμῶς ταῖς αἰσθήσεσι μαχομένας κατὰ τὴν τῶν πλείστων ὑπόληψιν, ὡς ἀστρολόγου τὸ διασῶσαι τὰς τῶν οὐρανίων κινήσεων ὑποθέσεις συμφώνους ταῖς τηρουμέναις παρόδοις, εἰλημμένας μὲν καὶ αὐτὰς ἀπὸ τῶν ἐναργῶν καὶ ὁλοσχερέστερον φαινομένων, εὑρούσας δὲ τῷ λόγῳ τὰ κατὰ μέρος ἐφ' ὅσον δυνατὸν ἀκριβῶς.

The instrument of this kind of method is called the harmonic *kanōn*, a term adopted out of common usage, and from its straightening (*kanonizein*) those things in sense perception that are inadequate to reveal the truth. The aim of the student of Harmonics must be to preserve in all respects the rational postulates of the *kanōn*, as never in any way conflicting with the perceptions that correspond to most people's estimation, just as the astronomer's aim is to preserve the postulates concerning the movements of the heavenly bodies in concord with their carefully observed courses, these postulates themselves having been taken from the obvious and rough and ready phenomena, but finding the points of detail as accurately as is possible through reason.[24]

Later in his treatise (II.12) Ptolemy defines his harmonic project as 'the enterprise of displaying with complete clarity the agreement of reason with perception',[25] and there too the *kanōn* is mentioned in the same sentence. Underpinning all the rational postulates which Ptolemy will put forward throughout the argument of the *Harmonics* is his basic scientific creed, stated here and elsewhere,[26] that nature is rational and ordered, not random:

ἐν ἅπασι γὰρ ἴδιόν ἐστι τοῦ θεωρητικοῦ καὶ ἐπιστήμονος τὸ δεικνύναι τὰ τῆς φύσεως ἔργα μετὰ λόγου τινὸς καὶ τεταγμένης αἰτίας δημιουργούμενα καὶ μηδὲν εἰκῇ, μηδὲ ὡς ἔτυχεν ἀποτελούμενον ὑπ' αὐτῆς καὶ μάλιστα ἐν ταῖς οὕτω καλλίσταις κατασκευαῖς, ὁποῖαι τυγχάνουσιν αἱ τῶν λογικωτέρων αἰσθήσεων, ὄψεως καὶ ἀκοῆς.

For in everything it is the proper task of the theoretical scientist to show that the works of nature are crafted with reason and with an orderly cause, and that

[24] 5.11–19, trans. Barker 1989: 278. I have quoted the passage already in a slightly different context (ch. 1 part 2).

[25] τὴν δι' ὅλης τῆς ἐναργείας ἔνδειξιν τῆς τοῦ λόγου πρὸς τὴν αἴσθησιν ὁμολογίας, 66.6–7, trans. Barker 1989: 340.

[26] E.g. 85.15–19.

nothing is produced by nature at random or just anyhow, especially in its most beautiful constructions, the kinds that belong to the more rational of the senses, sight and hearing.[27]

Thus the monochord, with its immediate appeal to reason and perception through sight and hearing, is – on a logical level – qualified to fill the requirement set out at the end of I.1. If beauty consists in order and proportion, then principles of order and proportion will be found to underlie the most beautiful natural phenomena. Ptolemy's insistence on the use of epimoric ratios is in part, as I suggested in chapter 1, a reflection of this thesis. Furthermore, his phrase αἱ τῶν λογικωτέρων αἰσθήσεων ('the more rational of the senses') blurs Ptolemaïs' sharp distinction between *logos* (reason) and *aisthēsis* (perception): there are two branches of *aisthēsis* with which *logos* can communicate directly, and the *kanōn* will be shown to communicate with both simultaneously. It is in this way that it becomes for Ptolemy an essential tool for demonstrating the rational beauty of the constructions of nature.

Ptolemy's logical defence for the use of the *kanōn* has now been made. His acoustical defence follows in I.3, and it can hardly be counted so successful. Because musical intervals are to be generated on the *kanōn* in accordance with ratios of numbers, he must show that musical sounds themselves are quantifiable; if he cannot, any attempt to defend arguments in harmonic theory by appeal to the monochord will be open to question. Like others in the Archytan tradition, he takes the fundamental causal constituent of pitch to be impact (*plēgē*, 6.21). This, we recall, was one of the postulates which Ptolemaïs said were studied by the *kanonikoi* 'beginning from the starting points given by perception'.[28] But because he must also account for the behaviour of strings, Ptolemy spends some time discussing such qualities as thickness and density, which apply particularly to strings (and not, for example, to auloi), and the effect of their variation on pitch (7.17–8.15). From his material examples (bronze and wood, gut and flax) he concludes that what is denser and finer is of higher pitch; and this,

[27] 5.19–24. On the special status of the ears and eyes, cf. Ptolemy's comments on hearing and sight at *Judic.* 16.4; cf. also *Harm.* 5.6–8 (quoted above).

[28] Ptolemaïs ap. Porph. *In Harm.* 23.18–19 (quoted more fully in ch. 5 part 1).

he argues, is due to its higher tension, 'since it is an attribute of things like this that they are tenser, while what is tenser is more vigorous in its impacts (*plēgai*): the more vigorous is more compacted, and the more compacted is sharper [or 'higher-pitched': ὀξύτερον]'.[29] Thus he is able to state his theory of sound as a quantifiable phenomenon in this way:

τάσις γάρ τίς ἐστι συνεχὴς τοῦ ἀέρος ὁ ψόφος, ἀπὸ τοῦ τοῖς τὰς πληγὰς ποιοῦσιν ἐμπεριλαμβανομένου διήκουσα πρὸς τὸν ἐκτός, καὶ διὰ τοῦτο, καθ' οἵαν ἂν δύναμιν εὐτονώτερον ἕκαστον ᾖ τῶν δι' ὧν αἱ πληγαί, ἐλάττων τε καὶ ὀξύτερος ἀποτελεῖται.

For sound is a sort of continuous tensing of the air, penetrating to the outer air from the air that immediately surrounds the things making the impacts, and for this reason, to whatever degree each of the things making the impacts is tenser, the sound is smaller and sharper to the same degree.[30]

This thesis is most similar to the one put forward in the *De audibilibus*, where series of impacts of varying speed were said to be transmitted through the stationary medium of the air. As we noted in chapter 3 (part 3), that theory could not properly account for the acoustic properties of strings; neither, in its own way, can this one. As he continues, Ptolemy digs himself deeper into the rut he has followed:

διά τε δὴ τούτων ἔοικεν ἡ κατὰ τὸ ὀξὺ καὶ βαρὺ τῶν ψόφων διαφορὰ ποσότητος εἶδος εἶναί τι, καὶ μᾶλλον ἐκ τῆς τῶν ἀποχῶν τοῦ τε πληττομένου καὶ τοῦ πλήττοντος ἀνισότητος. τῷ γὰρ ποσῷ τούτων ἐναργέστατα συνίσταται, ταῖς

[29] ὅτι τοῖς μὲν τοιούτοις εὐτονωτέροις εἶναι συμβέβηκεν, τὸ δὲ εὐτονώτερον ἐν ταῖς πληγαῖς γίνεται σφοδρότερον, τοῦτο δὲ ἀθρούστερον, τοῦτο δ' ὀξύτερον, 8.3–5, trans. Barker 1989: 281. Throughout the acoustical passages of Ptolemy's treatise, Barker regularly translates the words *oxys* and *barys* by 'sharp' and 'heavy' (1989: 281 n. 25); Solomon (2000) prefers 'high' and 'low'. It should be noted that Greek acoustical language did not distinguish between the two meanings. Words literally meaning 'up' and 'down' or 'height' and 'depth' are very rarely applied to notes in Greek and Roman musical discourse (cf. ἄνω and κάτω in Hippoc. *Vict.* I.18.3; [Arist.] *Pr.* XIX.37, 47; also *altitudines* and *altior* in Gell. *NA* XVI.18.4, quoted in ch. 5 part 1). Of course all such language is metaphorical: in terms of its pitch a sound is no more literally 'high' than it is 'sharp'. Where we employ the language of positional metaphor (assisted, certainly, by our notation), the Greeks applied that of the qualitative. One reason to prefer 'sharp' and 'heavy' here is that it preserves the strange apples-and-oranges pairing of the Greek: pitches sound like qualities, not like quantities. Anyone can measure 'height' and 'depth' with numbers. Ptolemy, like his predecessors, is going out on a limb to quantify perceptibles that are thought of and described in impressionistic terms.

[30] 8.12–15, trans. Barker 1989: 281.

μὲν ἐλάττοσι διαστάσεσιν ἑπομένης τῆς ὀξύτητος διὰ τὸ ἐκ τῆς ἐγγύτητος σφοδρόν, ταῖς δὲ μείζοσι τῆς βαρύτητος διὰ τὴν παρὰ τὸ ἀπώτερον ἔκλυσιν, ὥστε ἀντιπεπονθέναι ταῖς διαστάσεσι τοὺς ψόφους.

For these reasons it seems that difference of sounds in respect of sharpness and heaviness is a form of quantity, and arises more particularly from inequality in the distances between the striker and the thing struck. For it is in the quantity of these that they are most clearly constituted, sharpness following upon the smaller distances because of the vigour arising from proximity, heaviness upon the greater because of the relaxation that goes with being further away, so that the sounds are modified in the opposite way to the distances.[31]

That this acoustical model is retrogressive to that found in the introduction of the *Sectio canonis* will be clear. There, at least, the necessary components of an acoustic theory permitting the demonstrative use of strings were provided, even if they were not all set out in formally argued steps. Here, we are given a barely improved version of a fourth-century acoustics, which is then applied indiscriminately to a variety of instruments whose pitch variations are generated from some change in length, as though it would hold equally for them all. Ptolemy treats it as obvious that the theory is generally applicable, citing 'sounds that come about as the consequence of length, like those of strings and auloi and windpipes: for so long as other factors remain the same, in strings those produced with smaller distances between the bridges are invariably sharper than those produced with greater ones'.[32] He has, in fact, introduced two very different and incompatible sorts of distance,[33] and then equated them without further clarification. Since any such clarification would expose the inapplicability of his acoustic theory to the *kanōn*, it is hardly surprising that he leaves the matter as vague as he does. He then proceeds as though the matter were settled: 'Let this outline suffice to indicate how height

[31] 8.15–21, trans. Barker 1989: 281–2.

[32] 8.25–9.2, trans. Barker 1989: 282. It is curious that Ptolemy was satisfied with such a problematic acoustical account, especially since it appears from portions of I.5 that he knew the *Sectio canonis*. (He even takes on the flawed prop. 11 without complaint, 12.8–13.) It is possible, however, that he knew this document only through an intermediary compiler (such as Didymus), who may not have quoted it either faithfully or in its entirety.

[33] That 'between the striker and the thing struck' (8.17–18) and that 'between the bridges' (9.1–2).

and depth [or 'sharpness and heaviness'] of sound are constituted, and that their form is a kind of quantity.'[34]

This acoustical model underlies the arguments of the following chapter (I.4), in which Ptolemy divides notes into equal-toned and unequal-toned, and eventually into melodic and unmelodic, concordant and discordant. Having established a theory of pitch and interval, he can now proceed with his criticism of both the Pythagoreans and the Aristoxenians, whom he introduced in I.2. This will occupy much of the remainder of the book (I.5–7, 9–12). But not all of it can be completed without recourse to various types of *kanōn*, and so, having dealt with the Pythagoreans in I.5–7, Ptolemy must introduce the monochord in some detail before proceeding to a critique of Aristoxenian theory; this is the topic of I.8.

The black sheep revisited: Ptolemy's approach to the 8:3 interval

Before discussing this pivotal chapter, let us pause for a moment to consider the spearhead of Ptolemy's attack on the Pythagoreans. He takes aim at the well-known soft target of Pythagorean harmonics, the 8:3 octave plus fourth. We saw in chapter 5 that this interval seems to have posed problems for Adrastus, who argued on the one hand that it was a concord, and on the other that it is in instruments which have been prepared beforehand in accordance with the ratios (like the monochord) that reason and perception agree, and yet apparently omitted the octave plus fourth from his canonic division. Because a simple canonic division of the first six concords and the tone cannot be done without drawing attention

[34] πῶς μὲν οὖν ὀξύτης συνίσταται ψόφου καὶ βαρύτης καὶ ὅτι ποσότης τίς ἐστι τὸ εἶδος αὐτῶν, ὑποτετυπώσθω διὰ τούτων, 9.16–17, trans. Barker 1989: 282. On the insufficiencies of Ptolemy's acoustics for his harmonic project, see particularly Barker 2000: 50. Mathiesen (1999: 438) sees no difficulty, but his interpretation of the passage, that Ptolemy 'shares the view of earlier theorists who had noted that pitch is a matter of frequency of vibration' (citing the introduction of the *Sect. can.*, Nicom. *Harm.* 4, and [Arist.] *De audib.*), is both anachronistic and inaccurate. None of these texts presents a view of pitch which could be equated exactly with 'frequency of vibration', although the *Sect. can.* offers an acoustical model which is compatible with this (modern) definition. Ptolemy's, however, is not.

to the fact that 8:3 is the only epimeric ratio in the list, the connection between reason and perception claimed for the procedure must be modestly stated: what it can do is show that the intervals the ear recognises as concords conform to the ratios assigned to them, and that the combinations between each are parallel. What it cannot do, apparently, is show that intervals accepted as equally concordant by the ear conform to ratios judged equally special in status by reason. Adrastus, accordingly, did not overstate his goals for canonic division. But for Ptolemy this evidently will not do. He has already staked out the turf belonging to hearing and reason with greater care than any of his extant predecessors, and his project will be impossible without constant appeal to instruments on which the two faculties must be seen to agree.

The octave plus fourth, then, is the tip of an iceberg, and Ptolemy must deal with it before he introduces the monochord. If perception is competent to judge the concords, and only gets into trouble with the smaller intervals, and especially these in combination (as they occur in scales and melodies, which will shortly concern him directly), then it is in the realm of concords that perception must be treated as most trustworthy. As he put it in I.2, 'for the Pythagoreans did not follow the impressions of hearing even in those things where it is necessary for everyone to do so'.[35] He can therefore begin his account of what the Pythagoreans postulated about the concords in I.5 with a list of concords accepted by perception: these are the fourth, fifth, octave, octave plus fourth, octave plus fifth and double octave (11.1–4). But the Pythagoreans do not accept the octave plus fourth as a concord, he explains, 'because it makes the ratio of 8 to 3, which is neither epimoric nor multiple'.[36]

Ptolemy's acceptance of the octave plus fourth is quite similar to that of Adrastus, at least initially. The first criterion is perception: both at the outset of I.5 and later (13.22–3) he insists on its unequivocal testimony on the question. Like Adrastus, too, he

[35] οἱ μὲν γὰρ Πυθαγορικοὶ μηδὲ ἐν οἷς ἀναγκαῖον ἦν πᾶσι τῇ τῆς ἀκοῆς προσβολῇ κατακολουθήσαντες, 6.1–2, trans. Barker 1989: 279.

[36] διὰ τὸ ποιεῖν λόγον τὸν τῶν ὀκτὼ πρὸς τὰ τρία, μήτε ἐπιμόριον ὄντα μήτε πολλαπλάσιον, 12.5–7, trans. Barker 1989: 285.

takes the position that any interval put together from a combination of a simple concord and an octave is also concordant.

καθόλου γὰρ ἡ διὰ πασῶν συμφωνία, τῶν ποιούντων αὐτὴν φθόγγων ἀδιαφορούντων κατὰ τὴν δύναμιν ἑνός, ὅταν προσαφθῇ τινι τῶν ἄλλων, ἀπαράτρεπτον τὸ ἐκείνης εἶδος τηρεῖ, καθάπερ ἡ δεκὰς ἔχει, φέρε εἰπεῖν, πρὸς τοὺς ὑπ' αὐτὴν ἀριθμούς.

For it is always true of the concord of the octave, whose constituent notes do not differ in their function from a single note, that when it is attached to one of the others it keeps the form of the latter unaltered, just as does the number 10, for example, in relation to numbers smaller than itself.[37]

Therefore if the fourth is concordant, the octave plus fourth must also be concordant; and it follows 'that the impression of the fifth and an octave is related in the same way to that of the fourth and an octave as is that of the fifth alone to that of the fourth alone, which accords with what is found by plain perceptual experiment (*peira*)'.[38]

Ptolemy then brings out a further argument to account for the apparent discrepancy between mathematical and perception-based judgements of the interval (I.7). Here he begins from the same mathematical postulates as the Pythagoreans, but exchanges their sharp dichotomy between multiples and epimorics on the one hand and epimerics on the other for a more flexible approach, in which ratios can be graded on the criterion of nearness to equality (ἡ πρὸς τὰς ἰσότητας ἐγγύτης, 15.23). But the argument is not straightforward, and Ptolemy is not consistent in what he means by the word 'equality'. The result is that his claim to have saved the octave plus fourth is not convincingly defended.[39] His attempt, however, deserves attention for its uniqueness: his aim is to generate a spectrum on which to rank intervals by degree, not by kind. In this scheme the homophones and concords will be separated from the melodics and discords by a different kind of difference than

[37] Ptol. *Harm.* 13.3–7 (trans. Barker 1989: 287), with which cf. Adrastus ap. Theo. Sm. 52.1–9. The thesis goes back to Aristox. *El. harm.* 20.18–21.

[38] τὸν αὐτόν γε τρόπον ἔχειν τὴν τοῦ διὰ πέντε καὶ διὰ πασῶν ἀντίληψιν πρὸς τὴν τοῦ διὰ τεσσάρων καὶ διὰ πασῶν, ὅνπερ ἡ μόνου τοῦ διὰ πέντε πρὸς τὴν μόνου τοῦ διὰ τεσσάρων ἀκολούθως τοῖς ἀπὸ τῆς ἐναργοῦς πείρας καταλαμβανομένοις, 13.19–23, trans. Barker 1989: 287.

[39] See Barker 2000, ch. 5.

that postulated by the Pythagoreans.[40] The octave plus fourth then escapes condemnation because Ptolemy claims to have avoided the Pythagoreans' rigid assumption that all concords are either multiples or epimorics. The primacy of epimorics is nonetheless present in his argument,[41] though he avoids stating it explicitly, and rushes rather quickly to his conclusion on the octave plus fourth, whose 8:3 ratio is put together 'from the duple and the epitritic' (ἐκ τοῦ διπλασίου καὶ ἐπιτρίτου, 16.9–10): 'for the fact that this ratio is neither epimoric nor multiple will now be no embarrassment to us, since we have adopted no preliminary postulate of that sort'.[42]

Later, however (I.15), this interval proves to be enough of an embarrassment that like so many other theorists, he simply omits

[40] Barker notes that this view is novel, but also points out that Ptolemy does not attempt to remove completely the distinction between concord and discord (2000: 80–2). Ptolemy's argument is comparable in some respects with that of Schoenberg, whose *Harmonielehre* was first published nearly two decades before Düring's edition of Ptolemy's *Harmonics*. For Schoenberg, undermining the distinction between consonance and dissonance was an essential step, and he went about it using arguments similar to those we have seen in ancient authors. Like Panaetius and Adrastus, who attempted to identify consonance through the natural phenomenon of sympathetic vibration, Schoenberg constructs the notion of musical interval by appeal to the natural phenomenon of the overtone series. (It is noteworthy that all the intervals thereby generated will be multiples, epimorics, or their combinations.) What he does with the intervals in this series is very much like Ptolemy's spectrum approach, but unlike Ptolemy he follows it through to its logical conclusion. 'In the overtone series, which is one of the most remarkable properties of the tone, there appear after some stronger-sounding overtones a number of weaker-sounding ones. Without a doubt the former are more familiar to the ear, while the latter, hardly perceptible, are rather strange. In other words: the overtones closer to the fundamental [cf. Ptolemy's 'nearness to equality'] seem to contribute more or more perceptibly to the total phenomenon of the tone – tone as euphonious, suitable for art – while the more distant seem to contribute less or less perceptibly . . . But this relation is, to repeat, as follows: the more immediate overtones contribute *more*, the more remote contribute *less*. Hence, the distinction between them is only a matter of degree, not of kind. They are no more opposites than two and ten are opposites, as the frequency numbers indeed show; and the expressions "consonance" and "dissonance", which signify an antithesis, are false. It all simply depends on the growing ability of the analyzing ear to familiarize itself with the remote overtones, thereby expanding the conception of what is euphonious, suitable for art, so that it embraces the whole natural phenomenon. What today is remote can tomorrow be close at hand; it is all a matter of whether one can get closer. And the evolution of music has followed this course: it has drawn into the stock of artistic resources more and more of the harmonic possibilities inherent in the tone' (1911 = 1978: 20–1). A prime example is the major third (5:4), which immediately follows the fourth (4:3) in the overtone series. In antiquity it was considered a discord; by the Renaissance it was treated as a concord. Thus Schoenberg can refer to the dissonant intervals of the series as 'the more remote consonances' (21).

[41] 15.18–16.31, on which see Barker 2000: 77–8, 1991: 78–82 and 1994b.

[42] νῦν γὰρ οὐδὲν ἡμᾶς οὗτος οὐκ ὢν ἐπιμόριος ἢ πολλαπλάσιος δυσωπήσει μηδέν γε τοιοῦτο προϋποτεθειμένους, 16.10–12, trans. Barker 1989: 290.

it from his list. Here again, the importance of the epimoric and multiple forms is much in evidence, and even though what Ptolemy claims to give in this passage is a list of structurally important intervals – of which the octave plus fourth is only really so in the Lesser Perfect System[43] – his omission of it is rendered all the more noticeable by the fact that the way he sets out the other intervals would have highlighted the anomalous character of the 8:3 ratio. The octave ratio is the one 'in which the difference between the extremes is equal to the one that is exceeded'; the hemiolic fifth, 'in which the difference between the extremes contains a half of that which is exceeded'; and so on for the others.[44] It is a neat, and neatly closed, system. Had Ptolemy wished to include the octave plus fourth here, he would have had to insert it between the octave and the octave plus fifth, where its explanation ('in which the difference between the extremes is in 5:3 ratio to that which is exceeded, antithetically to none of the others', or something of the sort) would have caused some embarrassment indeed. For all his efforts, the 8:3 interval remained as difficult to accommodate as ever.

I have spent some time on Ptolemy's approach to the octave plus fourth, firstly because it was the principal stumbling-block for mathematical harmonics, and as such a crucial test-case in the wider question of how the roles of reason and perception were to be defined in harmonic science more broadly; secondly because Ptolemy's handling of it is indicative of his general approach to this wider question; and thirdly because it is in I.8, just after the monochord has been introduced, that we find what is surely Ptolemy's

43 Ptolemy treats this system, rather dismissively, in II.6. He views it as a derivative compound rather than an independent, autonomous system, since it is put together (through a modulation) from two parts of the Greater Perfect System. (An overview of these systems is given in the Introduction.)

44 33.12–15, trans. Barker 1989: 306. The list continues: the epitritic fourth, 'in which the difference between the extremes contains a third part of that which is exceeded' (καθ' ὃν ἡ ὑπεροχὴ τῶν ἄκρων τρίτον περιέχει μέρος τοῦ ὑπερεχομένου, 33.16–17); the triple-ratio octave plus fifth, 'in which the difference between the extremes makes up two of that which is exceeded, antithetically to the half' (καθ' ὃν ἡ ὑπεροχὴ τῶν ἄκρων δύο ποιεῖ τοὺς ὑπερεχομένους ἐν ἀντιθέσει τοῦ ἡμίσεος μέρους, 33.18–20); and the quadruple-ratio double octave, 'in which the difference between the extremes makes three of that which is exceeded, antithetically, once again, to the third part' (καθ' ὃν ἡ ὑπεροχὴ τῶν ἄκρων τρεῖς ποιεῖ τοὺς ὑπερεχομένους ἐν ἀντιθέσει πάλιν τοῦ τρίτου μέρους, 33.20–2).

most successful attempt to show that this awkward interval can be included among the concords in a way that will satisfy both reason and perception. First, however, we must examine his presentation of the instrument itself.

Introducing the kanōn

By the end of I.7 Ptolemy has got about as far as he can with arguments based on mathematics and 'rough and ready phenomena'.[45] He has defined the ratios of the concords and the tone,[46] and intends to treat the remaining melodic intervals in due course (16.27–8). 'But now it would be a good thing to demonstrate the clear truth of those that have already been set out, so that we may have their agreement with perception established beyond dispute, as a basis for discussion.'[47] Ptolemy is now following through with the requirements outlined in the first two chapters of his treatise, and he will not be able to proceed with the remainder of his critique of the work of his predecessors without such instrumental assistance. The first part of this critique (I.5–7) requires empirical confirmation; the second part (I.9–14, including the Archytas chapters) will rely heavily on the evidence of perception, evidence which will be thorougly meaningless in this context unless its reliability can be proven in advance. This chapter (I.8) is therefore a critical link in Ptolemy's treatise: without this demonstrable connection between *akoē* and *logos* all his arguments from perceptible evidence, by which the validity of so many of the larger constructions is to be confirmed, will not stand up.

From the beginning of the chapter Ptolemy shows his concern for the critical business of accuracy. The agreement between reason and perception will not be 'established beyond dispute' by anything but the most precise instruments; in substandard instruments unevenness and approximation give rise to quarrels (διαβολή,

45 5.17–18 (quoted more fully above).

46 The tone is the interval which, by his argument, constitutes the lower limit of the reliability of unassisted perception.

47 νῦν δὲ καλῶς ἂν ἔχοι τὴν ἐνάργειαν ἀποδεῖξαι τῶν ἤδη παραβεβλημένων, ἵνα τὸ πρὸς τὴν αἴσθησιν αὐτῶν ὁμολογούμενον ἀδιστάκτως ἔχωμεν ὑποτεθειμένον, 16.29–31, trans. Barker 1989: 290.

17.2) among investigators. From the outset, then, a number of inadequate instruments are removed from the harmonicist's workshop: auloi and *syringes* (one thinks of Archytas), weights hung from strings (one thinks of demonstrations attributed to Pythagoras), percussion instruments such as spheres,[48] discs (one thinks of Hippasus) and bowls. Ptolemy's explanations of why the ratios of the concords are not satisfactorily displayed in these instruments are not consistently accurate, but he does focus his attention on the main concern for stringed instruments: evenness of materials. This will become an important and recurring element of all the tests by which Ptolemy ensures the reliability of his instruments.

Having set these other instruments aside, he can focus his attention on the monochord. Later (II.12) he says that it had become something of a rarity in his day:

δοκεῖ μὲν γὰρ παρεληλυθέναι τὸ τοιοῦτον ὄργανον πρὸς χρῆσιν ἅμα χειρουργικὴν καὶ θεωρίαν τῶν ἀποτελεσματικῶν τοῦ ἡρμοσμένου λόγου, ἐπειδὴ τοῖς ἄλλοις οὐκ ἐφαίνετο τῶν εἰρημένων ἑκάτερον ὑπάρχον, ἀλλὰ τοῖς μὲν κανονικοῖς τὸ θεωρηματικὸν μόνον.

For this kind of instrument [the single-stringed *kanōn*] seems to have passed into disuse, both for practical performance and for the theoretical study of the things that produce the ratio of what is attuned, since the *kanonikoi* had as their domain only the study of theorems.[49]

This last statement is surprising, given what we have seen in chapter 5, unless Ptolemy means by it merely that what the *kanonikoi* did with the *kanōn* was limited to the demonstration of concords and scale-systems. That is, they studied their theorems, produced ratios, and made them audible; but at no point did they use the *kanōn* to allow perception to judge the results of their theorising.[50] The instrument was a one-way street, leading from reason to perception, but not back again.

But what Ptolemy actually says is not that the instrument was being used for purposes different from his own; that much is clear in any case. The key question is how to translate the verb

[48] 17.18: I follow Barker's reading (σφαῖραι), which is supported by the schol. to Ptol. *Harm.* 17.28. Düring prefers σφῦραι, 'hammers', which would fit the Pythagoras tradition better; see his apparatus ad loc. and Raasted 1979. Solomon (2000: 25) follows Düring's reading.

[49] 66.15–19, trans. Barker 1989: 340.

[50] This is Barker's view (1989: 340 n. 90).

παρεληλυθέναι (66.15). Düring took it in the sense 'fall into disuse'; both Barker and Solomon follow him in their translations.[51] Massimo Raffa has recently proposed that it should be understood in its second sense, 'be superior', 'stand out'; this was the interpretation of Wallis and Ruelle.[52] Both interpretations have merit: Ptolemy's concern in this passage is to point out the limitations of the single-stringed *kanōn* partly in order to showcase his own originality in proposing polychord *kanones* which address these limitations. He draws attention to the fact that no one has yet come up with anything better than the monochord for assessing mathematically derived attunements by ear (66.13–15). His concern to find a solution for the instrument's limitations is thus contrasted with either the neglect or the complacency of his immediate predecessors. They either neglected the monochord altogether because they were more concerned with theorems than with ensuring that the theorems were supported by the evidence of perception, or they found the instrument entirely satisfactory because it seemed to them to be superior to any others which might be used for the purpose (e.g. lyres, kitharas, auloi, *syringes*, and so on).

Ptolemy's claims for originality are founded partly on the detailed attention he gives the instrument, something which either reading of 66.15 supports. This begins with his introduction of the *kanōn* in I.8:

ἡ δὲ ἐπὶ τοῦ καλουμένου κανόνος διατεινομένη χορδὴ δείξει μὲν ἡμῖν τοὺς λόγους τῶν συμφωνιῶν ἀκριβέστερόν τε καὶ προχειρότερον, οὐ μὴν ὡς ἔτυχε λαβοῦσα τὴν τάσιν, ἀλλὰ πρῶτον μὲν μετά τινος ἀνακρίσεως πρὸς τὴν ἐσομένην ἂν ἐκ τῆς κατασκευῆς ἀνωμαλίαν, ἔπειτα καὶ τῶν περάτων τὴν προσήκουσαν λαμβανόντων θέσιν, ἵνα τὰ πέρατα τῶν ἐν αὐτοῖς ἀποψαλμάτων, οἷς ὁρίζεται τὸ πᾶν μῆκος, οἰκείας τε καὶ δήλας ἔχῃ τὰς ἀρχάς. νοείσθω δὴ κανὼν ὁ κατὰ τὴν ΑΒΓΔ εὐθεῖαν καὶ μαγάδες πρὸς τοῖς πέρασιν αὐτοῦ πανταχόθεν ἴσαι τε καὶ ὅμοιαι σφαιρικάς, ὡς ἔνι μάλιστα, ποιοῦσαι τὰς ὑπὸ τὴν χορδὴν[53] ἐπιφανείας, ἥ τε ΒΕ περὶ κέντρον τῆς εἰρημένης ἐπιφανείας τὸ Ζ, καὶ ἡ ΓΗ περὶ κέντρον ὁμοίως τὸ Θ, ληφθέντων τε τῶν Ε καὶ Η σημείων κατὰ τὰς διχοτομίας τῶν κυρτῶν

[51] LSJ s.v. παρέρχομαι I: Düring 1934: 241, Barker 1989: 340, Solomon 2000: 93.

[52] LSJ s.v. παρέρχομαι II: Wallis 1699: 84; Ruelle 1897: 310; Raffa 2002: 176, 405–8.

[53] τὴν χορδήν] Barker reads the singular here instead of the plural (τὰς χορδάς Düring). Solomon (2000: 26) follows Düring's reading, but cites Porphyry (*In Harm.* 121.24), who uses the singular in his close paraphrase of the passage; the MSS are unanimous in Porphyry's case (see Düring's apparatus ad loc.).

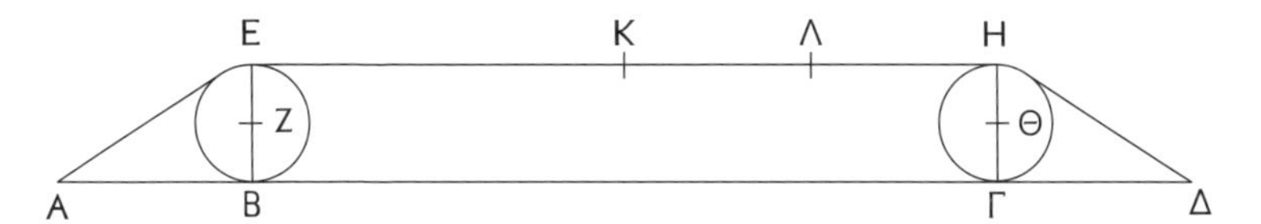

Fig. 6.1 Diagram accompanying Ptol. *Harm.* I.8 (after Düring 1930).

ἐπιφανειῶν. θέσιν ἐχέτωσαν τοιαύτην αἱ μαγάδες, ὥστε τὰς διὰ τῶν E καὶ H διχοτομιῶν καὶ τῶν Z Θ κέντρων ἐκβαλλομένας, τουτέστι τὴν EZB καὶ τὴν HΘΓ καθέτους εἶναι πρὸς τὴν ΑΒΓΔ. ἐὰν τοίνυν ἀπὸ τῶν A καὶ Δ διατείνωμεν χορδὴν σύμμετρον, ὡς τὴν ΑΕΗΔ, παράλληλός τε ἔσται τῇ ΑΒΓΔ, διὰ τὸ ἴσον ὕψος ἔχειν τὰς μαγάδας. καὶ λήψεται κατὰ τὰ E καὶ H σημεῖα τὰς ἀρχὰς τῶν ἀποψαλμάτων. ἐπ' αὐτῶν γὰρ ποιήσεται τὰς ἐπαφὰς τῶν κυρτῶν ἐπιφανειῶν, διὰ τὸ τὰς EZB καὶ HΘΓ καθέτους γίνεσθαι καὶ πρὸς αὐτήν.

But the string stretched over what is called the *kanōn* will show us the ratios of the concords more accurately and readily. It does not acquire its pitch in any random way, but in the first place it is equipped with a way of assessing any unevenness that might arise from the apparatus, and secondly its limits are appropriately placed so that the limits of the plucked sections between them, into which the whole length is divided, have suitable and clearly perceptible points of origin. Let us think of a *kanōn* on the straight line ΑΒΓΔ, and at its limits, bridges that are in all respects equal and similar, with the surfaces that lie under the string spherical, as nearly as is possible. Let one bridge, BE, have Z as the centre of the surface mentioned, and let the other, ΓH, have Θ, similarly, as the centre, where points E and H are found by bisection of the convex surfaces. Let the bridges be so placed that the lines drawn through the points of bisection E and H and through the centres Z and Θ are perpendicular to ΑΒΓΔ. If then from A and Δ we stretch a string of appropriate length, ΑΕΗΔ, it will be parallel to ΑΒΓΔ,[54] because the bridges have equal height; and at points E and H it will have the beginnings of its plucked sections. For it is at these points that it will touch the convex surfaces, since EZB and HΘΓ are perpendicular to it too.[55]

This is the fullest and most detailed description of the monochord we possess from an ancient author. Several things make it especially striking: firstly, its detail is both geometrical and practical, as we noted in chapter 1. Secondly, Ptolemy's concern for evenness and exactitude in the instrument's physical construction

[54] It is not strictly possible for ΑΕΗΔ to be parallel to ΑΒΓΔ, as Ptolemy has set them out, but the point is that the string segment EH will be parallel to the segment of the *kanōn* immediately below it (ΒΓ).

[55] 17.20–18.9. The translation is Barker's (1989: 291–2), but here and in subsequent quotations I have retained the Greek letters to facilitate comparison with the diagram.

is a persistent feature. The monochord is to be preferred to other instruments (even the ones Ptolemaïs said were sometimes used despite being 'not strictly canonic'[56]) because it will display the ratios of the concords 'more accurately and readily'. It has two further advantages: it can be used to detect its own flaws (an aspect to which Ptolemy returns), and the quantities of the intervals are presented in such a way as to make them 'clearly perceptible'.

Ptolemy sketches the *kanōn* in geometrical fashion first; we saw in chapter 1 that his use of the verb *noein* specifically invokes the language of mathematical discourse. But we also noted that by the end of the first sentence of his description (17.27–18.2) it is clear that he is thinking of a construction which is substantially under-represented by the accompanying diagram. 'At its limits', he says, let us imagine 'bridges that are in all respects equal and similar, with the surfaces that lie under the string spherical, as nearly as is possible.' The geometer does not need to resort to such language: he need only say 'let there be imagined two equal spheres', and he may then take for granted that they are 'in all respects equal and similar'. His accompanying diagram will under-represent them, but that is why he employs the verb *noein*, and however deficient the diagram actually is, it will not detract from the exact equality and regularity which the spheres possess ideally. The direct object of the verb *noein* in the key clause of Ptolemy's sentence, however, is not 'spheres', but 'bridges' (*magades*, 17.27); it is their contact surfaces that must be 'as spherical as possible'. This is a material construction described in geometrical terms, not a geometrical construction with material additions.

In this case, however, the word 'spherical' is somewhat problematic. If Ptolemy is really interested in the accuracy of his instruments, bridges with spherical surfaces would seem to be an awkward stipulation. Geometrically, of course, it makes no difference whether the bridges are spheres, cylinders or circles: all three meet their tangents at a single point; all three are represented simply by a circle in Greek mathematical diagrams.[57] Practically, however, the musicologist must be able to reassure himself that the string is

[56] Ptolemaïs ap. Porph. *In Harm.* 23.1–4. [57] Netz 1999a: 53.

raised to exactly the same height by each bridge; horizontal cylindrical bridges allow for this in a way that spherical ones (even hemispheres) do not. How could the theorist remain confident that the string was always sitting precisely on the highest point of each spherical surface throughout his operations with the instrument?

The problem has not gone unnoticed. Porphyry, whose commentary on the passage offers some additional construction details,[58] also adds the word 'cylindrical' to an account which otherwise follows Ptolemy's closely. But the bridges he suggests, interestingly, are not horizontally cylindrical: 'And let there be bridges made of horn, equal and alike on all sides, and let them be cylindrical up to a certain height, but spherical where their surfaces fall under the string.'[59] This is a slightly clarified image of what Ptolemy had described: the bridges are cylindrical at their bases, but spherical at their summits, where the string crosses them. Each is a vertical cylinder capped with a segment of a sphere, like the dome of an observatory.[60] This is not what the diagram of *Harmonics* I.8 shows (see fig. 6.1), but the diagram represents only geometrically relevant information, and does so in strict conformity with Ptolemy's text. Geometrically speaking, it is an accurate representation of Ptolemy's *kanōn*, because the string and the ruler are separated by a figure which makes contact with each at a single definable point: thus ΕΒ = ΗΓ. According to the diagram, the measurement on the ruler will be the measurement on the string (ΒΓ = ΕΗ), and this is what Ptolemy's instrumental procedures will require.

His description is ambiguous because it emphasises the geometry of the instrument. The geometrical model has bridges which are perfectly spherical in their entirety (not just in their contact surfaces). Porphyry interprets the passage with the practicalities of canonic division in mind: by not being completely spherical, his bridges cannot roll about on the surface of the *kanōn*. But what Porphyry's bridges gain in practicality they lose in geometrical

[58] The surface of the *kanōn* should be wooden (*In Harm.* 121.17) and the bridges of horn (121.23).

[59] καὶ μαγάδες ἔστωσαν ἀπὸ κεράτων πεποιημέναι πανταχόθεν ἴσαι τε καὶ ὅμοιαι μέχρι μέν τινος ὕψους κυλινδρικαὶ οὖσαι, σφαιρικὰς δ' ἔχουσαι τὰς ὑπὸ τὴν χορδὴν πιπτούσας ἐπιφανείας, 121.22–5.

[60] See Alexanderson 1969: 9, Barker 2000: 197.

precision: the exact equality of their heights must be stated simply as a desideratum, rather than as part of the definition of the figure. In two equal spheres, the fact that EB = HΓ is obvious; Porphyry's bridges could be described in such precise geometrical terms, but not nearly so efficiently.[61]

There remains the difficulty of ensuring that the string always crosses the bridge exactly at its highest point. This difficulty can be minimised by fixing Porphyry-style sphere-capped cylindrical bridges in place, aligned in such a way that the string bisects their bases exactly, and by keeping the string at high tension. But it can be surmounted more easily by substituting horizontal cylindrical contact surfaces for spherical ones. Horizontal cylindrical or semi-cylindrical bridges (dowel segments, in effect) have the advantage of permitting the string to return more easily to its 'true' position after accidental lateral displacement (as sometimes happens when the movable bridge is shifted in canonic division). This solution is not suggested by any of the ancient commentators, but one scholiast seems to have assumed that Ptolemy meant to say 'cylindrical' when he said 'spherical'.[62]

The remainder of Ptolemy's description of the *kanōn* presents a mixture of geometrical and practical information. Points, lettered as in the diagram, are located in geometrical fashion: by bisection of line segments. This is an essential step, since the theorist must be certain that the string, introduced as AEHΔ (18.5) will be parallel to the *kanōn* below it, ABΓΔ, so that division of the one may be taken to be the same as division of the other. (Strictly speaking, only EH is parallel to BΓ, but Ptolemy himself refers to AEHΔ and ABΓΔ and calls them parallel.) He can be confident of this because 'the bridges have equal height' (18.6), a fact established through the initial construction. Points E and H are the limits of the speaking length of the string, because EH is a tangent of both bridges, because EH and BΓ are parallel, and because EZB and HΘΓ are at right angles to these two parallels. The exactitude of

[61] Rather than saying that the bridges should be cylindrical 'up to a certain height', he would need to say that the contact surfaces should be equal sections of equal spheres fitted to the summits of equal cylinders whose bases are perpendicular to their sides.

[62] Schol. to Ptol. *Harm.* 17.28; see Solomon 2000: 26 n. 139.

the practical construction is assured through its proximity to the geometrical construction.

It is at this point that the theoretical problem which surfaced in the work of Adrastus (what I have called the P/Q discrepancy)[63] becomes relevant, since the physically constructed monochord must now comply exactly with the internal relationships of the geometrical figure, a requirement which is of course impossible. But Ptolemy draws attention to a more serious issue. The problem is not merely that the bridge has width, as Adrastus noted; that particular difficulty has already been minimised as far as it can be in Ptolemy's directions about the shape of the bridges. Now the problem is that unlike the purely mathematical division of the *kanōn*, the practical division must involve movable bridges (*hypagōgia* or *magadia*) which, if they are to stop the string, must be slightly higher than the end-bridges (*magades*). Geometrically we are now off the rails: ΕΚΗ is no longer a straight line, and hence no longer parallel to ΒΓ (see fig. 6.1). If we place one of these *magadia* at Κ, the string will now form an isosceles triangle with an extremely short altitude (ΕΚΗ). Suddenly the reassuring similarity between the arithmetical type of lettered diagram and the monochord, through which notes could be represented by simple line segments and intervals by the ratios between their lengths, breaks down because *real* canonic division turns out not to function by simple line division after all. When we place the bridge at Κ, we think we are simply bisecting ΕΗ, but instead we are generating the triangle ΕΚΗ. Because of the physical requirements of canonic division, the geometry of sound is no longer an arithmetical geometry.

Unlike the situation as Adrastus described it, where the width of the bridge took away from the overall length of the string, we now see that the height of the bridge adds to the overall length. But greater length only translates to lower pitch when all other factors are kept the same; the placing of the bridge will now have stretched the string to a higher tension than it had before. Furthermore, the acoustical effect of placing the movable bridge at Κ will be different to that of placing it at Λ, for the closer it is

[63] See ch. 5 part 2.

placed to one of the fixed bridges, the greater its distortion of the original tension of the string. Transferring canonic division from the geometrical to the physical therefore introduces irregularities of many sorts.[64] Because the notes sounded in canonic division must be quantifiable in order to be scientifically meaningful, these irregularities constitute a serious challenge to the harmonicist. For the upshot of all this is that strings no longer appear to be representable by lines. The fact that lines (as geometrical entities) have no thickness (Euc. *El.* I def. 2) can be circumvented by ensuring that the instrument's string is uniform in thickness, but the fact that lines cannot have tension now makes line segments incapable of representing musical distances. In geometry, segments of straight lines can differ only in length, but the string segment EK differs not only in length but also in tension according to the presence or absence of the bridge under K.

Ptolemy offers several responses to this issue. In I.11 he puts forward an argument by which it would appear that whatever sharpness is generated by the increase in tension is balanced exactly by an equal flattening from the increase in length. The argument is not applied specifically to this problem, and in fact it is incorrect.[65] But in III.2 Ptolemy does tackle the question of whether the greater height of the movable bridge skews the measurement of the string segment. By appeal to geometry, he argues that as long as the contact surfaces of the two bridges are segments of equal circles, the measurement will not be affected despite the height of the movable bridge. Thus it is by geometry that he attempts to reassure himself that the departure from arithmetical geometry required by the physical *kanōn* does not interfere with the fidelity of its evidence.

The geometrical reassurance, however, only goes so far. Ptolemy posits a *kanōn* AB with two bridges, one higher than the other (ΑΓΔ and BEZ in fig. 6.2). The highest point of the lower (fixed) bridge is Δ, that of the higher (movable) Z. Their surfaces are to be thought of as segments of circles (τμήματα κύκλων, 90.8). If the bridges were of equal height, the string would make contact with them at Δ and Z. But the second bridge must be higher for practical

[64] See particularly Barker 2000: 51, 198–203. [65] See Barker 2000: 199.

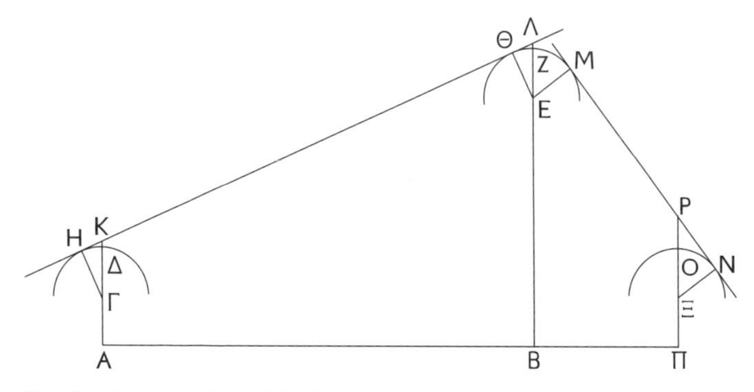

Fig. 6.2 An extension of the diagram accompanying the last part of Ptol. *Harm.* III.2 (p. 90 Düring). Ptolemy's diagram includes points A–Λ; points M–P and the structures on which they fall are my additions.

reasons, and so the true contact points are slightly lower than the summit of each bridge, at points H and Θ. Ptolemy's geometrical argument (90.6–91.19) is that because the bridges' surfaces are the circumferences of equal circles, the whole measurement has simply been shunted to one side (to the left, in the diagram), but is otherwise identical: the equality of the bridges' surfaces ensures that the distances HK and ΘΛ are equal, and therefore that HΘ and KΛ are also equal (90.18–91.6).

This argument has several limitations. First, it requires a new point of comparison: Ptolemy can reassure himself that the distances HΘ and KΛ are equal, but this now requires him to apply the *kanonion* directly to the string itself (90.16–18); HΘ and KΛ, though equal to one another, are each longer than AB. What this means in practice is that the harmonicist cannot simply read off the unit-numbers on the base of the *kanōn* (AB), even when the movable bridge has been placed precisely relative to the axes of the end-bridges. This introduces a practical complication that is even greater for his polychord *kanones* (as Ptolemy recognises, 91.8–19), on which the level of geometrical distortion (i.e. that of length irrespective of tension) will differ from one string to the next whenever the bridges need to be placed simultaneously at different positions, something which his procedures frequently require.

The second limitation of Ptolemy's geometrical argument is that it does not take account of the increase in tension which arises from stretching the string to a greater overall length. It is in his unconcern for this limitation that we can see most clearly the extent to which he was prepared to reduce the problem to geometry. In III.2 he focusses only on the string segment on one side of the movable bridge, but in fact his procedures are not always limited in this respect: his very first canonic division (I.8, 18.22–19.15) uses the segments of string on both sides of the movable bridge, as Panaetius and Didymus had done; his procedures with the *helikōn* and its variant (II.2) will require this technique also. If we imagine (see fig. 6.2) a continuation of Ptolemy's diagram to the other fixed bridge (ΠΞΟ), it is easy to see that although his geometrical argument will apply as well to MN as it did to HΘ, the string has now been stretched to accommodate not only the amount by which HΘ + MN is greater than ΔO (the length of the open string), but also by an amount equal to the arc ΘM. The level of acoustic distortion (i.e. that of length obtained by an increase in tension) this generates will vary not only according to the height and position of bridge BEZ, but also according to the amount of string left over between each of the end-bridges and the *kollaboi* (pegs) by which the string is fixed to the instrument.[66] The longer the total length of the string (including the portions beyond both end-bridges), the less the distortion caused by the placement of the movable bridge. Ptolemy's emphasis on controlling string length by appeal to geometry thus comes at the cost of controlling tension.

Ptolemy does not address the issue of tension beyond the brief and inadequate remarks of I.11. He was evidently satisfied that as long as the surfaces of the bridges made equal circles, and efforts were made to avoid sounding very short string segments, the distortion could be minimised. He was content to restore canonic division to the realm of arithmetic by a combination of geometrical argument and practical adjustment. To be fair, he had no means of calculating the acoustic effect of differential tension on his apparatus, and although his attempt to reduce the difficulty to

[66] That is, the portions of 'inactive' string to the left of H and to the right of N in fig. 6.2.

an inverse proportionality in I.11 is unsatisfactory, he is the first harmonicist to apply such an argument to the reliability of scientific instruments. Furthermore, the distortions introduced by the addition of length and tension are truly minuscule on a well-built *kanōn*. The importance of the way Ptolemy handles them is in what it tells us about his scientific priorities.

In I.8, Ptolemy's focus is on evenness and equality of materials; other factors are not considered. He offers a test by which the theorist is able to determine whether the instrument he has constructed gives results which can be trusted. If the instrument fails, he must correct the unevenness of its materials. If it passes, he may use it with confidence in subsequent procedures. (Designation-letters in the following quotation refer to the diagram of fig. 6.1.)

ἐφαρμόσαντες δὴ τῇ χορδῇ κανόνιον καὶ μεταλαβόντες ἐπ' αὐτοῦ τὸ ΕΗ μῆκος, ἵνα προχειρότερον ποιώμεθα τὰς παραμετρήσεις, πρῶτον μὲν ἐπὶ τὴν γινομένην τοῦ ὅλου μήκους διχοτομίαν, οἷον τὴν Κ, καὶ ἔτι τὴν τῆς ἡμισείας διχοτομίαν, ὡς τὴν Λ, καταστήσομεν ὑπαγώγια στενὰ εὖ μάλα καὶ λεῖα, ἢ καὶ νὴ Δία μαγάδια ἕτερα, ὑψηλότερα μὲν ἐκείνων βραχεῖ, ἀπαραλλάκτως δὲ ἔχοντα, θέσεως ἕνεκεν, ἰσότητος καὶ ὁμοιότητος κατὰ τῆς μέσης τοῦ κυρτώματος γραμμῆς, ἥτις ὑπ' αὐτὴν ἔσται τὴν τοῦ κανονίου διχοτομίαν ἢ πάλιν τὴν τῆς ἡμισείας διχοτομίαν, ἵνα ἐὰν μὲν τὸ ΕΚ τῆς χορδῆς μέρος ἰσότονον εὑρίσκηται τῷ ΚΗ καὶ ἔτι τὸ ΚΛ τῷ ΛΗ, δῆλον ἡμῖν αὐτῆς ᾖ τὸ κατὰ τὴν σύστασιν ἀπαράλλακτον. ἐὰν δὲ μή, μεταφέρωμεν τὴν δοκιμασίαν ἐπ' ἄλλο μέρος, ἤτοι χορδὴν ἄλλην, ἕως ἂν τὸ ἀκόλουθον διασωθῇ, τουτέστι τὸ ἐν τοῖς ὁμοίοις καὶ ἀναλόγοις καὶ ἰσομήκεσι καὶ μίαν ἔχουσι τάσιν ὁμότονον.

To the string we shall now fit a measuring-rod (*kanonion*), and use it to divide up the length of ΕΗ, so that we may make the comparative measurements more easily. First at the bisection of the whole length, Κ, and then at the bisection of the half, Λ, we shall place blades, very thin and smooth, or indeed other bridges, a little higher than the others, but no different from them in respect of their position, equality or similarity about the line through the middle of the convexity, which will be under the exact bisection of the measuring-rod or again under the bisection of the half. Then if part ΕΚ of the string is found to be of equal pitch to ΚΗ, and again ΚΛ to ΛΗ, the string's evenness of constitution will be evident to us. If they are not so, let us transfer the test to another part, or to another string, until the required consequence is preserved – that is, sameness of pitch in parts that are similar, corresponding, equal in length, and of single tension.[67]

[67] 18.9–22, trans. Barker 1989: 292. As in the preceding passage, I have retained the Greek letters.

The test is perceptibility. Ptolemy does not deny that the discrepancy between the ideal and physical exists, but it is relevant to the theorist only insofar as it distorts the results to a perceptible degree. If all reasonable attempts are made to ensure equality of materials, and no difference of pitch is detected between identical lengths of the same string, then the instrument will not lead the theorist astray. Under these considerations, the requirements for the bridges will be merely that whatever distortion they introduce is imperceptible. From our look at III.2, however, it is evident that the first type Ptolemy describes (those which are thin and smooth like blades, 18.13), will not satisfy his geometrical conclusions as well as the second type (those identical to the end-bridges except in their height). But the ultimate test of perception is the only one which matters in the end: an imperceptible error will not mislead the harmonic scientist; he has merely reached the limits of the capabilities of his own senses.

See no discord, hear no discord: rescuing the octave plus fourth

With this crucial test passed, Ptolemy can now move on to his first canonic division. This is in many ways the same procedure we have seen in many of his predecessors: a simple demonstration of the concord ratios. But by combining two of their procedures Ptolemy is able to produce a division which can include the controversial and mathematically difficult octave plus fourth without drawing attention to its ratio.[68]

ἔπειτα τοῦ τοιούτου καταληφθέντος καὶ καταδιαιρεθέντος τοῦ κανονίου τοῖς ἐκκειμένοις τῶν συμφωνιῶν λόγοις, εὑρήσομεν ἐκ τῆς ἐφ᾽ ἕκαστον τμῆμα τοῦ μαγαδίου παραγωγῆς ὁμολογουμένας ταῖς ἀκοαῖς ἐπὶ τὸ ἀκριβέστατον τὰς τῶν οἰκείων φθόγγων διαφοράς. τοιούτων μὲν γὰρ λαμβανομένης τῆς ΕΚ διαστάσεως τεσσάρων, οἵων ἐστὶν ἡ ΚΗ τριῶν, οἱ καθ᾽ ἑκάτερον αὐτῶν φθόγγοι ποιήσουσι τὴν διὰ τεσσάρων συμφωνίαν διὰ τὸν ἐπίτριτον λόγον. τοιούτων δὲ λαμβανομένης τῆς ΕΚ τριῶν, οἵων ἐστὶν ἡ ΚΗ δύο, ποιήσουσιν οἱ καθ᾽ ἑκάτερον φθόγγοι τὴν διὰ πέντε συμφωνίαν διὰ τὸν ἡμιόλιον λόγον. καὶ πάλιν ἐὰν μὲν οὕτως διαιρεθῇ τὸ πᾶν μῆκος, ὥστε τὴν μὲν ΕΚ γίνεσθαι δύο τμημάτων, τὴν δὲ ΚΗ τοῦ αὐτοῦ ἑνός, ἔσται τὸ διὰ πασῶν ὁμόφωνον παρὰ τὸν διπλάσιον λόγον. ἐὰν δὲ ὥστε τὴν μὲν ΕΚ συνάγεσθαι τμημάτων ὀκτώ, τὴν δὲ ΚΗ τῶν αὐτῶν

[68] Designation-letters in the following passage refer to the diagram of fig. 6.1.

τριῶν, ἡ διὰ πασῶν καὶ διὰ τεσσάρων ἔσται συμφωνία κατὰ τὸν ὀκτὼ πρὸς τὰ τρία λόγον. ἐὰν δὲ ὥστε τὴν μὲν ΕΚ τμημάτων εἶναι τριῶν, τὴν δὲ ΚΗ τοῦ αὐτοῦ ἑνός, ἡ διὰ πέντε καὶ διὰ πασῶν ἔσται συμφωνία κατὰ τὸν τριπλάσιον λόγον. ἐὰν δὲ ὥστε τὴν μὲν ΕΚ συνάγεσθαι τμημάτων τεσσάρων, τὴν δὲ ΚΗ τοῦ αὐτοῦ ἑνός, ἔσται τὸ δὶς διὰ πασῶν ὁμόφωνον παρὰ τὸν τετραπλάσιον λόγον.

When something of this kind [i.e. equality of length, tension and pitch] has been found, and the measuring-rod has been divided in the ratios of the concords that have been set out, by shifting the bridge to each point of division we shall find that the differences of the appropriate notes agree most accurately with the hearing. For if distance EK is constructed of four such parts as those of which KH is three, the notes corresponding to each of them will make the concord of a fourth through the epitritic ratio. If EK is constructed of three such parts as those of which KH is two, the notes corresponding to each will make the concord of a fifth through the hemiolic ratio. And again, if the whole length is so divided that EK is made up of two sections and KH of one that is the same, there will be the homophone of the octave, in accordance with the duple ratio. If it is divided so that EK is put together from eight sections and KH from three of the same, there will be the concord of an octave and a fourth, corresponding to the ratio of 8 to 3. If it is divided so that EK is of three sections and KH of one that is the same, there will be the concord of an octave and a fifth, corresponding to the triple ratio. And if it is divided so that EK is put together from four sections and KH from one that is the same, there will be the homophone of the double octave, in accordance with the quadruple ratio.[69]

In following the description, we realise that K is not a fixed point; it is the movable bridge, placed at a different position for each interval, and occupying six different points in the course of the division. We also note that both sides of the string are required: the comparisons are between segments of string on either side of K, not between one of these lengths and EH. This has affinities with the procedure of Panaetius' division, but there the comparison was between the length of the entire string and each of the two segments, not between the two segments. Thirdly, we note that the order of construction is much more transparent than in the divisions of either Panaetius or Adrastus.[70] Panaetius produced his concords in the order octave, fourth, double octave, fifth, octave plus fifth, tone; there are in fact many alternatives which use the same economy of bridge positions. The division attributed to Adrastus

[69] 18.22–19.15, trans. Barker 1989: 292–3.

[70] Panaetius' canonic division and that attributed to Adrastus are discussed in ch. 5 part 2.

proceeded in linear fashion, giving fourth, octave, double octave, fifth, octave plus fifth, tone: the same intervals in a different sequence.

Ptolemy's approach is novel. Firstly, his division gives the same number of intervals (six), but he omits the tone, and includes the octave plus fourth. This eliminates the distinction between concords and discords which was present in the divisions of Panaetius and Adrastus. Secondly, he constructs these six intervals in ascending order of magnitude, using one bridge position for each interval. A corollary of this is that K will be placed in a linear sequence of positions, a much neater sequence, in fact, than in the division of Adrastus, where we first encountered the procedure. If we construct the division on a 120-unit monochord, the positions will be as follows:[71]

EK:KH = 4:3	fourth	51 26′
EK:KH = 3:2	fifth	48
EK:KH = 2:1	octave	40
EK:KH = 8:3	octave plus fourth	32 44′
EK:KH = 3:1	octave plus fifth	30
EK:KH = 4:1	double octave	24

The octave plus fourth has been ingeniously accommodated. The explanation of its ratio is no different from that of the others (except for the unavoidable fact that in Greek the ratio requires twice as many words to articulate as any of the others[72]), and the interval itself can be placed between its neighbours without offending either reason or the ear. Nothing can be done about its ratio, which remains the only one on the list which is neither multiple nor epimoric, but in the practical context of the division, the position of the bridge has now become the more dominant numerical feature. Furthermore, the omission of the tone (which in this arrangement would have to come first, with a bridge position of 56 28′) avoids drawing attention to the formal similarities of its ratio to those of the other concords, as opposed to the anomalous

[71] As in fig. 6.1, the end-bridges are E and H, and the movable bridge is K. A complementary series of bridge positions will be found if the division is carried out at the other end of the string (68 34′, 72, 80, 87 16′, 90, 96). The bridge positions here and in the table are in sexagesimal notation (on which see below, and ch. 4 table 4.1).

[72] Cf. e.g. τὸν ἡμιόλιον λόγον (19.5–6) with τὸν ὀκτὼ πρὸς τὰ τρία λόγον (19.10–11).

ratio of the octave plus fourth. It is the best solution offered by any ancient theorist.[73]

Using the kanōn

The monochord, thus established by reassurances of its accuracy and reliability, can be put to work immediately. Firstly, it offers retroactive assistance. The canonic division of 18.22–19.15 comes as the confirmation of what was argued in the course of Ptolemy's critique of Pythagorean conceptions of concord, and of what he asserted in the course of his own account (I.5–7). The ratios he has specified have been proven by trustworthy apparatus to fit the intervals they are meant to fit; the one concord which embarrassed the Pythagoreans is no embarrassment to Ptolemy's division. But canonic division is, in the main, a vindication of what was essentially correct about Pythagorean analysis (19.16–17).

When he moves on to the Aristoxenians in I.9–12, Ptolemy has much recourse to the *kanōn*. It is only through the proof of the correspondence between ratio and interval offered in I.8 that he can both attack the Aristoxenian theorists for refusing to accept ratio-based harmonic analysis in the first place, and question the theoretical consequences of this refusal:

τοῖς Ἀριστοξενείοις δέ, ἐπεὶ μήτε τούτοις ἐναργῶς ἔχουσι συγκατέθεντο, μήτε εἴπερ ἠπίστουν αὐτοῖς, τοὺς ὑγιεστέρους ἐζήτησαν, εἴ γε θεωρητικῶς ὑπισχνοῦντο προσενηνέχθαι μουσικῇ. τὸ μὲν γὰρ τὰ τοιαῦτα πάθη ταῖς ἀκοαῖς παρακολουθεῖν ἐκ τοῦ πῶς ἔχειν τοὺς φθόγγους πρὸς ἀλλήλους ἀναγκαῖον αὐτοῖς ἐστιν ὁμολογεῖν, καὶ προσέτι τὸ τῶν αὐτῶν ἀντιλήψεων ὡρισμένας καὶ τὰς αὐτὰς εἶναι διαφοράς. πῶς δὲ ἔχουσι καθʼ ἕκαστον εἶδος οἱ ποιοῦντες αὐτὸ

[73] It has its limitations, of course. One might object that the presentation of the concords in this way could easily include discords without raising any more mathematical objections than those raised already about the 8:3 interval. If, for example, Ptolemy had wanted to include Philo's 5:2 octave plus major third (see ch. 5 n. 52), whose ratio differs in no mathematically significant way from that of the octave plus fourth, he could have constructed it between the octave and the octave plus fourth without disrupting his procedure; its bridge position in my scheme would be 34 17′. In this sense the procedure is blind: there is nothing about it which separates concords from discords. But it is not deaf, and that is perhaps Ptolemy's defence. The process does not highlight ratio form, but rather concentrates the attention on the perceptible evidence, which will be found to agree with the ratios previously established. For a more mathematical inclusion of the 8:3 interval, cf. the way Ptolemy sets it out in his account of the *helikōn* (II.2, discussed below; see n. 111).

δύο φθόγγοι πρὸς ἀλλήλους, οὔτε λέγουσιν οὔτε ζητοῦσιν, ἀλλ' ὥσπερ αὐτῶν ἀσωμάτων μὲν ὄντων, τῶν δὲ μεταξὺ σωμάτων, τὰς διαστάσεις τῶν εἰδῶν μόνας παραβάλλουσιν, ἵνα τι δόξωσιν ἀριθμῷ καὶ λόγῳ ποιεῖν.

But we should find fault with the Aristoxenians, since they neither accepted these ratios as clearly established, nor, if they really lacked confidence in them, did they seek more satisfactory ones – assuming that they were genuinely committed to the theoretical study of music. For they must necessarily agree that such experiences come to the hearing from a relation that the notes have to one another, and further that where the impressions are the same, the differences are determinate and the same. Yet in what relation, for each species [of concord], the two notes that make it stand, they neither say nor enquire, but as if the notes themselves were bodiless and what lie between them were bodies, they compare only the intervals belonging to the species, so as to appear to be doing something with number and reason.[74]

Ptolemy's jibe, 'assuming that they were genuinely committed to the theoretical study of music', illustrates the extent to which he was willing to equate rational analysis with music theory. The Aristoxenians' outright rejection of the mathematical language of canonics calls into question their very commitment to the theoretical investigation of music. This is a position which can only be sustained with confidence as a result of procedures such as those of I.8.

In I.10 Ptolemy begins to home in on specific issues; here it is the Aristoxenian assertion that the fourth consists of two and a half tones.[75] He summarises the argument of 'the Aristoxenians' in this way:

αὐτοὶ μὲν οὖν πειρῶνται τὸ προκείμενον δεικνύειν οὕτως. ἔστωσαν γὰρ δύο φθόγγοι διὰ τεσσάρων συμφωνοῦντες οἱ Α Β, καὶ ἀπὸ μὲν τοῦ Α δίτονον εἰλήφθω ἐπὶ τὸ ὀξὺ τὸ ΑΓ, ἀπὸ δὲ τοῦ Β ὁμοίως ἐπὶ τὸ βαρὺ δίτονον εἰλήφθω τὸ ΒΔ· ἑκάτερον ἄρα τῶν ΑΔ καὶ ΓΒ ἴσον ἐστίν, καὶ τηλικοῦτον ᾧ ἐλλείπει τὸ δίτονον τοῦ διὰ τεσσάρων. πάλιν δὴ ἀπὸ μὲν τοῦ Δ διὰ τεσσάρων εἰλήφθω ἐπὶ τὸ ὀξὺ τὸ ΔΕ, ἀπὸ δὲ τοῦ Γ ὁμοίως διὰ τεσσάρων ἐπὶ τὸ βαρὺ τὸ ΓΖ. ἐπεὶ τοίνυν ἑκάτερον τῶν ΒΑ καὶ ΓΖ διὰ τεσσάρων ἐστίν, ἴσον ἐστὶ καὶ τὸ ΒΓ τῷ ΑΖ, κατὰ τὰ αὐτὰ δὲ καὶ τὸ ΑΔ τῷ ΒΕ. ἴσα ἄρα τὰ τέσσαρα διαστήματά ἐστιν ἀλλήλοις. ἀλλ' ὅλον τὸ ΖΕ τὴν διὰ πέντε φασὶ ποιήσει συμφωνίαν, ὥστε ἐπεὶ τὸ μὲν ΑΒ διὰ τεσσάρων ἐστίν, τὸ δὲ ΖΕ διὰ πέντε, ὑπεροχὴ δ' αὐτῶν τὰ ΖΑ καὶ ΒΕ, συναμφότερα μὲν ταῦτα καταλείπεσθαι τόνου, ἑκάτερον δ' αὐτῶν,

[74] 19.18–20.9, trans. Barker 1989: 293.
[75] I have discussed this assertion in chapter 1 part 1.

τουτέστιν ἑκάτερον τῶν ΑΔ καὶ ΓΒ ἡμιτονίου, διτόνου δ' ὄντος τοῦ ΑΓ καὶ τὸ ΑΒ διὰ τεσσάρων δυσὶ καὶ ἡμίσει συντίθεσθαι τόνοις.

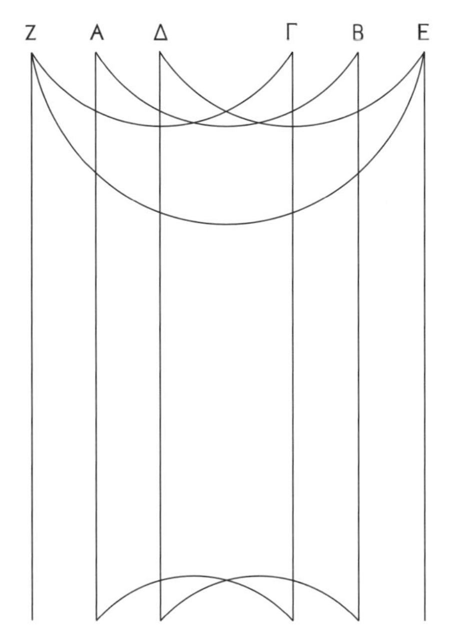

Fig. 6.3 Diagram accompanying Ptol. *Harm.* I.10, 22.2–16 (after Düring 1930).

They attempt to prove their proposition as follows. Let there be two notes, A and B, concordant at the fourth, and from A let a ditone, AΓ, be taken upwards; and from B, similarly, let a ditone BΔ be taken downwards. Then AΔ and ΓB are equal, and each is as great as is that by which the ditone is less than the fourth. Again, from Δ let a fourth be taken upwards, ΔE, and from Γ, similarly, let a fourth be taken downwards, ΓZ. Then since each of BA and ΓZ is a fourth, BΓ is equal to AZ, and on the same grounds AΔ is also equal to BE. The four intervals are therefore equal to one another. But, they say, the whole of ZE will make the concord of a fifth, so that since AB is a fourth and ZE is a fifth, and the difference between them is ZA plus BE, these differences taken together make up a tone, and each of them – that is [in effect], each of AΔ and ΓB – is a half-tone; and since AΓ is a ditone, the fourth AB is put together from two and a half tones.[76]

[76] 22.2–16, trans. Barker 1989: 295–6. The way in which Ptolemy represents this Aristoxenian argument hardly does justice to the hypothetical and (notably) empirical way in which Aristoxenus himself expressed it (*El. harm.* 56.14–58.6, discussed in ch. 1 part 1).

In this passage (as in some others), Ptolemy's letters signify something different from what they did in I.8. Here they represent not points on a line which can be equated with a string, but notes. Thus there are two systems of letter use in Ptolemy's treatise, as there are in the *Sectio canonis*:[77] the first, by which an interval is expressed as the ratio between two lengths, requires four letters (e.g. EK:KH) and is used elsewhere; the second, used here, requires only two (e.g. BΓ). The diagram accompanying the text of 22.2–16 gives six parallel lines of equal length, each labelled with a letter corresponding to one of the notes used in the construction (fig. 6.3); these are given in order of pitch, with lower notes at the left, and higher notes on the right: Z, A, Δ, Γ, B, E. They could be envisaged as six strings of a lyre, where pitch difference is not generated through difference in string length. This second use of letter and diagram belongs to a conception of notes which will not remind the reader of the monochord.

Ptolemy's objection to the Aristoxenian method he has recounted relies on the ratios of the intervals in question.

ὁ δὲ λόγος ἅπαξ τοῦ τόνου δειχθέντος ἐπογδόου καὶ τοῦ διὰ τεσσάρων ἐπιτρίτου δηλονότι αὐτόθεν ποιεῖ τὸ τὴν ὑπεροχήν, ᾗ ὑπερέχει τὸ διὰ τεσσάρων τοῦ διτόνου, καλουμένην δὲ λεῖμμα, ἐλάττονα εἶναι ἡμιτονίου.

But as soon as the tone has been shown to be epogdoic and the fourth epitritic, it is obvious from this very fact that reason entails that the difference by which the fourth exceeds the ditone, called the *leimma*, is smaller than a half-tone.[78]

He proves this by calculating the ratio of the *leimma* (a simple arithmetical argument). He posits three epogdoic tones in succession, the terms of whose bounding notes are 1536, 1728, 1944 and 2187. He then constructs a fourth downwards from the highest note, whose lower bounding note will turn out to be 2048 (2048:1536 = 4:3). The fourth is then the first two tones plus whatever is left between 2048 and 1944; this is of course 256:243 (= 2048:1944). That 256:243 is less than 2187:2048 is clear, and

[77] See ch. 1 part 1. [78] 22.17–23.3, trans. Barker 1989: 296.

so the Aristoxenian assertion that the fourth is composed of two and a half tones has been disproved.[79]

This argument relies entirely on the premise that the ratios 4:3 and 9:8 actually do correspond to the intervals perceived as fourth and tone respectively. Without proof of this connection it would not succeed.[80] Ptolemy can only bring out ratios at this point because he has already done the canonic division; the canonic division could only be used because he had first argued for its credentials with tests presided over by perception.

Next he invokes the *kanōn* explicitly; the numbers have been set out – 'now try it', he seems to say.

ἡ μὲν γὰρ αἴσθησις μονονοὺ κέκραγεν ἐπιγινώσκουσα σαφῶς καὶ ἀδιστάκτως τήν τε διὰ πέντε συμφωνίαν, ὅταν ἐπὶ τῆς ἐκτεθειμένης μονοχόρδου δείξεως κατὰ τὸν ἡμιόλιον ληφθῇ λόγον καὶ τὴν διὰ τεσσάρων, ὅταν κατὰ τὸν ἐπίτριτον.

For perception virtually shrieks its clear and unmistakable recognition of the concord of the fifth, when in the exposition of the monochord that has been set out it is constructed according to the hemiolic ratio, and of that of the fourth, when it is constructed according to the epitritic.[81]

The Aristoxenians will not try it, however, since as Ptolemy notes, they do not agree with the fundamental postulates of monochord use (24.1–4). He is clear that this is not a debate between reason and perception (23.19–21), but between uses of rational and perceptual evidence. The error of the Aristoxenians is that they trust perception in the smaller intervals where it cannot be trusted, and do not trust its judgements of the concords, for which it is eminently qualified.[82] The evidence of perception is there to be used, and the harmonicist's method must make use of it in those areas where it will speak reliably. This method is best exemplified, of course, in the construction, tests and divisions of the *kanōn*.

It is worth noting at this point that this is the first time since the *Sectio canonis* that we have seen an argumentative use of the monochord like this one. Ptolemy's predecessors may well have

[79] 23.3–18; cf. *Sect. can.* prop. 15.

[80] The same is true of the argument which follows it (24.8–19); the numbers of both procedures could be applied to a demonstration on the *kanōn*.

[81] 23.21–24.1, trans. Barker 1989: 297.

[82] 24.4–8; but Ptolemy has not been entirely fair to Aristoxenus (cf. *El. harm.* 55.3–12).

used the instrument argumentatively, but the only type of monochord use which survives in written accounts between the third century BC and the second century AD is simple demonstration, either of the concords and the tone, or of tetrachordal divisions and melodies in which to contextualise them (Didymus).

Additional strings

Ptolemy's second specific attack on Aristoxenian dogma is a refutation of the statement that the octave contains six tones (I.11). As the chapter title indicates,[83] the task is to show, once again through the Aristoxenians' prized faculty of perception (*aisthēsis*), that they are wrong in their view of this concord as well. This development is part of a crescendo of argument; an error about the octave is even graver than an error about the fourth – graver because the octave is the most basic concord (Ptolemy often refers to it as 'the octave homophone', τὸ διὰ πασῶν ὁμόφωνον, as at 19.8). The error follows logically from the misunderstanding of the fourth, so in this respect at least (but not in some others) the Aristoxenians' position is internally consistent.

Ptolemy's immediate appeal is to the musical ear.

ἐὰν δὲ ἐπιτάξωμεν τῷ μουσικωτάτῳ ποιῆσαι τόνους ἐφεξῆς καὶ καθ' αὑτοὺς ἕξ, μὴ συνεπιβαλλομένων μέντοι τῶν προηρμοσμένων φθόγγων, ἵνα μὴ καταφέρηται πρὸς ἄλλο τι τῶν συμφώνων, ὁ πρῶτος φθόγγος πρὸς τὸν ἕβδομον οὐ ποιήσει τὸ διὰ πασῶν. εἴτε δὴ μὴ παρὰ τὴν ἀσθένειαν τῆς αἰσθήσεως συμβαίνει τὸ τοιοῦτο, ψεῦδος ἂν ἀποφαίνοι τὸ τὴν διὰ πασῶν συμφωνίαν ἓξ εἶναι τόνων, εἴτε τῷ μὴ δύνασθαι λαμβάνειν αὐτὴν τοὺς τόνους ἀκριβῶς, πολὺ πλέον οὐκ ἔσται πιστὴ πρὸς τὴν τῶν διτόνων λῆψιν, ἀφ' ὧν εὑρίσκειν οἴεται τὸ διὰ τεσσάρων δύο καὶ ἡμίσεος τόνων.

If we instruct the most expert musician to construct six tones in succession, just by themselves, and without the aid of other notes attuned beforehand, so that he cannot refer to some other of the concords, the first note will not make an octave with the seventh. Now if this sort of result is not due to the weakness of perception, the claim that the concord of the octave consists of six tones would be shown to be false; but if it is because perception cannot construct the tones accurately, it will be much less reliable in the construction of ditones, from

[83] 'How it may be demonstrated through perception that the octave is less than six tones, by means of the eight-stringed *kanōn*' (πῶς ἂν καὶ διὰ τῆς αἰσθήσεως ἐπιδειχθείη τὸ διὰ πασῶν ἔλαττον ἓξ τόνων διὰ ὀκταχόρδου κανόνος, I.11).

which he[84] supposes that he can discover that the fourth consists of two and a half tones.[85]

Six tones constructed in succession by ratio, he notes, generate a magnitude greater than the octave; the exact ratio of this excess (the *komma*) had been known since at least the early third century BC.[86] Here, as he did with the difference between the *leimma* and the semitone in the last numerical argument of the preceding chapter (I.10), Ptolemy approximates to an epimoric interval, 65:64.[87]

Once again, the *kanōn* is the only way for the theorist to be sure either that the result obtained by the 'most expert musician' (25.5) was not due to the weakness of his perception, or that the result of calculation actually applies to the notes in question. It could be done on a single-stringed *kanōn*, but for reasons given later in his treatise (II.12), Ptolemy prefers to set out a larger instrument, on which each note will have a string to itself. As long as the potential causes of unevenness have been controlled by means of tests like that conducted in I.8, the eight-stringed *kanōn* will be as reliable as the single-stringed instrument.[88] Furthermore, it leaves less room for disagreement over its results: all eight notes of the demonstration will be available at once, something which the monochord cannot manage.

Ptolemy sets out the demonstration in this way:

ἔσται δ' ἡμῖν καὶ τὸ τοιοῦτον εὐκατανόητον συνάψασιν ἑπτὰ χορδὰς ἄλλας ἐν τῷ κανόνι τῇ μιᾷ κατὰ τὴν ὁμοίαν ἀνάκρισίν τε καὶ θέσιν. ἐὰν γὰρ ἰσοτόνους ἁρμοσώμεθα τοὺς ὀκτὼ φθόγγους ἐν ἴσοις τοῖς τῶν χορδῶν μήκεσιν ἀκριβῶς ὥστε τοὺς ΑΒΓΔΕΖΗΘ, ἔπειτα διὰ τῆς τοῦ κανονίου προσαγωγῆς εἰς ἓξ τοὺς

[84] Evidently Aristoxenus, though Ptolemy never mentions him by name in the chapter. Ptolemy has (again) misrepresented Aristoxenus, who constructs his tones and ditones 'through concords' (διὰ συμφωνίας, *El. harm.* 55.16).

[85] 25.5–13, trans. Barker 1989: 299.

[86] See *Sect. can.* prop. 9. The *komma* is not named there, nor is its value sought, but the terms of its ratio are (nearly) generated as a by-product of the proof that six epogdoics are greater than 2:1.

[87] It is a good approximation: the true *komma* is 531,441:524,288, while 65:64 is equal to 532,480:524,288. It is not the nearest epimoric, however: 74:73 is much closer. Ptolemy perhaps prefers approximation by epimoric to exactitude because one grasps their relative sizes much more readily; this is important in contexts such as the present one. (Compare also his use of 19:18 in place of the 256:243 *leimma* at 45.5.)

[88] See Ptolemy's reassurances about the theoretical identity of many-stringed and single-stringed demonstrations, 85.8–19.

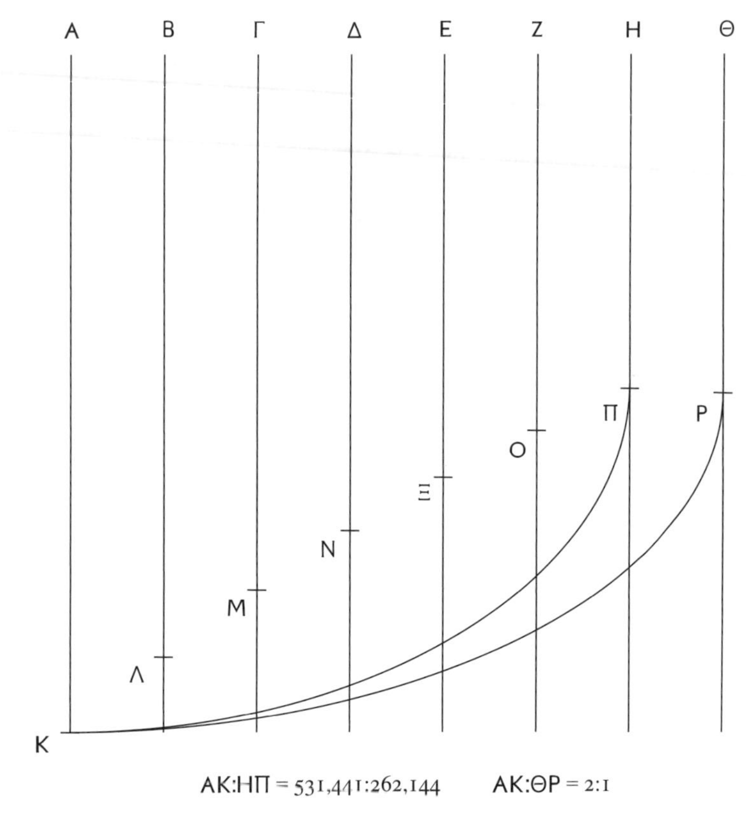

Fig. 6.4 Diagram accompanying Ptol. *Harm.* I.11, 26.3–14 (after Düring 1930).*

ἐφεξῆς ἐπογδόους λόγους διαιρεθέντος παραφέρωμεν καθ᾽ ἕκαστον φθόγγον τὸ παραπλήσιον ὑπαγώγιον ἐπὶ τὴν οἰκείαν τομήν, ἵνα ὡς ἐπόγδοος ᾖ ἥ τε AK διάστασις τῆς BΛ καὶ αὕτη τῆς ΓM καὶ αὕτη τῆς ΔN καὶ αὕτη τῆς EΞ καὶ αὕτη τῆς ZO καὶ αὕτη τῆς HΠ, ποιεῖ δὲ καὶ ἡ AK πρὸς τὴν ΘP τὸν διπλάσιον λόγον, οὗτοι μὲν ὁμοφωνήσουσιν ἀκριβῶς οἱ φθόγγοι κατὰ τὸ διὰ πασῶν, ὁ δὲ ΠH τοῦ ΘP βραχεῖ καὶ τῷ αὐτῷ πάντοτε ἔσται ὀξύτερος.

This sort of result [i.e. that found by calculation] will be easily demonstrated if we fasten seven more strings on the *kanōn*, in association with the one string,

* Düring's additions, 'KΠ $= \frac{531,441}{262,144}$' and 'KP $= \frac{2}{1}$', misrepresent Ptolemy's designation-letters as (arithmetical) quantities rather than (geometrical) points. In Solomon's version of the diagram (2000: 37) most of the designation-letters are either missing or incorrectly placed, with the result that Ptolemy's BΛ, ΓM, ΔN, EΞ, ZO, HΠ and ΘP cannot be located.

on the basis of the same kind of selection and placing. For if we accurately attune the eight notes at equal pitch in equal lengths of the strings, the notes ΑΒΓΔΕΖΗΘ, and then, by application of the measuring-rod, divided into six epogdoic [9:8] ratios in succession, we place an identical bridge at the proper division corresponding to each note, to make distance ΑΚ the epogdoic of ΒΛ, ΒΛ that of ΓΜ, ΓΜ that of ΔΝ, ΔΝ that of ΕΞ, ΕΞ that of ΖΟ, ΖΟ that of ΗΠ, while ΑΚ makes with ΘΡ the duple ratio, the latter notes will sound accurately the homophone of the octave, but ΠΗ will be slightly higher than ΘΡ, and always to the same degree.[89]

This last point is an important one. In his introduction of the single-stringed *kanōn* Ptolemy emphasised the fact that the instrument 'does not acquire its pitch in any random way' (οὐ μὴν ὡς ἔτυχε λαβοῦσα τὴν τάσιν, 17.22–3), and so we discover through the use of such instruments that the patterns governing the combinations of musical intervals are not random either. The amount by which six tones exceed an octave is constant, as the carefully controlled demonstration proves: the result will be no different if it is repeated, and the apparent aberration therefore cannot be explained away as observational error.

As though anticipating the objection that the eight-stringed *kanōn* may not give reliable evidence, Ptolemy offers two additional demonstrations to show that as long as each of the strings is even, differences of tension and density between them will not affect the notes sounded by the same ratios of their lengths (26.15–28.12). As we saw a little earlier, his physics is not always right, but his conclusion is correct insofar as it relates to the procedures of I.11.[90] The fact that he spends fully half of the chapter ensuring the trustworthiness of his eight-stringed *kanōn* illustrates the extent to which he considers this an essential part of his argument.

Having finished with the ratios of the concords, Ptolemy now moves on to deal with tetrachordal divisions (I.12–16). His approach is systematic; after establishing the concords and their

[89] 26.3–14, trans. Barker 1989: 300. In many respects this is simply an application of *Sect. can.* prop. 9 to an eight-stringed *kanōn*; Ptolemy even retains the same letters ΑΒΓΔΕΖΗ, which in his version no longer signify numbers standing to one another in determinate ratio, but the end-points of seven strings. The eighth letter, Θ, is Ptolemy's addition, by which (in the length ΘΡ) he can compare the true octave to the interval made up of six tones.

[90] See Barker 1989: 300 n. 102 and 2000: 202–3.

compositions and correcting the errors of his predecessors (I.12–14), he proceeds to build the larger structures, beginning with their constituent parts (I.15–II.2), until he has set out all the possible variations of the Greater Perfect System (II.3–15), and shown how these are related to the scales used by musicians in his day (II.16). Book III opens with two chapters on the use and trustworthiness of the instruments, and a final excursus on the wider implications of the harmonic principles established throughout the earlier part of the treatise ends the work (III.3–14).[91]

Ptolemy has introduced the eight-stringed *kanōn* in order to clear up a controversy over the octave homophone, but now it proves useful to him in the course of his analysis of scale-structures as well. Throughout his critique of the tetrachords proposed by his predecessors, as well as his explication of his own, he continues to rely not only on criteria such as ratio form (Archytas, for instance, incurs his criticism for using two non-epimoric ratios in his chromatic, 32.1–3), but also on what he calls 'the self-evidence of sense-perception' (ἡ ἀπὸ τῆς αἰσθήσεως ἐνάργεια, 32.3–4). To bring the two together Ptolemy will require his eight-stringed *kanōn* again.

He argues that Archytas' chromatic and enharmonic tetrachords are incorrectly formed, since (among other things) the musical ear demands that the smallest interval be placed at the bottom of the tetrachord, not in second position as it is in Archytas' enharmonic:

ταῦτα μὲν δὴ δοκεῖ τῷ λογικῷ κριτηρίῳ περιποιῆσαι τὴν διαβολήν, ὅτι κατὰ τοὺς ἐκτιθεμένους λόγους ὑπὸ τῶν προϊσταμένων αὐτοῦ γινομένης τῆς τοῦ κανόνος κατατομῆς οὐ διασῴζεται τὸ ἐμμελές. οἱ γὰρ πλεῖστοι τῶν τε προκειμένων καὶ τῶν τοῖς ἄλλοις σχεδὸν ἅπασι διαπεπλασμένων οὐκ ἐφαρμόζουσι τοῖς ὁμολογουμένοις ἤθεσιν.

These things, then, seem to set up a slanderous accusation against the rational criterion, since when the division of the *kanōn* is made according to the ratios set out by his proposals, that which is melodic is not preserved. For the majority of

[91] The end of the treatise has not survived as Ptolemy wrote it (if he finished it, that is). The text as we have it was completed (from the second sentence of III.14) by Nicephorus Gregoras in the fourteenth century. The final chapter (III.16) was not written by Gregoras, but transposed there by him from the text of III.9 in one MS, where it was clearly out of place. It is uncertain whether it should be attributed to Ptolemy or not.

those set out above, and of those that have been worked out by virtually everyone else, are not attuned to the characters generally agreed on.[92]

It is through the process of setting out the attunements on the *kanōn* that we are able to judge whether or not they preserve 'that which is melodic', and whether or not they retain the characters (*ēthē*) which most people agree on. There are two notions here; the first involves the ear's recognition of the intervals of an attunement as melodically acceptable, the second the recognition of a pattern of attunement as possessing the 'character' which it ought to have.[93] Both are presumably the result of one's being accustomed to hearing the attunements in melodies. When the enharmonic is set out properly, each interval will be accepted as melodic, and the character of the whole tetrachord will be immediately recognisable to the ear as that of the enharmonic.[94]

Nothing here requires absolutely that an eight-stringed *kanōn* (as opposed to a monochord) be used for this procedure. But because of the difficulties of using the monochord, explained in II.12, and because of the clear necessity expressed there of the theorist's being able to construct not just attunements but melodies on the instrument in order to judge the proposed attunements fairly, the *kanonos katatomē* of 32.12–13 will not be entirely satisfactory if the word *kanōn* in this passage means *monochordos kanōn*. As Barker has argued (2000: 236), the procedure described here is empirical: the only way to judge Archytas' tetrachords is to set them out on an instrument and listen to them. It is not merely demonstrative, as the concord-procedures were; here the reader is at liberty to follow the instructions and disagree with Ptolemy's judgement, and Ptolemy leaves himself open to this possibility.

Throughout the subsequent chapters there is a recurrent emphasis on the general agreement of musical perception. Thus the appeal to 'the characters generally agreed on' (32.15) is echoed a chapter later in a point argued 'from agreed perception' (παρὰ δὲ τῆς ὁμολογουμένης αἰσθήσεως, 33.22). The latter comes, in fact, from the

92 32.10–15, trans. Barker 1989: 305.
93 On Ptolemy's use of *ēthos* see Barker 2000: 235–6, Solomon 2000: 46 n. 224.
94 Something similar occurs with the plainchant psalm-tones. Given, for example, the intonation *a-g-a* followed by the reciting note *a*, the accustomed ear immediately recognises, by its 'character', tone IV.

same context as the former: the relative magnitudes of the intervals which make up the tetrachord in the three genera. In I.16 Ptolemy can say of his version of the tense chromatic that it 'seems most agreeable to the ears' because it makes a division into two nearly equal parts.[95] The rhetoric is more emphatic near the end of I.15, and it is hardly coincidental that the eight-stringed *kanōn* makes its reappearance in the same passage. The instrument itself, with its checks and tests completed (and these are mentioned explicitly at 37.7–9), is the foundation of such confident assertions as Ptolemy gives us here. He has just finished setting out his soft, tonic and tense diatonics, and now applies them to his eight-stringed *kanōn*:

ὅτι δὲ οὐ τὸ εὔλογον ἔχουσι μόνον αἱ προκείμεναι τῶν γενῶν διαιρέσεις, ἀλλὰ καὶ τὸ ταῖς αἰσθήσεσιν σύμφωνον, ἐξέσται πάλιν κατανοεῖν ἀπὸ τοῦ διὰ πασῶν περιέχοντος ὀκταχόρδου κανόνος, ἀκριβουμένων τῶν φθόγγων, ὡς εἴπομεν, κατά τε τὰς ὁμαλότητας τῶν χορδῶν καὶ τὰς ἰσοτονίας. ταῖς γὰρ γινομέναις τῶν παρατιθεμένων κανονίων κατατομαῖς ἀκολούθως τοῖς ἐφ' ἑκάστου γένους λόγοις συναποκαθισταμένων τῶν ὑπαγομένων μαγαδίων, οὕτως ἔσται τὸ διὰ πασῶν ἡρμοσμένον, ὡς μηδ' ἂν τὸ τυχὸν ἔτι παρακινῆσαι τοὺς μουσικωτάτους.

The fact that the divisions of the genera set out above do not contain only what is rational but also what is concordant with the senses can be grasped, once again, from the eight-stringed *kanōn* that spans an octave, once the notes are made accurate, as we have said, in respect of the evenness of the strings and their equality of pitch. For when the bridges set under [the strings] are aligned with the divisions marked on the measuring-rods placed beside them – the divisions that correspond to the ratios in each genus – the octave will be so tuned that the most musical of men would not alter it any more, not even a little.[96]

In I.16 Ptolemy also includes his own original tetrachord, the 'even diatonic' (10:9 × 11:10 × 12:11), with an explanation in which the requirement of an eight-stringed *kanōn* is clear not only because of the plural 'equal-pitched ⟨strings⟩' (ἰσοτονίων ⟨χορδῶν⟩, 38.30) but because of the necessity of trying out melodies in the genus to test its claim to inclusion alongside 'the genera that are more familiar to our ears':[97]

[95] προσφορώτατον φαίνεται ταῖς ἀκοαῖς, 38.13, trans. Barker 1989: 311. The ratios of this tetrachord are 7:6 × 12:11 × 22:21. The two nearly equal parts are the the upper interval (7:6) and the *pyknon* (8:7 = 12:11 × 22:21).

[96] 37.5–12, trans. Barker 1989: 310. [97] τὰ συνηθέστερα ταῖς ἀκοαῖς γένη, I.16.

τῆς δὲ δι᾽ αὐτῶν κατατομῆς ἐκλαμβανομένης ἐπὶ τῶν ἰσοτονιῶν ξενικώτερον μέν πως καὶ ἀγροικότερον ἦθος καταφανήσεται, προσηνὲς δ᾽ ἄλλως καὶ μᾶλλον συγγυμναζόμενον ταῖς ἀκοαῖς, ὡς μὴ δεόντως ἂν παροραθῆναι διά τε τὸ τοῦ μέλους ἰδιάζον καὶ διὰ τὸ τεταγμένον τῆς κατατομῆς. ἔτι δὲ ὅτι κἂν καθ᾽ αὑτὸ μελῳδῆται, οὐκ ἐμποιεῖ ταῖς αἰσθήσεσι προσκοπήν, ὃ μόνῳ σχεδὸν συμβέβηκε τῷ μέσῳ τῶν διατονικῶν τῶν ἄλλων.

When a division is taken in [strings] of equal pitch on the basis of these numbers [18, 20, 22, 24, 27, 30, 33, 36], the character that becomes apparent is rather foreign and rustic, but exceptionally gentle, and the more so as our hearing becomes trained to it, so that it would not be proper to overlook it, both because of the special character of its melody, and because of the orderliness of the division. Another reason is that when a melody is played in this genus by itself, it gives no offensive shock to the hearing, which is true, pretty well, only of the intermediate one of the other diatonics.[98]

By the end of book I Ptolemy has constructed all the tetrachords he will use in his treatise. He now changes approach; what he has done so far is to set out his tetrachords rationally and then submit them to the judgement of perception using the *kanōn*. In the first chapter of book II he reverses the procedure, requiring that a series of familiar attunements (which he lists by name) be constructed on the instrument, each of which he then matches to attunements previously laid out by ratio. His method throughout II.1 is consistent: first, 'let such and such an attunement be constructed' (πεποιήσθω . . .); then 'I say that what is contained by it is the genus of . . .' (λέγω ὅτι περιέχεται ὑπ᾽ αὐτοῦ τὸ . . . γένος), followed by the proof of the parallel. These paradigms correspond to what we have already encountered as the *ekthesis* and *diorismos* of a Greek mathematical proposition.[99] It is common in Greek mathematical texts for the *ekthesis* to refer to a diagram in which the proof will be carried out. Here the reference is to the instrument via the conventions of the lettered diagram. The diagrams are intermediary; the real *ekthesis* must be done on the *kanōn* or the demonstration will be incomplete. It is a point at which the modern reader can no longer participate as Ptolemy intends, since it relies on an intimate familiarity with a collection of second-century

[98] 38.29–39.1, trans. Barker 1989: 312. The 'intermediate' diatonic is the tonic diatonic, identical to Archytas' diatonic: 9:8 × 8:7 × 28:27.

[99] See ch. 1 part 1, Netz 1999a: 9–11 and 1999b: 284.

attunements: *tropoi*, *sterea*, *iastiaiolia* and *parhypatai*.[100] We can read his propositions backwards and arrive at his scales as it were by the back door, but we cannot follow his project in the correct order. This depends on the kind of 'agreed perception' (ὁμολογουμένη αἴσθησις) which only his ancient readers possessed.

Even Ptolemy's ancient readers needed more than mere familiarity with the common instrumental attunements of their day in order to read II.1 as he intended: the eight-stringed *kanōn* is central to the argument of the chapter, and the reader cannot follow it by merely imagining the instrument. Ptolemy's reversal of rational and perception-based procedure here is based on his assumption that the two are really in complete agreement about the intervals in music, for all the fact that they speak in different languages.[101] The *kanōn* is, in effect, simply the 'translator' between the two. Whichever faculty speaks first, the statements turn out to be the same.

Babylonios temptare numeros[102]

The diagrams of II.1 employ a numerical feature which marks a departure from Ptolemy's practices earlier in the *Harmonics*. Alongside each of the lines which represent strings are numbers indicating bridge positions. This in itself is not unusual, but the fact that they are not all integers is. The numbers themselves may well have been added by a later hand, but I draw attention to them here because the feature which distinguishes them is one they share with many of those found in the tables of II.14–15, all of which are derived according to a system introduced near the end of II.13.

In I.10 for the first time, and then in I.13 and elsewhere, Ptolemy supplements a series of ratios with a series of string-length numbers, worked out by putting all the ratios in their lowest common terms. This is mathematically useful, in that it allows the theorist to think of the intervals as they fall between a continuous sequence of whole numbers. To a certain degree it is also practically useful, in that a measuring-rod (*kanonion*) can be prepared in advance with

[100] 42.12, 43.11, 43.19, 44.15. [101] Cf. 66.6–7 (quoted above).
[102] With apologies to Horace (*Carm*. I.11.2–3).

these numbers marked on it, and then 'fitted to' the instrument.[103] Beyond the two instances already cited (I.10 and I.13), we find further examples of this means of expression in the remaining chapters of book I.[104]

The theorist must then have a set of differently marked *kanonia* in his laboratory toolbox, one for each division. This is perhaps reasonable if the total repertoire of divisions (and comparisons between them) is small, and it was evidently the method employed by many of Ptolemy's predecessors. As we saw in chapter 5, even those who may have had no intention of applying their divisions to the monochord still presented them in lowest terms; a prime example is the author of the *Timaeus Locrus*. Even Thrasyllus' division, which was certainly meant to be constructed on the monochord, required division numbers running up to 41,472. But by the time Ptolemy comes to set out his enharmonic, soft chromatic and tense chromatic in common terms in I.15 (35.7–12), the numbers are already running into hundreds of thousands: the lowest note of these three tetrachords is 141,680 units. It soon becomes a bulky and unworkable method, especially since Ptolemy intends to set out for comparison all the tetrachords of his predecessors along with his own, all in the context of one-octave scales (II.14). There are even more complicated constructions yet to come in the remainder of book II. By this point Ptolemy's project has outgrown the means of numerical expression traditional to harmonic science, and a new expedient must be found.

Ptolemy has one ready to hand. The astronomical practice (inherited from the Babylonians) of expressing fractional quantities in sexagesimals, to which he was thoroughly accustomed, provides an ideal solution to the problem.[105] By constructing his divisions in a 60-unit octave, and calculating bridge positions to the nearest sixtieth, he can now fit a standard all-purpose *kanonion* to his instruments. In the tables (*kanonia*) of II.14 no more than 120 units of string are required, though in several of the tables

[103] At 18.9 the verb is ἐφαρμόζειν; later Ptolemy describes the procedure as one of placing the *kanonion* beside or against the strings (37.9–10, 69.16–17).

[104] 35.7–12, 36.35–37.4, 38.27–9, 40.17–18.

[105] On the notation (which is slightly different in the *Almagest*), see Toomer 1984: 6–7. I have discussed its application to the tables of II.14 in ch. 4.

(*kanones*) of II.15 slightly longer lengths are needed; the size of the unit is constant.

The advantage of this system from Ptolemy's point of view is not simply that it eliminates the need for much painstaking calculation and a large supply of differently marked *kanonia*, although this is certainly a benefit. The principal importance of the development is one that it shares with the eight-stringed *kanōn*, and indeed the two are evidently meant to be used in conjunction: it allows for ease of comparison. From the beginning of his critique of Archytas' tetrachords in I.14 Ptolemy has been insisting on the construction and appraisal of tetrachords according to what he calls 'the self-evidence of sense-perception' (32.3–4), and in his criticisms of his predecessors in II.13 he attributes their errors to the fact that they were limited to the single-stringed *kanōn*; more strings are necessary in order for the theorist properly to contextualise his tetrachords, by playing melodies in them. No proof that Ptolemy's own tetrachords offer any improvement over those he has criticised would be convincing unless the reader could compare them all under the same conditions. Ptolemy is obliging:

καὶ ἵνα γε πρόχειρος ἡμῖν ἡ παραβολὴ ᾖ τῶν τε καθ' ἡμᾶς γενικῶν διαιρέσεων καὶ ἔτι τῶν ἄνωθεν παραδοθεισῶν, ὅσαις γοῦν ἐνετύχομεν, προεκθησόμεθα μερικήν τινα τούτων παράθεσιν ἐπὶ τοῦ μέσου καὶ δωρίου τόνου πρὸς ἔνδειξιν αὐτοῦ μόνον τῆς ἐκκειμένης διαφορᾶς. καθόλου μέντοι κεχρήμεθα ταῖς τῶν διαιρέσεων ἐφόδοις οὐ τὸν αὐτὸν τρόπον τοῖς παλαιοτέροις, τέμνοντες καθ' ἕκαστον φθόγγον τὸ ὅλον μῆκος εἰς τοὺς διασημαινομένους λόγους, διὰ τὸ ἐργῶδες καὶ δύσληπτον τῆς τοιαύτης καταμετρήσεως, ἀλλ' ἐξαρχῆς τοῦ προστιθεμένου ταῖς χορδαῖς κανονίου διαιροῦντες τὸ ἀπολαμβανόμενον μῆκος ἀπὸ τοῦ κατὰ τὸ ὀξὺ πέρας ἀποψάλματος μέχρι τῆς ὑπὸ τὸν βαρύτατον φθόγγον ἐσομένης σημειώσεως εἰς ἴσα καὶ σύμμετρα τῷ μεγέθει τμήματα, καὶ παρατιθέντες αὐτοῖς τοὺς ἀπὸ τῆς πρὸς τὸ ὀξὺ πέρας ἀρχῆς ἀριθμούς, δι' ὅσων ἂν ἐγχωρῇ μορίων, ἵνα τοὺς ἐν τοῖς οἰκείοις λόγοις συνισταμένους ἑκάστῳ τῶν φθόγγων ἀπὸ τοῦ εἰρημένου κοινοῦ πέρατος ἔχοντες ἐκτεθειμένους ὑπάγωμεν ἀεὶ προχείρως ἐπὶ τοὺς ἐκ τοῦ κανονίου διασημαινομένους τόπους τὰ ἀποψάλματα τῶν κινουμένων μαγαδίων. κἀπειδὴ τοὺς συνέχοντας ἀριθμοὺς τὰς κοινὰς τῶν γενῶν διαφορὰς εἰς μυριάδας ἐκπίπτειν συμβαίνει, μονάδων ὅλων ἀπαρτιζομένων συνεχρησάμεθα τοῖς ἐγγυτάτω μερισμοῖς μέχρι τῶν πρώτων τῆς μιᾶς μονάδος ἑξηκοστῶν, ὥστε μηδέποτε πλέον ἑνὸς ἑξηκοστοῦ τῆς ἐν τῇ κατατομῇ τοῦ κανονίου μιᾶς μοίρας διενεγκεῖν τὰς παραβολάς.

In order to make readily available the contrast between our divisions of the genera and those that have previously been handed down – those, at any rate, that we have come across – we shall set out a partial comparison of them, in the middle *tonos*, the Dorian, to display in just that case the difference that there is. In general, we have not undertaken our approach to the divisions in the same way as the older writers, dividing the whole length into the ratios indicated for each note, because of the laboriousness and difficulty of this sort of measurement. Instead, on the *kanonion* that is placed up against the strings, we have begun by dividing the length cut off, from the highest limit of the sounding length to the mark there will be to indicate the lowest note, into divisions that are equal and proportionate in size. We have placed numbers against these, beginning from the highest limit, through however many parts may be involved, so that now that we have got set out the numbers related in the ratios appropriate to each of the notes, starting from the common limit mentioned, we may always find it easy to bring the dividing points on the moveable bridges up against the positions indicated by the *kanonion*. And since it turns out that the numbers containing the differences shared by the genera run into tens of thousands, we have used the nearest sixtieth parts of complete whole units, down to the first sixtieths of a single unit, so that our comparisons are never in error by more than one sixtieth of one of the parts into which the *kanonion* is divided.[106]

Of the tables which follow in II.14–15, the last is perhaps the most indicative of the extent to which Ptolemy was concerned to encourage his reader to construct the attunements for himself. Here he collects all the bridge positions (from 56 11′ to 124 27′ units) assigned to each note of the octave, and sets them out in eight columns, one for each note. These could be used to place indicating marks on the *kanonion*, so as to make setting up the divisions less time-consuming. From this it is clear that Ptolemy intends the reader to construct the scales on an eight-stringed *kanōn*, and that he is willing to help by making the procedure as *kanōn*-friendly as possible.

Combining inheritances: the helikōn *and the* kanōn

In II.2 Ptolemy introduces an instrument which has not yet made an appearance in extant Greek musical texts,[107] although it is clear from what he says about it that it came to him from his predecessors.

[106] 69.8–29, trans. Barker 1989: 344–5.

[107] Or in any other context, for that matter. Ptolemy and Aristides Quintilianus are the first to mention it.

γίνοιτο δ' ἂν ἡ κατὰ τὸν ὀκτάχορδον κανόνα τοῦ διὰ πασῶν χρῆσις καὶ καθ' ἕτερον τρόπον παρὰ τὸ καλούμενον ὄργανον ἑλικῶνα, πεποιημένον τοῖς ἀπὸ τῶν μαθημάτων εἰς τὴν ἔνδειξιν τῶν ἐν ταῖς συμφωνίαις λόγων οὔτωσί πως.

It is possible to use the eight-stringed *kanōn* of the octave in a different way too, in conjunction with the instrument called the *helikōn*, which has been made by students of mathematics to display the ratios in the concords, in the following sort of way.[108]

He then proceeds to construct the *helikōn*. It consisted of a geometrical figure – a lettered diagram, in effect – the four vertical lines of which were meant to signify strings, and the one oblique one an extended bridge. Two horizontal lines indicated fixed end-bridges.

ἐκτίθενται τετράγωνον ὡς τὸ ΑΒΓΔ καὶ διελόντες δίχα τὰς ΑΒ καὶ ΒΔ κατὰ τὰ Ε καὶ Ζ ἐπιζευγνύουσι μὲν τὰς ΑΖ καὶ ΒΗΓ, διάγουσι δὲ παρὰ τὴν ΑΓ διὰ μὲν τοῦ Ε τὴν ΕΘΚ, διὰ δὲ τοῦ Η τὴν ΛΗΜ.

They construct a square, ΑΒΓΔ, and after dividing ΑΒ and ΒΔ in half at Ε and Ζ, they join up ΑΖ and ΒΗΓ, and draw parallel to ΑΓ the line ΕΘΚ through Ε, and the line ΛΗΜ through Η.[109]

Ptolemy then goes on to list the relative magnitudes of the line segments this procedure has generated. This could be summarised as follows:[110]

ΑΓ:ΒΖ::ΑΓ:ΖΔ = 2:1
ΒΖ:ΕΘ::ΖΔ:ΕΘ = 2:1, since ΑΒ:ΑΕ = 2:1
ΑΓ:ΕΘ = 4:1
ΑΓ:ΘΚ = 4:3
ΜΗ:ΗΛ = 2:1, since ΔΓ:ΓΜ::ΔΒ:ΗΜ and ΒΑ:ΑΛ::ΒΖ:ΛΗ
therefore ΒΔ:ΗΜ::ΒΖ:ΛΗ, and conversely ΒΔ:ΒΖ::ΜΗ:ΛΗ
ΑΓ:ΗΜ = 3:2 and ΑΓ:ΗΛ = 3:1

If strings are stretched in place of the lines ΑΓ, ΕΚ, ΛΜ and ΒΔ, he says, and if ΑΘΗΖ is replaced by a bridge, then the simple concords and the tone will be found through the differing lengths of ΑΓ (12 units), ΘΚ (9), ΗΜ (8) and ΖΔ (6). The lengths on the other side of the bridge are used as well, so that the larger concords

[108] 46.4–7, trans. Barker 1989: 319.
[109] 46.7–10, trans. Barker 1989: 319. Compare the much sketchier account of Aristides Quintilianus (*De musica* III.3).
[110] Summary based on 46.10–47.1.

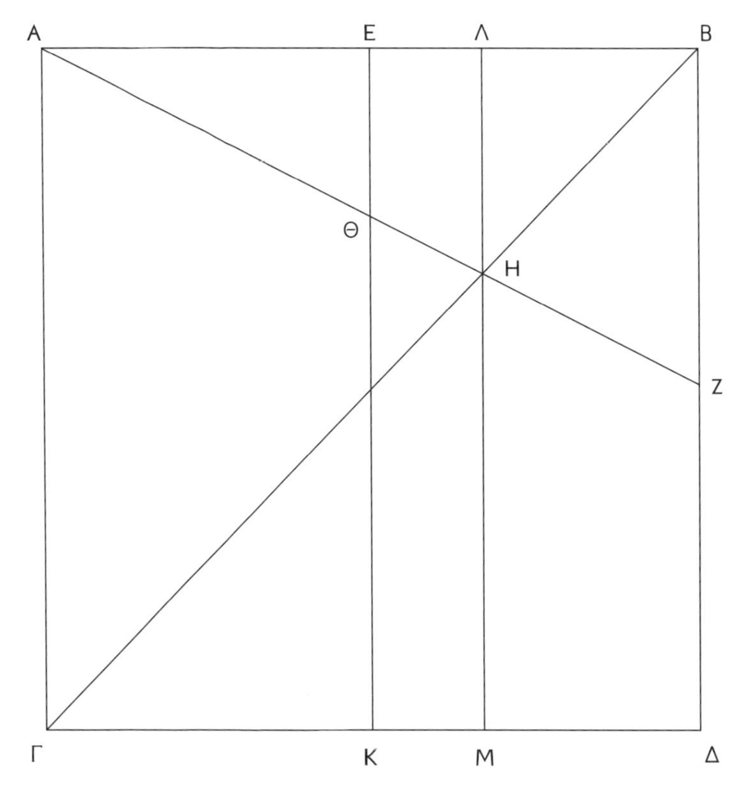

Fig. 6.5 The *helikōn* instrument: diagram accompanying Ptol. *Harm.* II.2, 46.2–47.17 (after Düring 1930).

can also be generated: the octave plus fourth (HM:ΘE = 8:3), the octave plus fifth (AΓ:ΛH = 3:1) and the double octave (AΓ:EΘ = 4:1).[111]

All this relies, though Ptolemy does not mention it here, on the assumption that the strings are even throughout their length and tuned in unison before the bridge has been placed. There is in fact remarkably little emphasis on the practical aspects of building or using the *helikōn*. Ptolemy presumably reports the instrument's description more or less as he has received it, and as

[111] 47.1–17. The interval HM:ΘE = 8:3 may or may not have been part of the division as Ptolemy inherited it. The part of the procedure which uses the string on both sides of the bridge may have been Ptolemy's addition.

Barker has noted it is not an account which seems to have practical construction as its main objective: it is a set of instructions for the geometer, not for the laboratory harmonicist.[112]

Ptolemy does very little with the *helikōn* other than to set it out and explain how it works; he then proceeds immediately into an account of what appears to be his own (unnamed) adaptation of the instrument. Because of the order in which the construction details of the *helikōn* are given, and because of his subsequent adoption of the most useful features of both instruments for his own more complex *kanones*, it appears that the *helikōn* is merely a starting point for Ptolemy.

The main advantage of the *helikōn* is that once its strings have been tested and tuned, the intervals between their pitches are determined not by the position of the bridge ΑΘΗΖ (which in Ptolemy's account acts like a fixed bridge), but by their lateral position relative to one another. This may not immediately appear advantageous, since moving a string (while maintaining its tension) is certainly more difficult than moving a bridge. But among the mathematicians who devised the instrument – if it was indeed ever constructed, rather than merely contemplated – it seems to have served simply to demonstrate the concords, which do not change.[113] This, at any rate, is the demonstration Ptolemy performs with it in his account. But as Aristides Quintilianus notes, the instrument could be made to sound not just the concords, but 'all the notes' (πάντας δὲ τοὺς φθόγγους) if more strings are added.[114] We should probably understand 'all the notes' to mean all those of a fixed scale; no change of genus and no modulation of *tonos* could be proposed without much fuss and bother.

Ptolemy's variations and adaptations of the *helikōn* retain the advantages of the original instrument while improving on some of its disadvantages. The result is – eventually – an instrument in which the attunement, once found, can be transposed with relative ease, and in which the practicalities of tuning and retuning are

[112] Barker 2000: 210.

[113] Ptolemy presumably inherited the *helikōn* in the form of a lettered diagram, with or without a written proof.

[114] *De musica* 99.9–12.

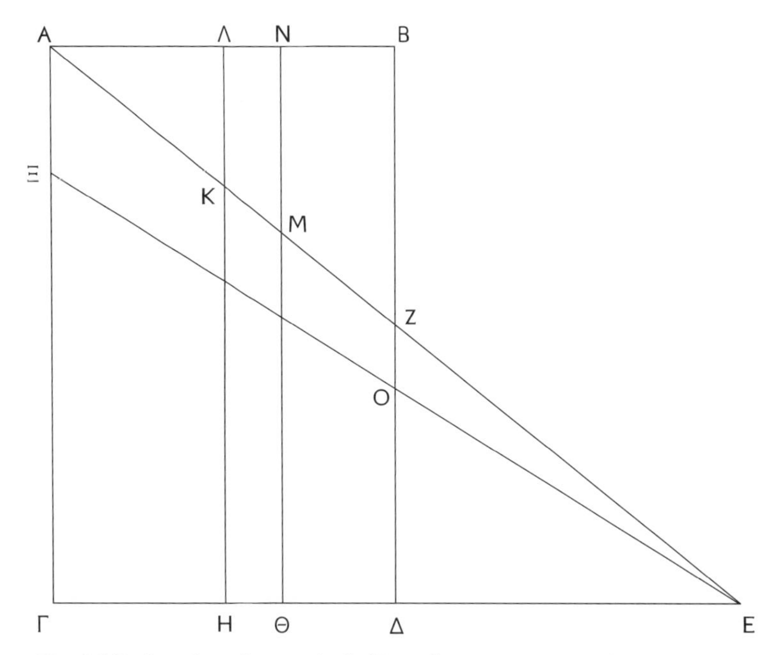

Fig. 6.6 Ptolemy's variant on the *helikōn*: diagram accompanying Ptol. *Harm.* II.2, 47.18–49.3 (after Düring 1930).

significantly improved. He proposes the first stage in this process immediately upon completing his description of the *helikōn*.

παρὰ δὴ τοῦτο τὸ ὄργανον ἐὰν ἐκθώμεθα παραλληλόγραμμον ἁπλῶς ὡς τὸ ΑΒΓΔ καὶ νοήσωμεν τὰς μὲν ΑΒ καὶ ΓΔ κατὰ τὰ ἀποψάλματα τῶν χορδῶν, τὰς δὲ ΑΓ καὶ ΒΔ κατὰ τοὺς ἄκρους φθόγγους τοῦ διὰ πασῶν, ἔπειτα προσεκβαλόντες τῇ ΓΔ ἴσην τὴν ΔΕ κατατέμωμεν ἀντὶ τῶν κανονίων τὴν ΓΔ πλευρὰν τοῖς οἰκείοις τῶν γενῶν λόγοις, ἐπὶ τοῦ Ε τὸ ὀξὺ πέρας ὑποτιθέμενοι, καὶ διὰ τῶν γινομένων ἐπ' αὐτῆς τομῶν τείνωμεν τὰς χορδὰς παραλλήλους τε τῇ ΑΓ καὶ ἰσοτόνους ἀλλήλαις, καὶ τούτου γενομένου τὸν κοινὸν ἐσόμενον ὑπαγωγέα τῶν χορδῶν ὑποβάλωμεν αὐταῖς κατὰ τὴν ὑποζευγνῦσαν τὰ ΑΕ σημεῖα θέσιν ὡς τὸν ΑΖΕ, ποιήσομεν πάντα μήκη τῶν χορδῶν ἐν τοῖς αὐτοῖς λόγοις, ὥστε ἐπιδέχεσθαι τὴν τῶν ἐφηρμοσμένων τοῖς γένεσι λόγων ἀνάκρισιν. ἐπειδήπερ, ὡς αἱ ἀπὸ τοῦ Ε λαμβανόμεναι κατὰ τὴν ΓΔ πρὸς ἀλλήλας ἔχουσιν, οὕτω καὶ αἱ διὰ τῶν περάτων αὐτῶν ἀναγόμεναι παρὰ τὴν ΑΓ μέχρι τῆς ΔΖ[115] ἕξουσι

[115] ΔΖ] Düring 1934: 18 (Berichtigungen); ΑΖ Düring 1930 and Barker 1989.

πρὸς ἀλλήλας, οἷον ὡς ἡ ΕΓ πρὸς τὴν ΕΔ, οὕτως ἡ ΓΑ πρὸς τὴν ΔΖ· διόπερ αὗται μὲν ποιήσουσι τὸ διὰ πασῶν, ὅτι διπλάσιος αὐτῶν ὁ λόγος.

Next to this instrument [i.e. the *helikōn*], suppose that we draw up a rectangle ΑΒΓΛ, and think of ΑΒ and ΓΔ as determining the vibrating lengths of the strings, and ΑΓ and ΒΔ as the extreme notes of the octave. Then we add ΔΕ, equal to and extending ΓΔ, and cut the side ΓΔ, by the application of rulers, in the ratios proper to the genera, making Ε the limit of high pitch. Through the resulting points of division on it we stretch strings parallel to ΑΓ and equal to one another in pitch, and when this is done we place under them what will be the bridge common to the strings in the position, ΑΖΕ, that joins the points Α and Ε. In this way we shall make all the lengths of the strings in the same ratios [as those marked on ΓΔ], so that it makes possible the assessment of the ratios that have been assigned to the genera. For as the lines taken from Ε along ΓΔ stand to one another, so will those drawn from the limits of these parallel to ΑΓ and as far as ΔΖ[115] stand to one another: for instance, as is ΕΓ to ΕΔ, so is ΓΑ to ΔΖ. Hence these lines will make the octave, since their ratio is the duple.[116]

Ptolemy goes on to set out the remaining parts of the instrument: Η is located a quarter of the way from Γ to Ε, and Θ at a third of this distance; on these points the strings ΗΚΛ and ΘΜΝ are set up. Certain relationships follow from this (48.15–23):

ΑΓ:ΗΚ = 4:3   ΑΓ:ΘΜ = 3:2   ΗΚ:ΘΜ = 9:8
ΘΜ:ΔΖ = 4:3   ΗΚ:ΔΖ = 3:2

The bridge, he observes, can be pivoted on point Ε, so as to transpose all the notes upward: bridge position ΞΟΕ will give all the strings a higher pitch than ΑΖΕ, while maintaining the same intervals between them (48.23–49.3). The strings are still as difficult to move from side to side as on the *helikōn*, however, and thus the principal advantage of this first improvement is simply that transposition is now possible. An attunement, once established, can be transposed to match the existing range of a performance instrument such as a lyre or kithara, perhaps in order to make the testing of the intervals of the proposed attunement even easier to judge: the kitharist can tune his instrument to the intervals of the *helikōn*-variant, and then play melodies in this attunement without sacrificing the familiarity of his usual instrument. But retuning the interval-series itself remains problematic, and this is an issue Ptolemy does not address until the end of book II.

[116] 47.18–48.14, trans. Barker 1989: 320–1 with one small amendment (see previous note).

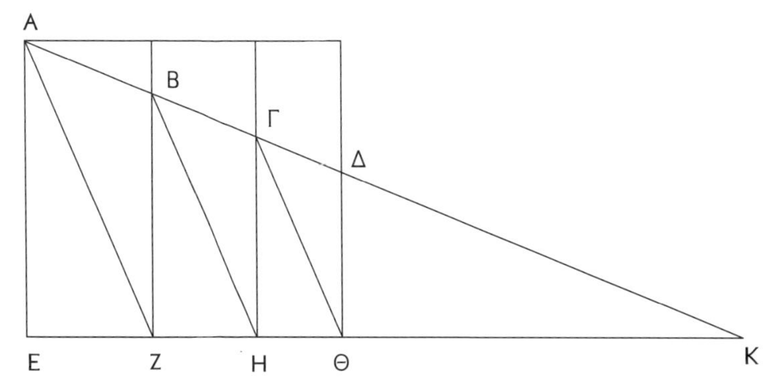

Fig. 6.7 Eratosthenes' *mesolabon* (after Thomas 1939).

Ptolemy's helikōn-*variant and Eratosthenes'* mesolabon

Before we proceed further, it seems appropriate to discuss the possible provenance of Ptolemy's improved *helikōn*. It is clear that this instrument is the real focus of II.2; Ptolemy seems only to have introduced the *helikōn* in order to bring out his adapted model of it. The *helikōn* is described for the geometer; Ptolemy's instrument is described in order of practical construction.[117] There is no reason to suspect that the *helikōn* improvement was not Ptolemy's own invention, but we may note all the same that it bears a striking resemblance to an instrument we have encountered already in a different context. The *mesolabon* of Eratosthenes (see chapter 4 and fig. 6.7) does not merely look similar to Ptolemy's instrument; it operates on similar mathematical principles, and performs a similar task.

The *mesolabon* finds means. Specifically, it finds two (or more) geometric means in continuous proportion. The relationships of the lengths of the vertical lines are key to this process, and because of the diagonal which crosses them (AK in Eutocius' diagram,[118] fig. 6.7) these relationships are paralleled by the distances between the vertical lines (AK:KB::EK:KZ and BK:KΓ::ZK:KH in fig. 6.7).

[117] See Barker 2000: 210.

[118] Eutocius is our immediate source for the diagram of fig. 6.7; see pp. 196–8.

Ptolemy's instrument, similarly, displays means. It does not find them, for in this case they are already known, and in any case mean-taking is not Ptolemy's objective. But in his description he does, effectively, set out the instrument in such a way as to locate the arithmetic (HK) and harmonic (ΘM) means between the lengths of the two outermost vertical lines (i.e. strings, AΓ and ΔZ in fig. 6.6). As in the *mesolabon*, the key relationships are those between the lengths of the vertical lines. Again, because of the diagonal which crosses the verticals (AZE in fig. 6.6), these length relationships are paralleled by the relationships of the distances between the strings (AE:EK::ΓE:EH and KE:EM::HE:EΘ in fig. 6.6). I have used the word 'paralleled', but of course in both cases it would be equally appropriate to say 'controlled'. In the *mesolabon* it is the lateral motion of the panels which governs both the distances between the vertical lines and the relationships between their lengths as cut by the diagonal. But here, of course, is the main difference between the two instruments. In the *mesolabon* (fig. 6.7) the diagonal AK is only a straight line (i.e. it only exists) when the vertical lines (BZ, ΓH and, in fact, ΔΘ as well) stand in a particular lateral relationship to one another, as determined by the placement of the instrument's panels. It has a conditional existence. Thus A must be the pivot-point, and K is not a fixed point at all; it is simply the point where the diagonal and the extension of EΘ happen to intersect.

Ptolemy's instrument is hinged the opposite way. His diagonal (i.e. bridge, AZE in fig. 6.6) pivots on E, rather than A, and so preserves the same ratios between the lengths of its verticals, even though the lengths themselves change. Yet it is still very close to the *mesolabon*; it uses the same mathematical relationships inherent in the same geometrical figures, applied to a musical requirement.[119]

Whether Ptolemy was in fact indebted to Eratosthenes for providing the inspiration for his modified *helikōn* is of course

[119] The instrument, in the proportions in which Ptolemy sets it out, will only accommodate scales of an octave span. This can be changed by a process similar to that of changing the two terms between which Eratosthenes' *mesolabon* will find two geometric means in continued proportion: that is, Ptolemy must start by pivoting the diagonal on A (in fig. 6.6), and situating E such that AΓ:ΔZ is the ratio of the bounding interval of the entire scale. At this point the diagonal can be re-hinged on E to transpose all the notes to a higher pitch without altering the intervals between them.

impossible to tell. Ptolemy may certainly have known of the *mesolabon*; he was no stranger to Alexandria, where Eratosthenes' column bearing the instrument and its dedication might still have been standing in the second century AD.[120] He could equally have come across the instrument in a compendium, or perhaps directly in a book by Eratosthenes; it was well known enough in the fourth and sixth centuries to have been reported by Pappus and Eutocius.[121] The parallel is useful insofar as it helps to provide a glimpse of the way in which Ptolemy's instrument-designing fits into the wider context of Greek mathematical activity. As we saw above, he was willing to adopt a numerical system from astronomy because it was ideally suited to the needs of his current project. Here too, the methods and materials of geometry are put to good use in harmonics.

Improving the eight-stringed kanōn

In II.16 we discover that some of the advantages of Ptolemy's modified *helikōn* can be combined with those of his expanded *kanones*. After suggesting the attachment of tuning pegs (*kollaboi*, 81.6) at both ends of the strings, he adds:

καὶ ἔτι τὸ ποιεῖν αὐτοὺς ἐν τῇ πελεκήσει κινουμένους ἐπὶ τοῦ πλάτους τοῦ κανόνος ἕνεκεν τῆς ἑτέρας τῶν χρήσεων, καθ' ἣν μιᾶς μαγάδος πλατείας ὑποβαλλομένης αἱ κατὰ πλάτος κινήσεις τῶν χορδῶν ποιοῦσι τὰς οἰκείας ἁρμογάς. διαιρουμένων γὰρ πάλιν δύο κανόνων ἴσων τῷ μήκει τῶν μενουσῶν μαγάδων εἰς τὰς ὑποκειμένας μεταξὺ τῶν ἄκρων φθόγγων μοίρας, καὶ προστιθεμένων ἑκατέρου τῶν μαγάδων ἑκατέρᾳ κατὰ τὴν ἐπὶ τὰ αὐτὰ ἀντίθεσιν τῶν ἴσων ἀριθμῶν, αἱ κατὰ πλάτος γε τῶν χορδῶν παραχωρήσεις ἀπ' αὐτῶν δειχθήσονται τοῖς ἁρμόζεσθαι δυναμένοις. τῶν δὲ κολλάβων συγκαθισταμένων μὲν καὶ αὐτῶν οἱ φθόγγοι τηρήσουσι τὰς αὐτὰς τάσεις, μενόντων δὲ συμβήσεται

[120] In *Alm.* III.1 Ptolemy mentions the inaccuracies of two astronomical instruments called equinoctial rings, situated 'in our palaestra', one of which had apparently been there for a sufficiently long period of time that Ptolemy blamed its deficiencies on the gradual shifting of its position over the course of many years (Taub 2002: 135–6). This suggests that it was not impossible for Eratosthenes' instrument to have survived in similar fashion for the three and a half centuries that separate his career from Ptolemy's.

[121] The immediate source (or sources) of Pappus' and Eutocius' accounts is not known (Cuomo 2000: 139).

τὰς χορδὰς ἐκ τῆς ἐπὶ τὰ πλάγια παραφορᾶς, ποτὲ μὲν χαλωμένας, ποτὲ δὲ ἐπιτεινομένας, δεῖσθαι πάλιν τῆς ἐξαρχῆς ἐπὶ τὰς ἰσοτονίας ἀποκαταστάσεως.

It will also be useful to make them moveable on their *pelekēseis* over the breadth of the *kanōn*, for the sake of a second form of usage, in which a single flat bridge is placed under the strings, and the sideways movements of the strings make the appropriate attunements. For when two *kanones*, equal to the length of the fixed bridge, are divided once again into the parts lying between the outermost notes, and when one of the two *kanones* is placed against each of the two bridges, in such a way that equal numbers are placed opposite one another against the same points, the sideways movements of the strings will be displayed by these numbers, for people who are capable of making an attunement. If the *kollaboi* themselves go along with them too, the notes will maintain the same pitches, but if the *kollaboi* stay still, the result will be that the strings, being sometimes slackened and sometimes tightened as a consequence of the sideways movement, will once again need to be restored to their original equality of pitch.[122]

The eight-stringed *kanōn* is now to be divided as the improved *helikōn* was – that is, the ratios of the attunements in the tables of II.14–15 are to be set out not in the lengths of the strings, but in the distances between them, using a 'single flat bridge' to divide them diagonally. In order to make this method of division workable, Ptolemy has had to find a way around the difficulty which vexed the *helikōn* and his variant of it: that the strings were not easily movable from side to side. His solution is to avoid anchoring the *kollaboi* directly to the *kanōn* itself, and to attach them instead to movable devices he calls *pelekēseis*. The word *pelekēsis* usually means simply 'wood-cutting', and is cognate with *pelekus*, 'axe'; other related words which give us a clue to its meaning here are *pelekēma*, 'wood-chip', 'splinter', and *pelekinos*, which sometimes indicates a 'dovetail' in carpentry, since it is shaped like an axe-head. Barker takes Ptolemy's use of *pelekēsis* here to mean a kind of axe-head-shaped wooden wedge with a hole drilled through it where an axe-head has a hole for the haft, so that it could slide from side to side on a rod fitted along the edge of the *kanōn* parallel to the end-bridges. A tuning-peg would be fixed to each *pelekēsis*, and its shape would prevent it from rotating on the rod under the tension of the string.[123] The strings can now be

[122] 81.9–21, trans. Barker 1989: 357.
[123] Barker 1989: 357 nn. 140–1 and 2000: 211–12.

moved sideways more easily and without increasing their tension. Additional rulers are used to ensure that the strings' new positions are measured accurately at both ends, and that they remain at right angles to the fixed bridges.

This development is the final stage in Ptolemy's series of adaptations of the *helikōn* which began in II.2. The advantage of the more flexible instrument described in II.16 is that transposition is now a much simpler process than it was with the basic version of the eight-stringed *kanōn*. In the case of the latter, to transpose an attunement up one tone, all eight movable bridges would have to be placed at new points so as to shorten the speaking length of each string by 9:8; this involves a separate calculation for each new bridge position. In Ptolemy's improved eight-stringed *kanōn*, the need for such calculations is done away with, and a single bridge can now be pivoted so as to bring the entire attunement evenly to precisely the desired pitch, whether this bears any rational relationship to the original one or not. Because of this, and because of its additional ease of retuning, the improved eight-stringed *kanōn* is compatible with the instruments of the real musical world in a way no other laboratory instrument has yet been.

This is surely not incidental to Ptolemy's project. In III.1 he is clear about his objectives:

μάλιστα μὲν παραστάσεως ἕνεκεν τῶν τῆς φύσεως δημιουργημάτων καὶ τῆς οὕτως ἀπαραβλήτου τέχνης, κατ' ἐπακολούθησιν δὲ καὶ τῆς δι' αὐτῆς χρήσεως, προϋποκεῖσθαι δεῖ τὸ τοιοῦτον εἴς τε τὴν εὕρεσιν καὶ τὴν ἔνδειξιν τῶν ἀκριβούντων τὸ ἡρμοσμένον λόγων.

For the sake, most importantly, of exhibiting the quite incomparable skill with which the works of nature are crafted, and for the sake, secondly and in consequence, of promoting the practice that makes use of it, it is essential that this sort of method be adopted as a foundation, for the discovery and the exhibition of the ratios that make attunement accurate.[124]

The sort of method Ptolemy means is one which uses carefully constructed and thoroughly tested apparatus, through which the ratios of the intervals can be most fairly judged by perception. This necessitates the melodic capabilities of many-stringed *kanones* – and not only the eight-stringed model, but a larger version as well.

[124] 85.15–19, trans. Barker 1989: 365.

The fifteen-stringed kanōn

Book III opens with a new consideration about the best way of displaying to perception the ratios of the attunements constructed in the course of book II. The tables of II.14–15 gave only one octave, and for theoretical purposes, Ptolemy says, this is sufficient.

εἰ δέ τις ἐκ περιουσίας βούλοιτο συμπληροῦν τῷ κανόνι τὸ δὶς διὰ πασῶν σύστημα τῆς παντελοῦς ποικιλίας ἕνεκεν, ὥστε προσποιεῖν τοῖς ὀκτὼ φθόγγοις τοὺς ἑπτὰ τοὺς λείποντας εἰς τοὺς ἐν τῇ λύρᾳ δεκαπέντε τοῦ δὶς διὰ πασῶν μεγέθους, ἐνέσται μεθοδεύειν καὶ τὴν τοιαύτην προσθήκην, ἵνα μήτε βραχέα καταλειπόμενα τὰ τῶν ὀξυτάτων φθόγγων δυσήχους αὐτοὺς ποιῇ, μήτε τὰ προενεχθησόμενα κανόνια μέχρι τοῦ δὶς διὰ πασῶν λαμβάνῃ τὰς διαιρέσεις.

But suppose that someone's enthusiasm should make him want to fill out on the *kanōn* the double-octave *systēma*, to achieve a complete repertoire of complexity, so as to add to the eight notes the remaining seven, adding up to the fifteen of the magnitude of the double octave on the *lyra*. A method can be found of achieving this objective too, in such a way that the short lengths left for the highest notes do not make them lack sonority, and the *kanonia* that are to be introduced do not have to be marked with divisions up to the double octave.[125]

A complete two-octave division does indeed present challenges even to Ptolemy's most advanced eight-stringed *kanōn*, and the issue of the strings' sonority is not negligible: the ear must be able to make a fair judgement of the intervals regardless of how high they are pitched.[126] His solution is to construct a fifteen-stringed *kanōn*, with eight fine strings and seven thicker ones, to be tuned an octave apart according to their thicknesses (III.1). (By the end of the chapter it is clear that this fifteen-stringed *kanōn* can be made with sliding *pelekēseis* as well.[127]) He includes a demonstration to prove that this tuning will not affect the results, as long as the bridge positions are kept the same in each section of the instrument (83.22–84.10).

[125] 83.5–12, trans. Barker 1989: 362.

[126] Barker suggests that the instrument's use in public demonstrations (harmonic *epideixeis* such as those discussed in ch. 3 part 1), where the volume of the instrument is an important factor, may have been one of Ptolemy's considerations (1989: 362 n. 2).

[127] 85.19–34. This method has its limitations, as Ptolemy explains. Not all the attunements can be constructed, as the outermost string of one octave must occasionally be placed too close to the projecting end of the diagonal bridge used to divide the other octave. It is a purely practical difficulty, which suggests that Ptolemy had actually constructed this most complex *kanōn* for himself, and tested its efficacy.

This largest *kanōn*, he declares, is useful for the nearly tone-deaf: as long as one is capable of tuning strings in unison (and, by implication, in octaves), the bridge positions can be found by measurement on the *kanonion* and the correct attunement will be heard. One wonders whether Ptolemy considered the use to which this instrument could be put as an expedient for the lyre-student who had difficulty tuning his own instrument, like the modern electronic tuners used by string-players who cannot tune their fifths by ear.

For the more capable and enthusiastic harmonicist, the instrument can be used to generate the entire Greater Perfect System in both ways: that is, beginning from the ratios and producing the notes, or the opposite. (The less capable harmonicist must content himself with the former.) The absolute pitch of the strings is unimportant, since it is the intervals of the notes relative to one another which constitute the attunement.

The issue of the investigator's confidence in his apparatus (which must, in some cases, exceed his confidence in his own sense-perception) is an important feature of this chapter (III.1), and Ptolemy devotes some effort to establishing beyond doubt the complete trustworthiness of this most complex of his instruments. To his reader who is capable of making an attunement by ear, he suggests the following test (84.11–20). Tune the fifteen strings in any random intervals. Next, place the bridges according to one of the attunements given in book II. Then use the *kollaboi* to tune the strings so that their pitches match the genus and *tonos* of the chosen attunement. Thereafter any of the other attunements will be faithfully reproduced by the bridge positions in Ptolemy's tables, and when the bridges are removed the strings will sound in unison. Further tests are offered in addition to this one (84.21–85.8). These are simple confidence-building demonstrations, to strengthen the scientist's trust not only in the apparatus itself, but in Ptolemy's general thesis that the faculties of reason and perception, which are connected in the construction and use of the instruments, are indeed in fundamental agreement with one another.

The following chapter (III.2) is devoted to a number of ingenious alternative methods of dividing the eight-stringed *kanōn* into the fifteen notes of the Greater Perfect System. The procedures illustrating this poor-man's fifteen-stringer show Ptolemy at his most creative and most attentive to practical detail. The problem of poor resonance in shorter string lengths is his main concern, and he offers several expedients for maximising high-note sonority while maintaining the instrument's accuracy.[128] The requirements of reason and perception must now be met in a half-sized instrument. It is an economising task; Ptolemy devotes a significant amount of energy to helping his reader avoid the necessity of building his largest instrument. If one has got as far as making an eight-stringed *kanōn*, these suggestions offer a way to construct the full two-octave system on it. The fact that the discussion is theoretically irrelevant illustrates just how important the purely practical aspects of canonic division were to Ptolemy. What he has taken time over is, he says, a description of the 'more reliable' methods of performing the division of the two-octave system (89.25–6). It is part of his project insofar as encouraging accuracy and good scientific practice in harmonics is also a part of his project.[129]

He concludes the strictly harmonic portion of his treatise with a summary in which the *kanōn* holds pride of place:

ὅτι μὲν οὖν καὶ μέχρι τῶν ἐμμελειῶν οἰκείους ἔχει λόγους ἡ τοῦ ἡρμοσμένου φύσις καὶ τίνας ἑκάστων αὐτάρκως καὶ διὰ πλειόνων ἀποδεδεῖχθαί μοι δοκῶ, ὡς μηδ' ἂν ἐπιδιστάσαι τοὺς ἅμα τῷ εὐλόγῳ τῶν ὑποτεθειμένων καὶ πρὸς τὴν διὰ τῆς ἐνεργείας δοκιμασίαν, τουτέστι τὰς μεμεθοδευμένας ἡμῖν κανονικὰς χρείας, φιλοτιμησαμένους ἕνεκεν τοῦ διὰ πάντων τῶν εἰδῶν κατανοῆσαι τὰς ἀπὸ τῶν αἰσθήσεων συγκαταθέσεις.

It seems to me, then, that we have demonstrated adequately and in several ways that the nature of attunement possesses its own proper ratios right down to the melodics, and that we have shown which ratio belongs to each of them, in such a way that those who strive eagerly to master both the rational grounds of the principles laid down and their assessment in practice – that is, the methods of using the *kanōn* that we have expounded – can be in no doubt that they conform, throughout all the species [of *systēma*], to what we accept on the basis of the senses.[130]

[128] They involve the double-bridging of each string, the replacement of thick strings with thinner ones, and the retuning of the latter to higher pitches (86.1–89.27).

[129] See 85.15–19 for Ptolemy's expression of these priorities.

[130] 91.22–92.1, trans. Barker 1989: 371.

Ptolemy has accomplished his project through the careful development of experimental and demonstrative instruments, and through the rigorous application of a scientific method which allowed a role for the ear that was unprecedented in mathematical harmonic science. His approach is both critical and creative: he has brought with him to harmonics the tools of other sciences – sexagesimal notation from the astronomers, the *helikōn* from the mathematicians – and has modified them to suit his project. But his principal debt to the past is surely the *kanōn*, without which his project could never have been undertaken. Like the abacus, it allows the manipulation of perceptible quantities; like the ruler and compasses, it can be used to construct such perceptible quantities through distances; like the geometrical diagram, it mediates between the realm of number and the realm of perception – the realm of being and the realm of becoming. Unlike any of these, however, it speaks plainly, directly and simultaneously to both sight and hearing, the two senses which Ptolemy regards as 'most especially the servants of the theoretical and rational part of the soul' (*Harm.* 5.7–8).

The celestial monochord

I shall end with a brief look at the way the *kanōn* is involved in the big picture, as Ptolemy sketches it. In the final chapters of his treatise (beginning at III.3), he aims to show that the conclusions of his harmonic arguments are not limited to the realm of music, but apply equally to a number of others also. It turns out that harmonics is simply the science in which the ratios and proportions which also govern the structures of the human soul and the *kosmos* are apprehended most readily and accurately. The significance of this particular science is therefore much broader than is suggested by its focus on the correct attunement of musical scales: harmonics, it seems, is the key to the universe.

One of the first stages by which Ptolemy develops this broader argument is a return to the parallel he introduced at the beginning of the treatise (in I.1–2) between sight and hearing as 'the more rational of the senses' (5.23–4). Each is related to a science concerned with the movements of things that are either only seen or only heard (94.13–16): these are astronomy and harmonics, which 'employ both arithmetic and geometry, as instruments of indisputable authority, to discover the quantity and quality of the primary movements'; they are, he adds, 'as it were cousins, born of the sisters, sight and hearing, and brought up by arithmetic and geometry as children most closely related in their stock'.[131]

The task of displaying the closeness of this relationship begins in III.8.[132] Ptolemy's first concern is to establish circular motion and cyclical recurrence as characteristic features of the type of movement both sciences study. Harmonic motion appears to be in straight lines, through determinate pitches of scales, but in fact it is cyclical, like the movements of the heavenly bodies. To put it anachronistically, harmonic space may appear linear, but in fact it is curved. In terms of their function (*dynamis*), the notes of the *systēmata* form a closed cycle: there is no objective starting point (100.32–101.6), only an eventual recurrence of the same interval-series at a new pitch. A musical scale is thus as circular as the orbits of the heavenly bodies. Ptolemy gives two brief examples. Suppose one were able to take the circle of the ecliptic (the zodiac) and cut it at one of the equinoctial points, open it up, and lay it flat alongside the double-octave *systēma*: one would then see that significant points on each structure corresponded to one another.[133] Suppose, next, that one were to take the double-octave scale and bend it into a circle by joining up the highest and lowest notes: one would see a similar correspondence of structures.[134] By extension

[131] χρώμεναι μὲν ὀργάνοις ἀναμφισβητήτοις ἀριθμητικῇ τε καὶ γεωμετρίᾳ πρός τε τὸ ποσὸν καὶ τὸ ποιὸν τῶν πρώτων κινήσεων, ἀνεψιαὶ δ' ὥσπερ καὶ αὗται, γενόμεναι μὲν ἐξ ἀδελφῶν ὄψεως καὶ ἀκοῆς, τεθραμμέναι δὲ ὡς ἐγγυτάτω πρὸς γένους ὑπ' ἀριθμητικῆς τε καὶ γεωμετρίας, 94.16–20, trans. Barker 1989: 373.

[132] The intervening chapters argue that the power of *harmonia* is in everything which has a source of movement in itself, especially the soul and the heavens (III.4), and pair the concords with the parts of the soul (III.5), the genera with the primary virtues (III.6), and harmonic modulations (*metabolai*) with changes (*metabolai*) in the human soul due to crises in life (III.7).

[133] At the middle of the 'cut' ecliptic is the other equinoctial point, and at the middle of the double-octave *systēma* is the note *mesē*; the two ends of the 'cut' ecliptic, which had been joined at the first equinoctial point, correspond to the notes at the two ends of the double-octave *systēma*, namely *proslambanomenos* and *nētē hyperbolaiōn* (101.6–12).

[134] *Mesē* would then lie at the point on the circumference exactly opposite to the join, just where the equivalent equinoctial point lies on the circle of the ecliptic; this is a significant relationship from an astronomical and astrological point of view (101.12–26). This image is perhaps an oblique reference to *Timaeus* 35b–36b.

(though Ptolemy does not extend the analogy this far), the particular semicircle of the ecliptic visible in the sky at a given moment will be analogous to a single-octave segment of the double-octave *systēma*, and just as the arrangement of constellations changes within this visible semicircle as one constellation sets in the west and another rises in the east, so too the arrangement of intervals within an octave may be changed by a similar process (one note disappearing from one end of the octave-scale, to be replaced at the other end by a new note).[135]

This is fairly simple, and vague enough not to involve any of the details of the harmonic arguments of the treatise in any meaningful way. But in the following chapter (III.9) Ptolemy sets out a structure for harmonic and astronomical comparison which applies the musical proportions directly to the zodiac.

ἐὰν γὰρ ἐκθώμεθα κύκλον τὸν ΑΒ καὶ διέλωμεν αὐτὸν ἀπὸ τοῦ αὐτοῦ σημείου, οἷον τοῦ Α, εἰς μὲν δύο ἴσα τῇ ΑΒ, εἰς δὲ τρία ἴσα τῇ ΑΓ, εἰς δὲ τέσσαρα ἴσα τῇ ΑΔ, εἰς δὲ ἓξ ἴσα τῇ ΓΒ, ἡ μὲν ΑΒ περιφέρεια ποιήσει τὴν διάμετρον στάσιν, ἡ δὲ ΑΔ τὴν τετράγωνον, ἡ δὲ ΑΓ τὴν τρίγωνον, ἡ δὲ ΓΒ τὴν ἑξάγωνον. καὶ περιέξουσιν οἱ λόγοι τῶν ἀπὸ τοῦ αὐτοῦ σημείου, τουτέστι πάλιν τοῦ Α, λαμβανομένων περιφερειῶν, τούς τε τῶν ὁμοφώνων καὶ τοὺς τῶν συμφώνων καὶ ἔτι τὸν τονιαῖον, ὡς ἐξέσται σκοπεῖν ὑποθεμένοις τὸν κύκλον τμημάτων ιβʹ, διὰ τὸ πρῶτον εἶναι τὸν ἀριθμὸν τοῦτον τῶν ἥμισυ καὶ τρίτον καὶ τέταρτον ἐχόντων μέρος.

Let us draw a circle, AB, and divide it [i.e. its circumference], starting from some one point, A, into two equal parts by means of line AB, into three equal parts by means of line ΑΓ, into four equal parts by line ΑΔ, and into six equal parts by line ΓΒ. Then arc AB will make the configuration of diametrical opposition, ΑΔ that of a square, ΑΓ that of a triangle, and ΓΒ that of a hexagon. And the ratios of the arcs, starting from the same point, that is, from A, will include those of the homophones and the concords, and that of the tone besides, as we can see if we suppose the circle to consist of twelve segments, since this is the first number to have a half, a third and a fourth part.[136]

[135] In fact, Ptolemy is more interested in another feature of the comparison, namely the consequent diametrical opposition of notes an octave apart when the double-octave *systēma* is bent into a circle. When he does develop comparisons between astronomical and harmonic motion, he will make the lateral motion of the heavenly bodies (that is, toward the north or south) parallel to modulations between *tonoi* (III.12; *tonoi* are two-octave scales). As is clear from his different treatments of the zodiac and the double-octave *systēma* in III.8–9, he was keen to consider more than one way of envisaging the comparisons between astronomical and harmonic structures.

[136] 102.4–13, trans. Barker 1989: 381–2.

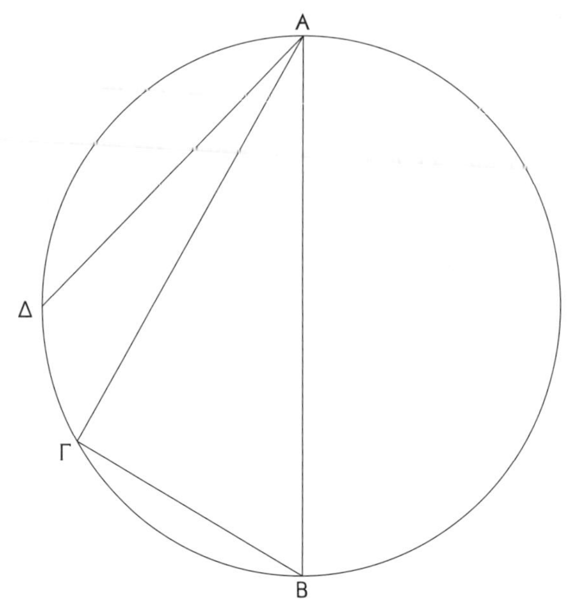

Fig. 6.8 Diagram accompanying Ptol. *Harm.* III.9 (after Düring 1930).

The figure could perhaps have been set out more clearly: it is the circumference of the circle which is being divided, and what Ptolemy means by dividing it into two, three and four equal parts by means of the straight lines AB, AΓ and AΔ is that these lines cut off arcs equal to a half, a third and a quarter of the circumference; the arc subtended by the chord ΓB is a sixth of the circumference. Likewise, chord AΔ is the side of a square inscribed in the circle, AΓ that of an (equilateral) triangle, and ΓB that of a hexagon.

The point of this operation is to show that if one divides a circle in this way, the ratios of the lengths of the arcs so defined will correspond to musically significant ratios. The easiest way to display these ratios, Ptolemy says, is to assign the circumference a length of 12 units: the parallel with the circle of the ecliptic and its twelve 30-degree divisions will now be clear. On this measurement, arc ABΔ = 9, arc ABΓ = 8 and arc AB = 6 (102.13–15). He then proceeds to list the ratios in their various manifestations within the figure thus quantified (to which I shall return shortly).

The purpose of the parallel, then, is to show that the primary division of the zodiac conforms in some way to numerical relationships whose significance in harmonics has already been established. Unlike the previous chapter, in which the ecliptic was mapped on to the non-proportional line of the double-octave scale, here the parallel is made through ratio and proportion. The only way to connect the harmonic and the astronomical in this operation is through the 'rational postulates of the *kanōn*' demonstrated earlier in the treatise. Distances of arc are not musical distances, unless they are reconceived as string lengths. To do this, we must cut the circle at point A and lay it flat; it now becomes a sort of *kanōn*, and the points originally found on the circle through geometry (Β, Γ, Δ) acquire arithmetical and harmonic significance as bridge positions by which the primary concords are sounded (see fig. 6.9).

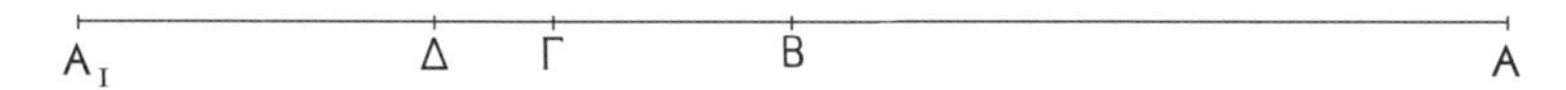

Fig. 6.9 The circumference of Ptolemy's divided circle (fig. 6.8), cut at point A (now A and A_I) and laid flat as a *kanōn*. The other points are located as Ptolemy defines them.

Ptolemy says nothing quite like this. He merely lists the distances of arc in which the harmonically significant ratios will be found in his divided circle. But his reader knows that these ratios have no harmonic significance as distances of arc; they have acquired their meaning through canonic division. The ecliptic-cutting and *systēma*-joining of III.8 has suggested already that the difference between the straight and curved lines of harmonics and astronomy is incidental, and only masks a greater structural unity between the two sciences. Now in III.9 a proportional view of the comparison reveals something similar. Ptolemy sets out the ratios and their intervals at 102.15–103.4; I arrange them below as he does, except that I designate them according to the positions they would occupy if his circle were opened as in fig. 6.9. Thus where Ptolemy says 'the 12 ⟨segments⟩ of the whole circle' (τὰ τοῦ ὅλου κύκλου ιβ′) I give the length AA_I, and where he gives arc ΑΓ I give length $A_I\Gamma$. (The direction of his division is therefore preserved in my scheme.)

the octave (2:1) in three ways:	$AA_I:AB$	= 12:6
	$AB\Gamma:A_I\Gamma$	= 8:4
	$A_I\Gamma B : A_I\Delta$	= 6:3
the fifth (3:2) in three ways:	$AA_I:AB\Gamma$	= 12:8
	$AB\Delta:AB$	= 9:6
	$AB:A_I\Gamma$	= 6:4
the fourth (4:3) in three ways:	$AA_I:AB\Delta$	= 12:9
	$AB\Gamma:AB$	= 8:6
	$A_I\Gamma:A_I\Delta$	= 4:3
the octave plus fifth (3:1) in two ways:	$AA_I:A_I\Gamma$	= 12:4
	$AB\Delta:A_I\Delta$	= 9:3
the double octave (4:1) in one way:	$AA_I:A_I\Delta$	= 12:3
the octave plus fourth (8:3) in one way:	$AB\Gamma:A_I\Delta$	= 8:3
the tone (9:8) in one way:	$AB\Delta:AB\Gamma$	= 9:8

Several things are happening in Ptolemy's division of the ecliptic. First, the twelve units of arc remind his reader of the twelve zodiacal signs which had become the standard measure of solar months: they are units of temporal as well as of astronomical measurement. Second, the division is carried out in two different directions (clockwise and anticlockwise), just as the segments of string on both sides of the bridge are used in the division of I.8. Thus with the single 'bridge position' Γ, the homophone of the octave is demonstrated by using the distances of arc on either side of it (8:4). This illustrates, in a way similar to the curling and uncurling images of III.8, that in some broader sense the fixed end-bridges of the *kanōn*, though they constitute the points of origin and termination of a string's vibrating length, are only so because of the practical necessities of canonic division. When the *kanōn* is bent into a circle, they are no longer extremities but a single point, and the locations of the harmonically significant bridge positions which lay between them now have geometrical significance demonstrable by means of inscribed polygons, and astronomical significance because of the zodiacal constellations which lie on them.[137]

[137] The remainder of III.9 (103.5–104.17) offers further connections between the harmonic and astronomical structures of the figure, but they contain so many 'uncharacteristic errors, misleading claims and confusions of expression' that Barker was tempted to doubt Ptolemy's authorship of the section (1989: 384 n. 70). On the poor state of the text at the end of book III, see n. 91 above.

The instruments are therefore central to Ptolemy's conception of harmonic science in ways that go beyond the strictly harmonic parts of his treatise. Even in his broader approach to the subject the *kanōn* continues to be relevant, by lending itself to the shaping of the mathematical structure of the universe. In this way Ptolemy can present the universe itself as a kind of perfectly tuned *kanōn*.

CONCLUSION

One of the central aims of this book has been to show not only the advantages, but also the complications and difficulties that arose from the introduction of a geometrical instrument into an arithmetical science, and the ways in which these were addressed by harmonicists in the mathematical tradition down to the time of Claudius Ptolemy. Mathematical harmonics, because it continued to express intervals exclusively as ratios of numbers, never left the realm of arithmetic, but the monochord, when it was introduced, brought with it the challenge of constraining the science within the bounds of arithmetic while making use of an instrument whose operation could not be described (not even mathematically) solely within the bounds of arithmetic.[1] The geometricisation of harmonics, in its various aspects, was the inevitable result of attempts to make compelling mathematical arguments for the use of numbers and ratios as a language and analytical tool for the investigation of harmonic structures with the assistance of an instrument which represented numbers and notes as measurable visible distances.

This geometricisation is at the mathematical level similar to that which attends the Euclidean expression of arithmetic by means of lettered diagrams (*El.* V, VII–IX), and the earliest harmonic text which mentions the monochord introduces it via the conventions of the lettered diagram (*Sect. can.* prop. 19). There are, however, important differences between this type of geometricisation and that of Euclid's *Elements*. The most far reaching of these is that the diagram of the canonic division does not represent only arithmetical entities, but also the relative physical dimensions of a vibrating string which must be replicated exactly in reality if the instrument is to fulfil its apodeictic function. This is simply a manifestation

[1] *Sect. can.* 19 begins with the words ἔστω τοῦ κανόνος μῆκος, 'let there be a length of the *kanōn*'; the word μῆκος, 'length', signals the shift toward geometry.

of the difference between arithmetic as an abstract science and harmonics as a science with perceptible objects, but its consequences are important. Minor imprecision in the rendering of the diagram is mathematically and apodeictically irrelevant, but in the construction of the instrument that it represents, such imprecision could result in the reader's inability to perceive (or be certain of) the connection between the musical phaenomena (notes within a recognisable scale-system) and the arithmetical relationships that are argued to exist between them.

This is a difficulty which does not attend the abacus (where the exact dimensions of the pebbles, or for that matter their uniformity, do not affect their ability to represent units), because the abacus does not represent arithmetical entities geometrically, and dot- or pebble-figures, when they eventually appear in Greek mathematical texts, do not rely on geometrical precision for their apodeictic value in the way that the monochord does. But the instruments of geometrical sciences with perceptible objects (like Ptolemy's armillary sphere or Diocles' parabolic mirror) share the monochord's problem of representation even though the monochord differs from some of them (e.g. the armillary sphere) by doing its representation without modelling the phaenomena under investigation.

It was the monochord's introduction of geometry into harmonics that brought with it the problems of representation and exactitude. In a sense this difficulty must have attended all instrumental attempts to demonstrate the relationship between musical intervals and ratios of numbers, for even Hippasus' discs would have failed to show the correspondence if they had been so inexpertly constructed as to be badly out of tune with one another. It is the high standard set by the argumentative rigour of Euclidean mathematics that seems to have brought the issue to the attention of some harmonic scientists in antiquity, and of course a solution was impossible: no concrete object can approach the perfection of a mathematical object. Adrastus apparently got only so far as recognising that the width of the monochord's bridge must take up some of the string's length, but he did not apply this discrepancy to his own account of canonic division. Ptolemy realised that the problem was a three-dimensional one, not merely one of length, and that the bridge must also raise the string in order to divide it into two segments which would vibrate independently of one another. He was also aware that this extra height must add both to the length and to the tension of the string: in combination an acoustical problem whose components are reducible to geometry on the one hand and physics on the other. Because he could not measure the acoustical effect of the additional tension caused by the height of the bridge, Ptolemy sought to reduce the entire problem to geometry in order to claim that as long as the bridges' surfaces made the circumferences of equal circles, the additional length would cancel out the effect of the additional tension exactly (a claim which, as we have seen, is not true).

What this shows is that the problem which arose in mathematical harmonics because of the introduction of a geometrical instrument, itself initially part of an attempt to furnish the science with incontrovertible arguments, was eventually dealt with (imperfectly) by the introduction of yet more geometry.

One of the other aims of the book has been to show how important the ancient discussions of instruments in mathematical harmonics can be to our understanding of Greek scientific method. Particularly significant in this respect are those of Ptolemy, because he not only extended and developed the theoretical and mathematical analysis of his predecessors in the field, but also added rigour through empirical verification on instruments (many of them improved in ways that went beyond the addition of strings) designed especially for the purpose. Here the application of advances made in other mathematical sciences is especially relevant. Ptolemaïs had already suggested an analogy between what reason discovers through the science of canonics and the *parapēgma*, a device which shares some aspects of both instruments and tables in its presentation of astronomical and meteorological information. Ptolemy was to go much further, not only applying the tabularising habit of his astronomical writings to harmonics in tables of bridge positions, but also employing, apparently for the first time, the sexagesimal notation used in astronomy to standardise and simplify his presentation of these bridge positions according to a common numerical measure on the *kanōn*. His more complex instruments, too, show the influence of other sciences. The *helikōn* was already a more geometrical instrument than the monochord even in the hands of Ptolemy's predecessors; his improvement on it, which may just possibly owe something to Eratosthenes' *mesolabon* (an instrument devised as a solution to a problem in geometry), appears to have suggested to Ptolemy at least one element of the design of his improved eight-stringed *kanōn*: that is, generating the intervals between strings by their lateral relationship to one another rather than by the placement of a separate bridge for each string.

What all of these advances brought to the student of mathematical harmonics was the possibility of greater certainty that the fleeting beauty of music did not lie beyond the reach of analysis, and that if melody was the audible expression of a series of arithmetical relationships, then musical beauty itself could be described and demonstrated mathematically. Arithmetic alone could not make the connection between numbers and notes. The sum of developments in what I have called the geometry of sound both furnished that connection more securely than had previously been possible, and also raised doubts about the capacity of instruments to reveal the numerical basis of music. It was in their attempts to minimise these uncertainties that Greek mathematical harmonicists rose to one of the central challenges that faces nearly all sciences that employ instruments.

BIBLIOGRAPHY

Abbreviations used in the Bibliography

AJPh	*American Journal of Philology*
AncPhil	*Ancient Philosophy*
ASMG	*Atti e memorie della Società Magna Grecia*
BCH	*Bulletin de correspondance hellénique*
BJHS	*British Journal for the History of Science*
CIMAGL	*Cahiers de l'Institut du moyen âge grec et latin*
CPh	*Classical Philology*
CQ	*Classical Quarterly*
CRAI	*Comptes rendus de l'Académie des inscriptions et belles-lettres*
HM	*Historia Mathematica*
ICS	*Illinois Classical Studies*
JHA	*Journal for the History of Astronomy*
JHAS	*Journal for the History of Arabic Science*
JHS	*Journal of Hellenic Studies*
OSAPh	*Oxford Studies in Ancient Philosophy*
PBA	*Proceedings of the British Academy*
PCPhS	*Proceedings of the Cambridge Philosophical Society*
QUCC	*Quaderni urbinati di cultura classica*
REG	*Revue des études grecques*
RPh	*Revue de philologie, de littérature et d'histoire anciennes*
TAPhA	*Transactions and Proceedings of the American Philological Association* (1897–1972); *Transactions of the American Philological Association* (1974–)
TAPhS	*Transactions of the American Philosophical Society*
ZPE	*Zeitschrift für Papyrologie und Epigraphik*

Acerbi, F. 2008. 'In what proof would a geometer use the ποδιαία?', *CQ* 58: 120–6.

Adkins, C. 1963. 'The theory and practice of the monochord', Ph.D. diss. State University of Iowa.

1967. 'The technique of the monochord', *Acta Musicologica* 39: 34–43.

Adler, A. (ed.) 1928–35. *Suidae Lexicon* (4 vols.). Leipzig.

Alexanderson, B. 1969. *Textual Remarks on Ptolemy's Harmonica and Porphyry's Commentary*. Gothenburg.

Anderson, W. D. 1994. *Music and Musicians in Ancient Greece*. Ithaca.
Anton, J. P. (ed.) 1980. *Science and the Sciences in Plato*. New York.
Arnim, H. von (ed.) 1903–24. *Stoicorum veterum fragmenta* (4 vols.). Leipzig.
Asmis, E. 1984. *Epicurus' Scientific Method*. Cornell Studies in Classical Philology, vol. 42. Ithaca and London.
Avezzù, G. 1994. 'Papyrus Hibeh I, 13: "anonymi fragmentum" de musica', *Musica e Storia* 2: 109–38.
Baird, D. 2004. *Thing Knowledge: A Philosophy of Scientific Instruments*. Berkeley.
Baird, D. and Faust, T. 1990. 'Scientific instruments, scientific progress and the cyclotron', *British Journal for the Philosophy of Science* 41: 147–75.
Baltes, M. 1972. *Timaios Lokros Über die Natur des Kosmos und der Seele*. Leiden.
Barbera, A. 1977. 'Arithmetic and geometric divisions of the tetrachord', *Journal of Music Theory* 21: 294–323.
1984a. 'The consonant eleventh and the expansion of the musical tetraktys: a study of ancient Pythagoreism', *Journal of Music Theory* 28: 191–223.
1984b. 'Placing *Sectio Canonis* in historical and philosophical contexts', *JHS* 104: 157–61.
1991. *The Euclidean Division of the Canon: Greek and Latin Sources*. Lincoln and London.
Barker, A. 1977. 'Music and mathematics: Theophrastus against the number-theorists', *PCPhS* 23: 1–15.
1978a. 'ΟΙ ΚΑΛΟΥΜΕΝΟΙ ΑΡΜΟΝΙΚΟΙ: the predecessors of Aristoxenus', *PCPhS* 24: 1–21.
1978b. 'ΣΥΜΦΩΝΟΙ ΑΡΙΘΜΟΙ: a note on *Republic* 531c1–4', *CPh* 63: 337–42.
1981. 'Methods and aims in the Euclidean *Sectio Canonis*', *JHS* 101: 1–16.
1984. *Greek Musical Writings* vol. I: *The Musician and his Art*. Cambridge Readings in the Literature of Music. Cambridge.
1988. 'Che cos' era la "mágadis"?', in *La musica in Grecia*, eds. B. Gentili and R. Pretagostini. Bari: 96–107.
1989. *Greek Musical Writings* vol. II: *Harmonic and Acoustic Theory*. Cambridge Readings in the Literature of Music. Cambridge.
1991. 'Three approaches to canonic division', in *ΠΕΡΙ ΤΩΝ ΜΑΘΗΜΑΤΩΝ*, ed. I. Mueller. *Apeiron* 24: 49–83.
1994a. 'Greek musicologists in the Roman Empire', in *The Sciences in Greco-Roman Antiquity*, ed. T. Barnes. *Apeiron* 27: 53–74.
1994b. 'Ptolemy's Pythagoreans, Archytas, and Plato's conception of mathematics', *Phronesis* 39: 113–35.
2000. *Scientific Method in Ptolemy's 'Harmonics'*. Cambridge.
2001. 'Diogenes of Babylon and Hellenistic musical theory', in *Cicéron et Philodème: la polémique en philosophie*, eds. C. Auvray-Assayas and D. Delattre. Paris: 353–70.

2002. 'Words for sounds', in Tuplin and Rihll 2002: 22–35.
2006. 'Archytas unbound', *OSAPh* 31: 297–321.
2007. *The Science of Harmonics in Classical Greece*. Cambridge.
Barker, A. and Creese, D. E. 2001. 'Eratosthenes', in *Die Musik in Geschichte und Gegenwart*, Kassel 1999–, Personenteil 6: 399–400.
Barnes, J. 1975. 'Aristotle's theory of demonstration', in Barnes *et al.* 1975–9, 1: 65–87.
Barnes, J., Schofield, M. and Sorabji, R. (eds.) 1975–9. *Articles on Aristotle* (4 vols.). London.
Bekker, I. (ed.) 1831–70. *Aristotelis opera* (5 vols.). Berlin.
Bélis, A. 1985. 'À propos de la construction de la lyre', *BCH* 109: 201–20.
1986. *Aristoxène de Tarente et Aristote: le traité d'harmonique*. Études et commentaires, 100. Paris.
Benoît, P., Chemla, K. and Ritter, J. (eds.) 1992. *Histoire de fractions, fractions d'histoire*. Berlin.
Bethe, E. (ed.) 1900–37. *Pollucis Onomasticon*. Leipzig.
Bowen, A. C. 1978. 'The minor sixth (8:5) in early Greek harmonic science', *AJPh* 99: 501–6.
1982. 'The foundations of early Pythagorean harmonic science: Archytas, fragment 1', *AncPhil* 2: 79–104.
(ed.) 1991a. *Science and Philosophy in Classical Greece*. New York.
1991b. 'Euclid's *Sectio canonis* and the history of Pythagoreanism', in Bowen 1991a: 164–87.
2007. 'The demarcation of physical theory and astronomy by Geminus and Ptolemy', *Perspectives on Science* 15: 327–58.
Bowen, A. C. and Bowen, W. R. 1997. 'The translator as interpreter: Euclid's *Sectio canonis* and Ptolemy's *Harmonica* in the Latin tradition', in *Music Discourse from Classical to Early Modern Times: Editing and Translating Texts*, ed. M. R. Maniates. Toronto: 97–148.
Bower, C. M. 1989. *Boethius: Fundamentals of Music*. New Haven.
Braund, D. and Wilkins, J. (eds.) 2000. *Athenaeus and his World: Reading Greek Culture in the Roman Empire*. Exeter.
Britton, J. P. 1992. *Models and Precision: The Quality of Ptolemy's Observations and Parameters*. New York.
Brockett, C. W., Jr. 1981. 'A comparison of the five monochords of Guido of Arezzo', *Current Musicology* 32: 29–42.
Bromley, A. G. 1986. 'Notes on the Antikythera mechanism', *Centaurus* 29: 5–27.
Brown, M. 1975. 'Pappus, Plato and the harmonic mean', *Phronesis* 20: 173–84.
Brunet, P. and Mieli, A. 1935. *Histoire des sciences* vol. I: *Antiquité*. Paris.
Bundrick, S. D. 2005. *Music and Image in Classical Athens*. Cambridge.
Burkert, W. 1972. *Lore and Science in Ancient Pythagoreanism*, translated by E. L. Minar, Jr. Cambridge, Mass. Originally published as *Weisheit und*

Wissenschaft: Studien zu Pythagoras, Philolaos und Platon. Nuremberg, 1962.

Burnet, J. (ed.) 1901. *Platonis opera* vol. I. Oxford.

(ed.) 1902. *Platonis opera* vol. IV. Oxford.

(ed.) 1903. *Platonis opera* vol. III. Oxford.

1920. *Greek Philosophy*. (2 vols.) London.

Burnett, C. *et al.* (eds.) 1991. *The Second Sense: Studies in Hearing and Musical Judgement from Antiquity to the Seventeenth Century*. Warburg Institute Surveys and Texts XXII. London.

Cassio, A. C. 1988. 'Nicomachus of Gerasa and the dialect of Archytas, fr. 1', *CQ* 38: 135–9.

Chailley, J. 1979. *La musique grecque antique*. Paris.

Clarke, M. L. 1971. *Higher Education in the Ancient World*. London.

Clement, P. A. and Hoffleit, H. B. 1969. *Plutarch's Moralia* vol. VIII. Loeb Classical Library. Cambridge, Mass.

Colson, F. H. and Whitaker, G. H. 1929–53. *Philo* (10 vols.). Loeb Classical Library. Cambridge, Mass.

Comotti, G. 1983. 'Un'antica arpa, la magadis, in un frammento di Teleste (fr. 808 P.)', *QUCC* n.s. 15 (44): 57–71.

1991. 'Pitagora, Ippaso, Laso e il metodo sperimentale', in Wallace and MacLachlan 1991: 20–9.

Cooper, J. M. (ed.) 1997. *Plato: Complete Works*. Indianapolis.

Cornford, F. M. 1922–3. 'Mysticism and science in the Pythagorean tradition', *CQ* 16: 137–50; 17: 1–12.

Coulton, J. J. 2002. 'The dioptra of Hero of Alexandria', in Tuplin and Rihll 2002: 150–64.

Creese, D. E. 2006. 'Music', in *The Edinburgh Companion to Ancient Greece and Rome*, eds. E. Bispham, T. Harrison and B. A. Sparkes. Edinburgh: 413–22.

2008a. 'Dionusios', in Keyser and Irby-Massie 2008: 263.

2008b. 'Hērakleidēs of Hērakleia on the Pontos, Junior', in Keyser and Irby-Massie 2008: 369–70.

Crocker, R. L. 1963–4. 'Pythagorean mathematics and music', *Journal of Aesthetics and Art Criticism* 22: 189–98, 325–35.

Cuomo, S. 2000. *Pappus of Alexandria and the Mathematics of Late Antiquity*. Cambridge.

2001. *Ancient Mathematics*. London and New York.

Dalby, A. 1991. 'The curriculum vitae of Duris of Samos', *CQ* 41: 539–41.

Decorps-Foulquier, M. 1999. 'Sur les figures du traité des *Coniques* d'Apollonios de Pergé édité par Eutocius d'Ascalon', *Revue d'histoire des mathématiques* 5: 61–82.

Delattre, D. (ed.) 2007. *Philodème de Gadara: Sur la musique, livre IV*, with tr. and comm. (2 vols.). Paris.

De Young, G. 2005. 'Diagrams in the Arabic Euclidean tradition: a preliminary assessment', *HM* 32: 129–79.

Diehl, E. (ed.) 1904. *Procli Diadochi in Platonis Timaeum commentaria.* (3 vols.) Leipzig.

Diels, H. (ed.) 1882–95. *Simplicii in Aristotelis physicorum libros octo commentaria* (2 vols.). Berlin.

(ed.) 1922. *Die Fragmente der Vorsokratiker*. Berlin.

Diels, H. and Kranz, W. (eds.) 1951. *Die Fragmente der Vorsokratiker*, 6th edn. Berlin.

Diggle, J. (ed.) 2004. *Theophrastus: Characters*, with tr. and comm. Cambridge.

D'Ooge, M. L. 1926. *Nicomachus of Gerasa: Introduction to Arithmetic*. New York.

Dupuis, J. 1892. *Théon de Smyrne, philosophe platonicien: Exposition des connaissances mathématiques utiles pour la lecture de Platon*. Text and translation. Paris.

Düring, I. (ed.) 1930. *Die Harmonielehre des Klaudios Ptolemaios*. Gothenburg.

(ed.) 1932. *Porphyrios Kommentar zur Harmonielehre des Ptolemaios*. Gothenburg. Reprinted Hildesheim, 1978.

(ed.) 1934. *Ptolemaios und Porphyrios über die Musik*. Gothenburg. Reprinted New York, 1980.

Einarson, B. and De Lacy, P. H. (eds.) 1967. *Plutarch's Moralia* vol. XIV. Loeb Classical Library. Cambridge, Mass.

Evans, J. 1998. *The History and Practice of Ancient Astronomy*. Oxford.

1999. 'The material culture of Greek astronomy', *JHA* 30: 237–307.

Evans, J. and Berggren, J. L. 2006. *Geminos's Introduction to the Phenomena: A Translation and Study of a Hellenistic Survey of Astronomy*. Princeton.

Feldman, L. H. 1996. 'Philo's views on music', in *Studies in Hellenistic Judaism*, ed. L. H. Feldman. Leiden, New York and Cologne: 504–28. Originally published in *Journal of Jewish Music and Liturgy* 9 (1986–7): 36–54.

Fletcher, N. H. and Rossing, T. D. 1998. *The Physics of Musical Instruments*, 2nd edn. New York.

Fortenbaugh, W. *et al.* (eds.) 1992. *Theophrastus of Eresus: Sources for His Life, Writings, Thought and Influence* (2 vols.). Philosophia Antiqua 54. Leiden and Cologne.

Fowler, D. H. 1991. 'Ratio and proportion in early Greek mathematics', in Bowen 1991a: 98–118.

1992. 'Logistic and fractions in early Greek mathematics: a new interpretation', in Benoît, Chemla and Ritter 1992: 133–47.

1999. *The Mathematics of Plato's Academy: A New Reconstruction*, 2nd edn. Oxford.

Fowler, D. H. and Taisbak, C. M. 1999. 'Did Euclid's circles have two kinds of radius?' *HM* 26: 361–4.

Fowler, H. N. 1914. *Plato* vol. I: *Euthyphro, Apology, Crito, Phaedo, Phaedrus*. Loeb Classical Library. Cambridge, Mass.

Fraser, P. M. 1970. 'Eratosthenes of Cyrene', *PBA* 56: 175–207.
Freeth, T., Bitsakis, Y. *et al.* 2006. 'Decoding the ancient Greek astronomical calculator known as the Antikythera mechanism', *Nature* 444: 587–91.
Freeth, T., Jones, A. *et al.* 2008. 'Calendars with Olympiad display and eclipse prediction on the Antikythera mechanism', *Nature* 454: 614–17.
Friedlein, G. (ed.) 1867. *Anicii Manlii Torquati Severini Boetii De institutione arithmetica libri duo; De institutione musica libri quinque*. Leipzig.
(ed.) 1873. *Proclus In primum Euclidis librum commentarii*. Leipzig.
Gentili, B. and Perusino, F. (eds.) 1995. *Mousike: Metrica Ritmica e Musica Greca in Memoria di Giovanni Comotti*. Pisa.
Gera, D. L. 2000. 'Two thought experiments in the *Dissoi Logoi*', *AJPh* 121: 21–45.
Geus, K. 1995. 'Anmerkungen zur "Musiktheorie" des Eratosthenes', in *Antike Naturwissenschaft und ihre Rezeption*, eds. K. Döring, B. Herzhoff and G. Wörle. Trier: 49–62.
2002. *Eratosthenes von Kyrene: Studien zur hellenistischen Kultur- und Wissenschaftsgeschichte*. Münchener Beiträge zur Papyrusforschung und antiken Rechtsgeschichte, Heft 92. Munich.
Gibson, S. 2005. *Aristoxenus of Tarentum and the Birth of Musicology*. New York and London.
Gingerich, O. 1980. 'Was Ptolemy a fraud?', *Quarterly Journal of the Royal Astronomical Society* 21: 253–66.
Godt, I. 1984. 'New voices and old theory: early music theorists on the monochord and singing, and recent advances in vocal technique in the performance of early music', *Journal of Musicology* 3: 312–19.
Goldstein, B. R. and Bowen, A. C. 1999. 'The role of observation in Ptolemy's lunar theories', in Swerdlow 1999: 341–56.
Goldstein, S. J. 1982. 'Problems raised by Ptolemy's lunar tables', *JHA* 13: 195–205.
Gottschalk, H. B. 1968. 'The *De Audibilibus* and Peripatetic acoustics', *Hermes* 96: 435–60.
1980. *Heraclides of Pontus*. Oxford.
Grattan-Guinness, I. 1996. 'Numbers, magnitudes, ratios, and proportions in Euclid's *Elements*: how did he handle them?' *HM* 23: 355–75.
Grenfell, B. P. and Hunt, A. S. (eds.) 1906. *The Hibeh Papyri*. Part I. London and Oxford.
Gulick, C. B. 1927–41. *Athenaeus: The Deipnosophists* (7 vols.). Loeb Classical Library. Cambridge, Mass.
Guthrie, K. S. 1988. *The Pythagorean Sourcebook and Library: An Anthology of Ancient Writings which Relate to Pythagoras and Pythagorean Philosophy*. Grand Rapids.
Guthrie, W. K. C. 1962. *A History of Greek Philosophy* vol. I: *The Earlier Presocratics and the Pythagoreans*. Cambridge.

Hamilton, E. and Cairns, H. (eds.) 1961. *The Collected Dialogues of Plato*. New York.

Hannah, R. 2002. 'Euctemon's *parapēgma*', in Tuplin and Rihll 2002: 112–32.

Havelock, E. A. 1963. *Preface to Plato*. Oxford.

Heath, T. L. 1921. *A History of Greek Mathematics* (2 vols.). Oxford. Reprinted New York, 1981.

Heiberg, J. L. (ed.) 1883–8. *Euclidis Elementa* (4 vols.). Leipzig.

(ed.) 1895. *Euclidis opera omnia* vol. VII. Leipzig.

(ed.) 1898–1903. *Claudii Ptolemaei Syntaxis mathematica* (2 vols.). Leipzig.

(ed.) 1903. *Heronis Alexandrini opera quae supersunt omnia* vol. IV: *Definitiones, cum variis collectionibus*. Leipzig.

Heiberg, J. L. and Stamatis, E. S. (eds.) 1915. *Archimedis opera omnia cum commentariis Eutocii* vol. III. Leipzig.

Heilmann, A. 2007. *Boethius' Musiktheorie und das Quadrivium: eine Einführung in den neuplatonischen Hintergrund von 'De institutione musica'*. Göttingen.

Heinze, R. 1892. *Xenokrates: Darstellung der Lehre und Sammlung der Fragmente*. Leipzig. Reprinted Hildesheim, 1965.

Herlinger, J. 1987. *Prosdocimo de' Beldomandi: Brevis summula proportionum quantum ad musicam pertinet and Parvus tractatulus de modo monacordum dividendi*. Lincoln and London.

2002. 'Medieval canonics', in *The Cambridge History of Western Music Theory*, ed. T. Christensen. Cambridge: 168–92.

Hett, W. S. 1936. *Aristotle: Problems, Books I–XXI*. Loeb Classical Library. Cambridge, Mass.

Higgins, R. A. and Winnington-Ingram, R. P. 1965. 'Lute-players in Greek art', *JHS* 85: 62–71.

Hiller, E. 1870. 'Der Πλατωνικός des Eratosthenes', *Philologus* 30: 60–72.

(ed.) 1872. *Eratosthenis carminum reliquiae*. Leipzig.

(ed.) 1878. *Theonis Smyrnaei philosophi Platonici Expositio rerum mathematicarum ad legendum Platonem utilium*. Leipzig.

Hoche, R. (ed.) 1866. *Nicomachi Geraseni Pythagorei Introductionis arithmeticae libri II*. Leipzig.

Holford-Strevens, L. 2003. *Aulus Gellius: An Antonine Scholar and his Achievement*, revised edn. Oxford.

Hubert, C. (ed.) 1954. *Plutarchi Moralia* vol. VI, fasc. 1. Leipzig.

Huby, P. and Neal, G. (eds.) 1989. *The Criterion of Truth*. Liverpool.

Huffman, C. A. 1985. 'The authenticity of Archytas fr. 1', *CQ* 35: 344–8.

1988. 'The role of number in Philolaus' philosophy', *Phronesis* 33: 1–30.

1993. *Philolaus of Croton: Pythagorean and Presocratic*. Cambridge.

1999. 'The Pythagorean tradition', in *The Cambridge Companion to Early Greek Philosophy*, ed. A. A. Long. Cambridge: 66–87.

2005. *Archytas of Tarentum: Pythagorean, Philosopher and Mathematician King*. Cambridge.

Hughes, A. 1969. 'Ugolino: the monochord and musica ficta', *Musica disciplina* 23: 21–39.

Hultsch, F. (ed.) 1876–8. *Pappi Alexandrini Collectionis quae supersunt* (3 vols.). Berlin.

Hussey, E. 2002. 'Aristotle and mathematics', in Tuplin and Rihll 2002: 217–29.

Ilievski, P. H. 1993. 'The origin and semantic development of the term harmony', *ICS* 18: 19–29.

Irby-Massie, G. L. and Keyser, P. T. 2002. *Greek Science of the Hellenistic Era: A Sourcebook*. London and New York.

Jacoby, F. (ed.) 1923–58. *Die Fragmente der Griechischen Historiker*. Berlin.

Jan, K. von (ed.) 1895. *Musici Scriptores Graeci*. Leipzig. Reprinted Stuttgart and Leipzig, 1995.

Joly, R. 1967. *Hippocrate: Du régime*. Paris.

Jones, A. 1999. 'A classification of astronomical tables on papyrus', in Swerdlow 1999: 299–340.

Kahn, C. H. 1974. 'Pythagorean philosophy before Plato', in Mourelatos 1993: 161–85.

2001. *Pythagoras and the Pythagoreans: A Brief History*. Indianapolis.

Kenney, J. P. (ed.) 1995. *The School of Moses: Studies in Philo and Hellenistic Religion*. Atlanta.

Keyser, P. T. and Irby-Massie, G. L. (eds.) 2008. *Encyclopedia of Ancient Natural Scientists: The Greek Tradition and its Many Heirs*. London.

Kirk, G. S., Raven, J. E. and Schofield, M. 1983. *The Presocratic Philosophers: A Critical History with a Selection of Texts*, 2nd edn. Cambridge.

Kleingünther, A. 1933. *Πρῶτος Εὑρετής: Untersuchungen zur Geschichte einer Fragestellung. Philologus* Supplementband 26.1. Leipzig.

Knorr, W. 1993. '*Arithmētikē stoicheiōsis*: on Diophantus and Hero of Alexandria', *HM* 20: 180–92.

Koller, H. 1960. 'Das Modell der griechischen Logik', *Glotta* 38: 61–74.

Landels, J. G. 1999. *Music in Ancient Greece and Rome*. London.

Landucci Gattinoni, F. 1997. *Duride di Samo*. Rome.

Lawergren, B. 1984. 'The cylinder kithara in Etruria, Greece, and Anatolia', *Imago Musicae* 1: 147–74.

Lehoux, D. 2005. 'The parapegma fragments from Miletus', *ZPE* 152: 125–40.

2007. *Astronomy, Weather, and Calendars in the Ancient World: Parapegmata and Related Texts in Classical and Near-Eastern Societies*. Cambridge.

Levarie, S. 1991. 'Philo on music', *Journal of Musicology* 9: 124–30.

Levin, F. R. 1975. *The Harmonics of Nicomachus and the Pythagorean Tradition*. American Classical Studies, no. 1. The American Philological Association. University Park, Pa.

1990. 'Unity in Euclid's *Sectio Canonis*', *Hermes* 118: 430–43.

1994. *The Manual of Harmonics of Nicomachus the Pythagorean*, with tr. and comm. Grand Rapids.

Lewis, M. J. T. 2001. *Surveying Instruments of Greece and Rome*. Cambridge.
Lindley, M. 1980. 'Pythagorean intonation and the rise of the triad', *Royal Musical Association Research Chronicle* 16: 4–61.
Lippman, E. A. 1964. *Musical Thought in Ancient Greece*. New York.
Lloyd, G. E. R. 1973. *Greek Science after Aristotle*. London.
1978. 'Saving the appearances', *CQ* 28: 202–22, reprinted with new introduction in Lloyd 1991: 248–77.
1979. *Magic, Reason and Experience*. Cambridge.
1987. *The Revolutions of Wisdom: Studies in the Claims and Practice of Ancient Greek Science*. Berkeley.
1990a. 'Plato and Archytas in the Seventh Letter', *Phronesis* 35: 159–74.
1990b. *Demystifying Mentalities*. Cambridge.
1991. *Methods and Problems in Greek Science*. Cambridge.
1992. 'Methods and problems in the history of ancient science: the Greek case', *Isis* 83: 564–77.
2002. *The Ambitions of Curiosity: Understanding the World in Ancient Greece and China*. Cambridge.
Lloyd-Jones, H. and Parsons, P. (eds.) 1983. *Supplementum Hellenisticum*. Berlin and New York.
Lorch, R. 1981. 'A note on the technical vocabulary in Eratosthenes' tract *On mean proportionals*', *JHAS* 5: 166–70.
Lukinovich, A. 1990. 'The play of reflections between literary form and the sympotic theme in the *Deipnosophistae* of Athenaeus', in *Sympotica: A Symposium on the Symposion*, ed. O. Murray. Oxford: 263–71.
Maas, M. and Snyder, J. M. 1989. *Stringed Instruments of Ancient Greece*. New Haven.
MacLachlan, B. 1991. 'The harmony of the spheres: dulcis sonus', in Wallace and MacLachlan 1991: 7–19.
Macran, H. S. (ed.) 1902. *ΑΡΙΣΤΟΞΕΝΟΥ ΑΡΜΟΝΙΚΑ ΣΤΟΙΧΕΙΑ: The Harmonics of Aristoxenus*. Oxford. Reprinted Hildesheim, 1990.
Mansion, S. (ed.) 1961. *Aristote et les problèmes de méthode*. Communications présentées au Symposium Aristotelicum tenu à Louvain du 24 août au 1er septembre 1960. Louvain.
Marrou, H. I. 1977. *A History of Education in Antiquity*, 2nd edn, translated by G. Lamb. New York.
Mathiesen, T. J. 1999. *Apollo's Lyre: Greek Music and Music Theory in Antiquity and the Middle Ages*. Lincoln and London.
McKirahan, R. D. 1978. 'Aristotle's subordinate sciences', *BJHS* 11: 197–220.
1992. *Principles and Proofs: Aristotle's Theory of Demonstrative Science*. Princeton.
Meineke, A. 1843. *Analecta Alexandrina*. Berlin. Reprinted Hildesheim, 1964.
Mendell, H. 1998. 'Making sense of Aristotelian demonstration', *OSAPh* 16: 161–225.

Meyer, C. 1996. *Mensura monochordi: la division du monocorde, IXe–XVe siècles*. Paris.

Michel, P.-H. 1950. *De Pythagore à Euclide: contribution à l'histoire des mathématiques préeuclidiennes*. Paris.

Moehring, H. R. 1978. 'Arithmology as an exegetical tool in the writings of Philo of Alexandria', in *Society of Biblical Literature 1978 Seminar Papers* vol. I, ed. P. J. Atchemeier (Missoula), reprinted in Kenney 1995: 141–76.

Morrow, G. R. 1970. *Proclus: A Commentary on the First Book of Euclid's Elements*. Princeton.

Mountford, J. F. 1917. 'The Greek modes', MA thesis. University of Birmingham.

1923. 'The musical scales of Plato's *Republic*', *CQ* 17: 125–36.

1926. 'The *Harmonics* of Ptolemy and the lacuna in *ii*, 14', *TAPhA* 57: 71–95.

Mourelatos, A. P. D. (ed.) 1993. *The Pre-Socratics: A Collection of Critical Essays*, 2nd edn. Princeton.

Mueller, I. 1969. 'Euclid's *Elements* and the axiomatic method', *British Journal for the Philosophy of Science* 20: 289–309.

1980. 'Ascending to problems: astronomy and harmonics in *Republic* VII', in Anton 1980: 103–21.

1981. *Philosophy of Mathematics and Deductive Structure in Euclid's Elements*. Cambridge, Mass.

1997. 'Greek arithmetic, geometry and harmonics: Thales to Plato', in *Routledge History of Philosophy* vol. I: *From the Beginning to Plato*, ed. C. C. W. Taylor. London and New York: 271–322.

Müller, C. (ed.) 1841–70. *Fragmenta Historicorum Graecorum*. Paris.

Muwafi, A. and Philippou, A. N. 1981. 'An Arabic version of Eratosthenes *On mean proportionals*', *JHAS* 5: 147–65.

Najock, D. 1995. 'A canon for Ptolemy', *Skutala Moisan* 2: 3–7.

Nauck, A. (ed.) 1860. *Porphyrii Philosophi Platonici Opuscula Tria*. Leipzig.

Netz, R. 1999a. *The Shaping of Deduction in Greek Mathematics*. Cambridge.

1999b. 'Proclus' division of the mathematical proposition into parts: how and why was it formulated?' *CQ* 49: 282–303.

2002. 'Greek mathematicians: a group picture', in Tuplin and Rihll 2002: 196–216.

2004a. *The Transformation of Mathematics in the Early Mediterranean World: From Problems to Equations*. Cambridge.

2004b. *The Works of Archimedes* vol. I: *The Two Books On the Sphere and the Cylinder*. Cambridge.

Neubecker, A. J. (ed.) 1986. *Philodemus, Über die Musik IV. Buch: Text, Übersetzung und Kommentar*. Naples.

Neugebauer, O. 1957. *The Exact Sciences in Antiquity*, 2nd edn. Providence. Reprinted New York, 1969.

1975. *A History of Ancient Mathematical Astronomy*. Berlin and New York.

Newton, R. R. 1985. *The Origins of Ptolemy's Astronomical Tables*. Baltimore.

Norman, J. 2006. *After Euclid: Visual Reasoning and the Epistemology of Diagrams*. Stanford.

Omondi, W. A. 1984. 'Tuning of the thum, the Luo lyre: a systematic analysis', *Selected Reports in Ethnomusicology* 5: 263–81.

Oppel, H. 1937. 'ΚΑΝΩΝ', *Philologus* Suppl. vol. 30. Leipzig.

Owen, G. E. L. 1961. 'Τιθέναι τὰ φαινόμενα', in Mansion 1961: 83–103.

Page, D. L. (ed.) 1962. *Poetae Melici Graeci*. Oxford.

(ed.) 1981. *Further Greek Epigrams*. Cambridge.

Paquette, D. 1984. *L'Instrument de musique dans la céramique de la Grèce antique*. Paris.

Patterson, R. 2007. 'Diagrams, dialectic, and mathematical foundations in Plato', *Apeiron* 40: 1–33.

Pédech, P. 1989. *Trois historiens méconnus: Théopompe, Duris, Phylarque*. Paris.

Pesce, D. 1999. *Guido d'Arezzo's Regule Rithmice, Prologus in Antiphonarium, and Epistola ad Michahelem: A Critical Text and Translation*. Ottawa.

Pistelli, H. (ed.) 1894. *Iamblichi in Nicomachi Arithmeticam introductionem*. Leipzig.

Plumley, G. A. 1966. *El Tanbur: The Sudanese Lyre or the Nubian Kissar*. Cambridge.

Pöhlmann, E. and West, M. L. 2001. *Documents of Ancient Greek Music: The Extant Melodies and Fragments Edited and Transcribed with Commentary*. Oxford.

Powell, J. U. (ed.) 1925. *Collectanea Alexandrina*. Oxford.

Price, D. J. de Solla. 1957. 'Precision instruments to 1500', in *A History of Technology*, eds. C. Singer, E. J. Holmyard, A. R. Hall and T. I. Williams (New York and Oxford), III: 582–619.

1974. 'Gears from the Greeks: the Antikythera mechanism, a calendar computer from ca. 80 BC', *TAPhS* 64: 1–70.

Raasted, J. 1979. 'A neglected version of the anecdote about Pythagoras' hammer experiment', *CIMAGL* 31a: 1–9.

Raffa, M. 2002. *La Scienza Armonica di Claudio Tolemeo: Saggio critico, traduzione e commento*. Messina.

Riedweg, C. 2005. *Pythagoras: His Life, Teaching, and Influence*, translated by S. Rendall. Ithaca. Originally published as *Pythagoras: Leben, Lehre, Nachwirkung. Eine Einführung*. Munich, 2002.

Rihll, T. E. 1999. *Greek Science*. New Surveys in the Classics no. 29. Oxford.

Riley, M. T. 1995. 'Ptolemy's use of his predecessors' data', *TAPhA* 125: 221–50.

Rios, R. da (ed.) 1954. *Aristoxeni Elementa Harmonica*. Rome.

Roberts, H. 1981. 'Reconstructing the Greek tortoise-shell lyre', *World Archaeology* 12: 303–12.

Rocconi, E. 2003. *Le parole delle Muse*. Rome.

Rome, A. 1927. 'L'Astrolabe et le météoroscope d'après le commentaire de Pappus sur le 5e livre de l'Almageste', *Annales de la Société Scientifique de Bruxelles* 47: 77–102.

(ed.) 1931–43. *Commentaires de Pappus et de Théon d'Alexandrie sur l'Almageste* (3 vols.). Vatican City.

Ross, W. D. 1924. *Aristotle's Metaphysics: A Revised Text with Introduction and Commentary*. (2 vols.) Oxford.

(ed.) 1964. *Aristotelis Analytica Priora et Posteriora*. Oxford.

Ruelle, C.-É. 1897. 'Le monocorde, instrument de musique', *REG* 10: 309–12.

Rusten, J., Cunningham, I. C. and Knox, A. D. 1993. *Theophrastus Characters, Herodas Mimes, Cercidas and the Choliambic Poets*. Loeb Classical Library. Cambridge, Mass.

Sandbach, F. H. 1969. *Plutarch's Moralia*. (16 vols.) Loeb Classical Library. Cambridge, Mass.

De Santillana, G. and Pitts, W. 1951. 'Philolaos in limbo, or: what happened to the Pythagoreans?' *Isis* 42: 112–20.

Sarton, G. 1993. *Hellenistic Science and Culture in the Last Three Centuries BC*. New York.

Schaarschmidt, C. M. W. 1864. *Die angebliche Schriftstellerei des Philolaus und die Bruchstücke der ihm zugeschriebenen Bücher*. Bonn.

Schlesinger, K. 1933. 'Further notes on Aristoxenus and musical intervals', *CQ* 27: 88–96.

1939. *The Greek Aulos*. London.

Schoenberg, Arnold. 1911. *Harmonielehre*. Vienna, translated by R. E. Carter as *Theory of Harmony*. London, 1978.

Schönberger, P. L. 1914. *Studien zum 1. Buch der Harmonik des Claudius Ptolemaeus*. Augsburg.

Schöne, H. (ed.) 1903. *Heronis Alexandrini opera quae supersunt omnia* vol. III. Leipzig.

Scholz, H. 1975. 'The ancient axiomatic theory', in Barnes *et al.* 1975–9, I: 50–64.

Shiloah, A. 1979. *The Theory of Music in Arabic Writings (c. 900–1900)*. Répertoire international des sources musicales B 10. Munich.

2003. *The Theory of Music in Arabic Writings (c. 900–1900)*. Répertoire international des sources musicales B 10 A (supplement to B 10). Munich.

Sicking, C. M. J. 1998. 'Theophrastus on the nature of music', in *Theophrastus: Reappraising the Sources*, eds. J. M. van Ophuijsen and M. van Raalte. New Brunswick: 97–142.

Sidoli, N. 2004. 'On the use of the term *diastēma* in ancient Greek constructions', *HM* 31: 2–10.

Slings, S. R. (ed.) 2003. *Platonis Respublica*. Oxford.

Solmsen, F. 1942. 'Eratosthenes as Platonist and poet', *TAPhA* 73: 192–213.

Solomon, J. 2000. *Ptolemy Harmonics: Translation and Commentary*. Leiden.

Stahl, W. H. 1952. *Macrobius' Commentary on the Dream of Scipio*. New York.

Striker, G. 1996. *Essays on Hellenistic Epistemology and Ethics*. Cambridge.

Strunk, Oliver. 1952. *Source Readings in Music History* vol. I: *Antiquity and the Middle Ages*. London and Boston.

Swerdlow, N. M. (ed.) 1999. *Ancient Astronomy and Celestial Divination*. Cambridge, Mass.

Szabó, Á. 1978. *The Beginnings of Greek Mathematics*, translated by A. M. Ungar. Hingham, Mass.

Tannery, P. 1902. 'Sur les intervalles de la musique grecque', *REG* 15: 336–52, reprinted in Tannery 1915: 97–115.

1904a. 'Inauthenticité de la "Division du Canon" attribuée à Euclide', *CRAI* 4: 439–45, reprinted in Tannery 1915: 213–19.

1904b. 'À propos des fragments philolaïques sur la musique', *RPh* 28: 233–49, reprinted in Tannery 1915: 220–43.

1913. 'L'Évolution des gammes antiques', published posthumously in Tannery 1915: 312–20.

1915. *Mémoires scientifiques* vol. III: *Sciences exactes dans l'antiquité*. Toulouse and Paris.

Tarán, L. 1975. *Academica: Plato, Philip of Opus and the Pseudo-Platonic Epinomis*. Philadelphia.

Tarrant, H. 1993. *Thrasyllan Platonism*. Ithaca and London.

Taub, L. 1993. *Ptolemy's Universe: The Natural Philosophical and Ethical Foundations of Ptolemy's Astronomy*. Chicago.

2002. 'Instruments of Alexandrian astronomy: the uses of the equinoctial rings', in Tuplin and Rihll 2002: 133–49.

Taylor, A. E. 1928. *A Commentary on Plato's Timaeus*. Oxford.

Taylor, C. A. 1965. *The Physics of Musical Sounds*. London.

Thesleff, H. 1961. 'An introduction to the Pythagorean writings of the Hellenistic period', *Acta Academiae Aboensis*, ser. A (Humaniora), 24.3.

1965. 'The Pythagorean texts of the Hellenistic period', *Acta Academiae Aboensis*, ser. A (Humaniora), 30.1.

1989. 'Platonic chronology', *Phronesis* 34: 1–26.

Thomas, I. 1939–41. *Greek Mathematical Works* (2 vols.). Loeb Classical Library. Cambridge, Mass.

Tobin, T. H. 1985. *Timaeus of Locri, On the Nature of the World and the Soul: Text, Translation and Notes*. Chico.

Toomer, G. J. 1976. *Diocles on Burning Mirrors*. New York.

1984. *Ptolemy's Almagest*. London.

Triadafillidis, T. A. 1995. 'Circumventing visual limitations in teaching the geometry of shapes', *Educational Studies in Mathematics* 29: 225–35.

Tuplin, C. J. and Rihll, T. E. (eds.) 2002. *Science and Mathematics in Ancient Greek Culture*. Oxford.

Van Brummelen, G. 1994. 'Lunar and planetary interpolation tables in Ptolemy's *Almagest*', *JHA* 25: 297–311.

Van der Waerden, B. L. 1943. 'Die Harmonielehre der Pythagoreer', *Hermes* 78: 163–99.

Ver Eecke, P. 1948. *Proclus de Lycie: Les commentaires sur le premier livre des Éléments d'Euclide*. Bruges.

Wallace, R. W. 1995. 'Music theorists in fourth-century Athens', in Gentili and Perusino 1995: 17–39.

Wallace, R. W. and MacLachlan, B. (eds.) 1991. *Harmonia Mundi: Musica e filosofia nell'antichità*. Biblioteca di Quaderni Urbinati di Cultura Classica 5. Rome.

Wallis, J. (ed.) 1682. *ΚΛΑΥΔΙΟΥ ΠΤΟΛΕΜΑΙΟΥ ΑΡΜΟΝΙΚΩΝ ΒΙΒΛΙΑ Γ*. Oxford.

(ed.) 1699. *Opera mathematica* vol. III. Oxford.

Wantzloeben, S. 1911. *Das Monochord als Instrument und als System*. Halle.

Warner, D. J. 1990. 'What is a scientific instrument, when did it become one, and why?' *BJHS* 23: 83–93.

Wegner, U. 1984. *Afrikanische Saiteninstrumente*. Berlin.

Wehrli, F. (ed.) 1945. *Die Schule des Aristoteles: Texte und Kommentar* vol. II: *Aristoxenos*. Basle.

(ed.)1953. *Die Schule des Aristoteles: Texte und Kommentar* vol. VII: *Herakleides Pontikos*. Basle.

West, M. L. 1992. *Ancient Greek Music*. Oxford.

1997. 'When is a harp a panpipe? the meanings of πηκτίς', *CQ* 47: 48–55.

Wheeler, E. L. 1983. 'The *hoplomachoi* and Vegetius' Spartan drillmasters', *Chiron* 13: 1–20.

Winnington-Ingram, R. P. 1932. 'Aristoxenus and the intervals of Greek music', *CQ* 26: 195–208.

1956. 'The pentatonic tuning of the Greek lyre: a theory examined', *CQ* 6: 169–86.

(ed.) 1963. *Aristidis Quintiliani De Musica libri tres*. Leipzig.

Wohl, V. 1998. 'Plato avant la lettre: authenticity in Plato's epistles', *Ramus* 27: 60–93.

Wolfer, E. P. 1954. *Eratosthenes von Kyrene als Mathematiker und Philosoph*. Groningen.

Zancani Montuoro, P. 1974–6. 'Francavilla Marittima: A. Necropoli', *ASMG* 15–17: 9–106.

Zhmud, L. J. 1997. *Wissenschaft, Philosophie und Religion im frühen Pythagoreismus*. Berlin.

2006. *The Origin of the History of Science in Classical Antiquity*, translated by A. Chernoglazov. Berlin and New York.

INDEX LOCORUM

For discussion of texts without the citation of passages, see the general index.
† before a title = spurious or dubious; after a passage citation = authorship uncertain; * = synopsis; lac. = lacuna; test. = testimonium

GENERAL INDEX